C0-AWZ-777

Practical Applications of Infrared Techniques

Practical Applications of Infrared Techniques

A NEW TOOL IN A NEW DIMENSION
FOR PROBLEM SOLVING

RICCARDO VANZETTI

A WILEY-INTERSCIENCE PUBLICATION

JOHN WILEY & SONS New York • London • Sydney • Toronto

Library of Congress Cataloging in Publication Data:

Vanzetti, Riccardo.
Practical applications of infrared techniques.

Bibliography: p.
1. Infra-red technology. I. Title.

TK4500.V35 621.36'2 72-6441
ISBN 0-471-90360-4

Printed in the United States of America

10 9 8 7 6 5 4 3 2 1

Preface

During the last million years, man has been successful in increasing his abilities and in enlarging his understanding of the world where he lives.

The tool responsible for such progress is an outstanding computer (usually called the brain), which receives the outside information through five inputs (the five senses), compares it against the stored information (experience and/or knowledge), and makes decisions and orders their implementation by feeding appropriate output signals to the body tools: the voice for communications; the limbs for action; and the reproductive system for perpetuation.

Contrary to the general behavior of living beings, man has shown no patience and no willingness to keep within the original limitations of the human system as described above. In his unrelenting drive for progress, he has been constantly striving to widen the range of his capabilities and performance.

First came the improvements in the action area: with his bare hands, man made tools, and with the tools he made devices capable of expanding greatly the range of physical performance.

Then came the effort to improve the decision-making mechanism: from maternal education to philosophy, man kept striving to establish clearer and wider standards of understanding and judgment.

Only recently man has attacked the third area: expanding the input information acquisition. It seems logical that the widening of the limits of our senses, or the addition of new ones, should enable the brain to receive more information for a more complete evaluation of the outside world.

Among the latest advances in this area has been the development of sensors to detect and measure infrared radiation and the capability of presenting this information in a way that is understandable to the human mind.

This is as if man increased the number of his senses from five to six:

now man can perceive the infrared radiation emitted by all physical matter. This might not sound impressive but its great importance will be illustrated by an example. Let us suppose that a farmer was born blind. He tills his land and selects the seeds by shape, size, and weight. One day, through a successful operation, he acquires eyesight. Now he realizes that the seeds have a color, too. Perhaps there is a correlation between color and the chances of growth; perhaps the black ones are those that will never sprout. Thus the farmer must start to find out the meaning of the color of every visible object.

The same thing is true with infrared. Man has just acquired this new sense. Suddenly darkness is no more. Even the blackest night is filled with the glow of infrared radiation emitted constantly by all physical matter. However, man must learn to understand the message that this radiation is broadcasting. When man succeeds, he will be one step further on the road to the knowledge of the universe and control of the physical world.

It is very long, probably unending, this road that man has been walking ever since. Remember the first recording of it?

> " . . . let us build ourselves a city and a tower with its top in the heavens; let us make a name for ourselves. . . ."

The tower that men started to build to conquer the heavens and stand close to God was the Tower of Babel. What is the tower that will lift man to superhuman heights? Is it not really the tower of science, which will give him wider knowledge and greater power under the heavens than ever before? However, the construction becomes more and more difficult as the tower rises, because of the confusion of the languages: the chemist, the mathematician, the geologist, the doctor, the engineer, the biologist, and the physicist, all speak a different language. They cannot cooperate in the construction of the tower. Should these seven men be locked together in a room, they could only carry out nonconstructive conversation, such as discussions of the weather, hobbies, and television. But they would have extreme difficulty in coordinating a joint effort for the further construction of the tower: this is the history of today, as written 5000 years ago by Moses in the Genesis.

In this book I avoid the difficulty of the different languages. Infrared radiation is such a general phenomenon that its knowledge cannot be restricted to a limited class of scientists. This is why I write in a language that will be understood by men such as the seven mentioned above, and by all the people who are interested in the construction of the Tower.

This book carries a message of common interest, namely how to use the information yielded by the infrared radiation to improve the design and construction of today's hardware, especially that of electronic hardware. This improvement can be defined as an increase in the reliability of the equipment, since by definition, reliability is the probability that a piece of equipment will perform within specifications under specified environmental conditions for a given length of time.

Better design and better construction will automatically increase such probability. However, my approach will be far from the statistical concept so dear to the majority of our reliability experts. I feel that we must try to go beyond the statement that "during the next 1000 hours of work only an average of three transistors and an average of five diodes will fail in the average computer of Type X." I want to identify, among the 2000 transistors and the 8000 diodes wired inside a specific computer of Type X, those few transistors (not necessarily three) and those few diodes (not necessarily five) that will fail during the required operating life of the computer. I believe that this can be achieved with the help of infrared techniques.

In other words, this book will stress the practical approach and point out how the infrared radiation can be used for the solution of technical problems and how it can supply information that will lead to the design and manufacture of better products.

The use of infrared radiation for reliability is now in its very early infancy. This is the first book written on this subject. It can only describe the little progress that has taken place so far and point out the findings indicating the tremendous potential of this new approach. The road leading to the full implementation of these techniques is long and not easy, and until now too few people have been walking along it. In full humility, I acknowledge the inadequacy of my effort, and with deep gratitude I welcome the help of any proselyte.

To whomever will elect to join in this effort, I make no promises of raises, longer vacations, promotions, awards, or decorations. I only promise harder and longer work and—perhaps some day in the future—the deep satisfaction when looking earnestly at each other in the eyes to know that we helped add one more step to the construction of the tower of knowledge, with the goal of revealing God and not replacing Him.

RICCARDO VANZETTI

February 1972
Walpole, Massachusetts

Acknowledgments

Sincere gratitude and appreciation is hereby expressed to those people who helped me either in the preliminary phase of infrared development or in the actual work connected with the present book.

One asterisk indicates indirect contribution of work and effort; two asterisks indicate direct encouragement and support of my work; three asterisks denote direct contribution to the text of this book or to the work connected with its preparation and publication.

* Dr. Frederick E. Alzofon
Lockheed
Houston, Texas

*** Dr. Richard Barbera
Department of Education
California State Colleges
Los Angeles, California

** Stephen N. Bobo
U. S. Department of Transportation
Cambridge, Mass.

* Neville Burrowes
U. S. Navy Underwater Sound Lab.
New London, Conn.

* Dr. Peter Debye
Raytheon Company
Waltham, Mass.

** E. F. Dertinger
ITT-TELECOM
Milan, Tenn.

*** A. S. Dostoomian
Vanzetti Infrared & Computer Systems, Inc.
Canton, Mass.

*** Kathleen Flynn
Raytheon Company
Norwood, Mass.

*** Augustus G. Grace
Vanzetti Infrared & Computer Systems, Inc.
Canton, Mass.

** Fritz A. Gross
Raytheon Company
Waltham, Mass.

** Leon C. Hamiter
National Aeronautics & Space Administration
Huntsville, Ala.

** Ruth A. Herman
Consultant
Dayton, Ohio

*** Anthony J. Intrieri
Dynarad, Inc.
Norwood, Mass.

* Edward J. Kubiak
General American Transportation Corp.
Niles, Ill.

* Dale Maley
Martin-Marietta
Orlando, Florida

** Joseph L. Murphy
National Aeronautics & Space Administration
Washington, D. C.

* R. J. Simms
Agricultural Control Systems
Redwood City, Calif.

** Theo Tsacoumis
National Aeronautics & Space Administration
Washington, D. C.

*** Nelly-Aimee Vanzetti
Walpole, Mass.

* Paul E. J. Vogel
Army Materials & Mechanics Research Center
Watertown, Mass.

*** John P. Ward
Computer Systems Engineering, Inc.
Billerica, Mass.

** W. T. Welsh
Riker-Maxson Corp.
New York, New York

* David K. Wilburn
U. S. Army Physical Sciences Lab.
Warren, Michigan

Contents

PART II APPLICATIONS
Who cares to listen?

Practical Applications of Infrared Techniques

Part I Infrared Radiation and Detection

Chapter 1 Fundamentals

IN THE DARK OF THE NIGHT

"Hello, sir," said the voice at the other end of the line, "this is the Security Officer speaking. Sorry to report patient 89156 missing from the criminal ward. . . ."

"Patient 89156 . . . you mean Jim, the Boston strangler?"

"Yes, sir. We are going out now with a patrol to canvass the Black Woods area, where he must be hiding. Hope we catch him before the sun sets."

However, this search did not succeed. The thick underbrush provided a perfect cover for the fugitive, and by nightfall the search party returned empty-handed. About 9:00 P.M. the hospital's director and his junior assistant started a new kind of search. The overcast sky made the night pitch black, and they were carrying neither weapons nor lights. Nevertheless, their stride was fast and sure while they crisscrossed the Black Woods.

"Down there, sir. Can you see a faint light?" called the assistant with excitement.

"It might very well be him," agreed the director. "Let's go." (See Figure 1)

"I bet he'll be surprised," commented the assistant ironically.

As they stood close to the fugitive, the director spoke severely. "Jim, what are you doing here in the middle of the night? Shame on you!"

"You can't see me, and you can't catch me, Doctor," answered Jim defiantly.

"Jim, don't you know that doctors can see in the dark? Just put out some fingers, I'll tell you how many they are . . . three . . . don't cheat, Jim. Now there are two . . . now five . . . you know you can't fool me, Jim. Let's go."

Figure 1. Hiding in the dark. (Courtesy of Night Vision Laboratory, U.S. Army Electronics Command.)

"I might be crazy," Jim said plaintively, "but I'm not stupid, and I know when I am beat. Sorry, Doc, forgive me. But if you want me to move, better lead me by the hand; I can't see as far as the tip of my nose."

Back at the hospital, the director and his assistant took off their goggles. "What a gadget!" exclaimed the latter, admiring the pair in his hand, "I guess these will soon bring artificial illumination to an end."

"Yes," nodded the director, "for the infrared seeing eye, darkness is no more. Even in the blackest night the whole world glows by what used to be known as *invisible light*."

INVISIBLE LIGHT

We have just taken a glance into the near future. Let us now look back in the past.

A man in a darkened laboratory is investigating the nature of light, the nature of heat, and their correlation. The only window of the room is covered by a thick wooden panel (Figure 2). In this panel there is a small rectangular opening, and in this opening a prism separates the sun's light into the colors of the spectrum. The man is measuring the heat content of each color of the spectrum with a thermometer. To compensate for changes in the ambient temperature, a few control thermometers are scattered around the work area. But what is happening to one of these controls? It shows a temperature much higher than any other thermometer in the room, and this high temperature area is clearly located out of the colors of the spectrum, way beyond the outer edge of the red band.

The year is 1800. Napoleon Bonaparte is crossing the Alps on his way to conquer Italy. Ludwig van Beethoven is composing the Second Symphony. And Frederik William Herschel has just accidentally discovered the *infrared radiation.*

In his address to the Royal Society of London (Figure 3) on March 27, 1800, he states: " . . . the radiant heat from the sun will at least partly consist—if I may be permitted the expression—of *invisible light.* . . ."

Strange as it may sound, *invisible light* is absolutely correct. Maybe we

Figure 2. F. W. Herschel discovering infrared. (Courtesy Elmer Reese.)

Figure 3. Sir William Herschel, discoverer of the infrared. (Crown Copyrighted reserved. Reproduced by permission of Sir J. C. W. Herschel.)

should specify—*invisible to the human eye;* the limits of the visible radiation band appear to vary for the different species of living beings. Larger animals seem to be less sensitive to the lower frequencies, as indicated by the case of the bull: yes, the bull of the Corridas, the daring show starred by the Matador with his red cape that makes the bulls so furious. Do you know how the red cape looks to the bull? Just plain black. In the red range, the bull is color-blind!

Conversely, for some smaller animals, the visible bandpass seems to shift in the other direction, that is, some of them are not able to see the colors at the upper end of the spectrum: this is probably the reason why few flowers are blue or violet: most insects can not see them, and the recent discovery that mosquitoes can see the invisible radiation of the human body is not surprising; through bitter personal experience I found out that on the darkest of the nights I look to the mosquitoes like a lighthouse looks to seamen.

Today this invisible light is called "infrared radiation." It is an electromagnetic oscillation of the same nature of the visible light and of the radiowaves. As a matter of fact, it is located below (hence the term "infra") the red band of the visible light and above the millimeter radio waves in the electromagnetic spectrum.

THE ELECTROMAGNETIC SPECTRUM

Figure 4 shows the spectral distribution of the electromagnetic radiation as known today. Starting at the right end with the so-called audio frequencies, we progress through the radio waves (long, medium, short, VHF, UHF, XHF, and millimeter waves) until we reach the infrared region. There is an overlap between the lowest frequencies of the infrared spectrum and the high frequency end of the millimeter waves, an overlap in name only, because it is the same radiation that can be designated by either name. We will not even mention the submillimeter radio waves that occupy a large portion of the far-infrared spectral area.

At the high frequency end of the infrared region is the visible light, next ultraviolet, and then as the energy content increases, x-rays, γ-rays, and cosmic rays. Here the human knowledge stops for the time being.

Figure 4 shows three scales that are the key to the classification of the electromagnetic radiation: frequency, wavelength, and energy content. They are related by the equations:

$$f = \frac{c}{\lambda}$$

$$E = hf$$

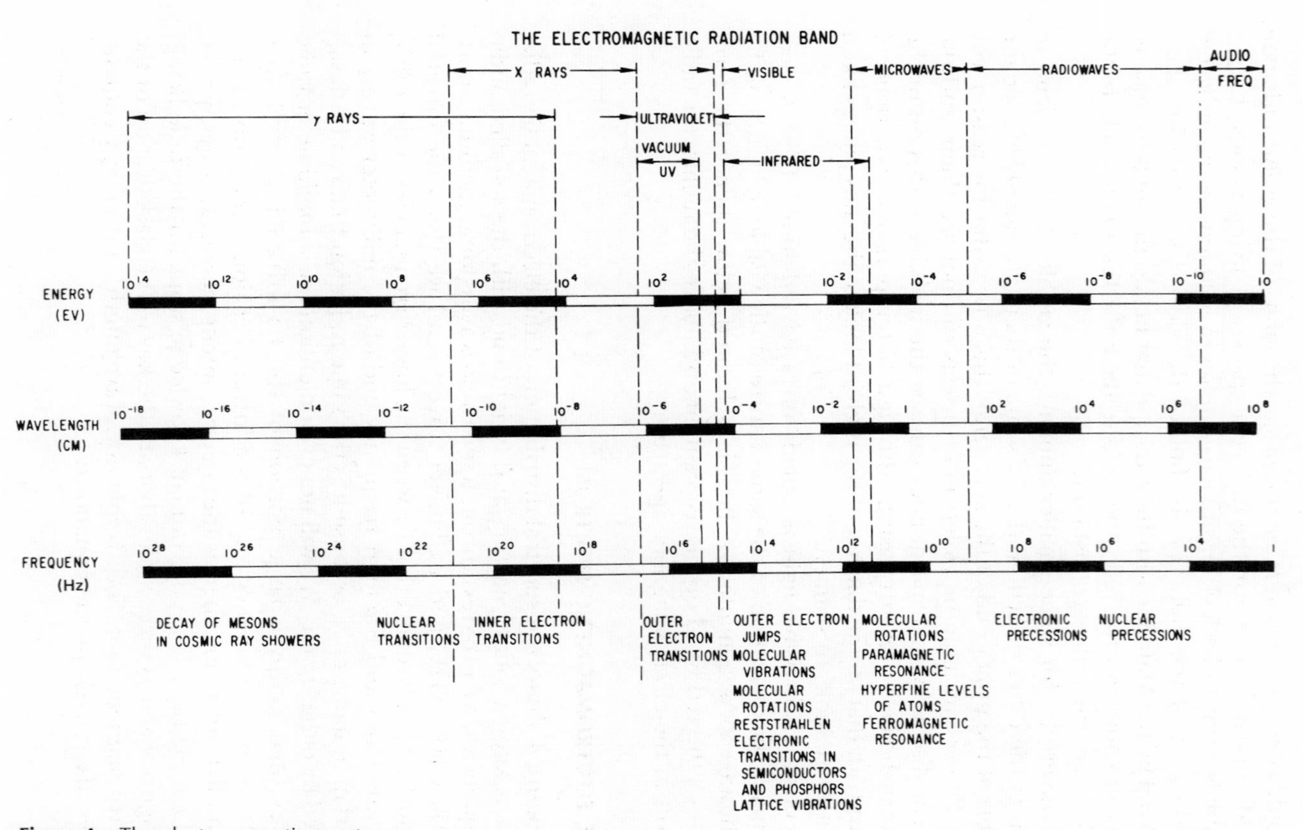

Figure 4. The electromagnetic spectrum.

where f = frequency (Hz)

c = velocity of light (2.99793×10^{10} cm/sec)

λ = wavelength (cm)

E = radiation quantum energy (J)

h = Planck's constant (6.6252×10^{-34} W/sec^2)

From the chart we can see that they cover an extremely wide range. However, in spite of tremendous differences in frequency, wavelength, and energy content, all these radiations are identical in that they all transport energy and always travel with the same speed: the speed of light, which, in the vacuum, is close to 300,000 km/sec.

These radiations are different in the way they are generated: by electromechanical generators (alternators), at the low-frequency end, through resistive-capacitive electronic networks, all the way to nuclear transitions and meson decay at the high-frequency end. Infrared radiation is produced by the rotational and vibrational movements of the atomic and subatomic particles of which physical matter is made. Since atomic and subatomic particles are always in motion, infrared radiation is always emitted by physical matter, at frequencies corresponding to the resonant constants of the oscillators. Because there is a very great variety of different particles, different frequencies are generated in great variety, covering the range of what is known as the infrared spectrum.

THE INFRARED SPECTRUM

Figure 5 shows the infrared radiation band and also the band of the visible light right next to it. We can see that the infrared region is divided in three areas: the near infrared, the intermediate, and the far infrared. These divisions are rather arbitrary, as are their upper and lower limits.

The different detection techniques used in the past were responsible for the boundaries shown in the chart, although at the present time there does not seem to be real need for dividing the infrared band in any number of subareas. However, the sources of infrared radiation can be divided in three groups, based on their mass:

1. Subatomic particles whose "jumps" supply most of the radiation in the near infrared.
2. Atomic particles whose movements produce radiation in the intermediate infrared.
3. Molecules whose vibration and rotations generate radiation in the far infrared.

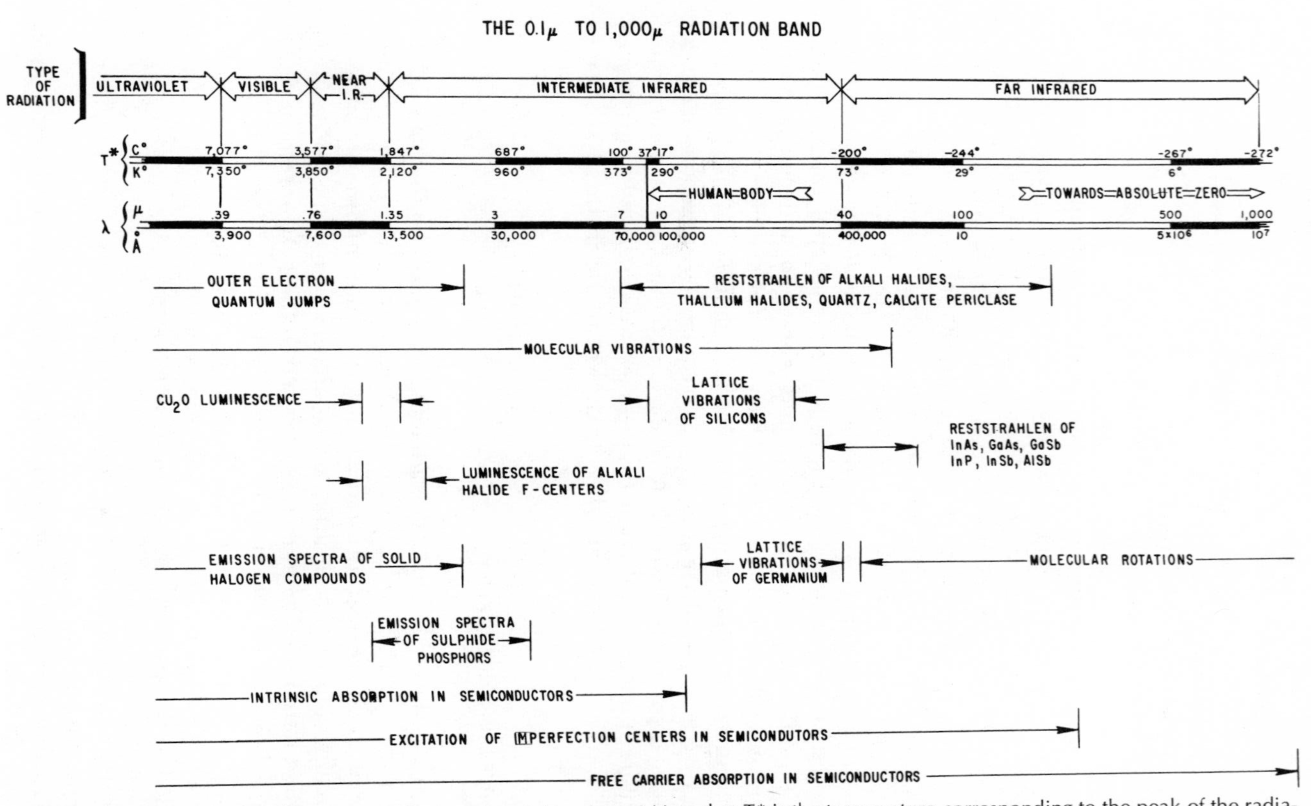

Figure 5. Visible and infrared spectrum. (Courtesy Leo J. Neuringer.) Note that T* is the temperature corresponding to the peak of the radiation band. For every T, radiation occurs (at lower levels) at all other λ.

The one thing that stands out at first glance is the great disproportion between the area covered by the visible light and the area taken by infrared—approximately 1 octave, as opposed to more than 10 octaves. One could wonder about the beautiful sights that an infrared-seeing eye would enjoy, being able to see approximately seventy new colors, in addition to the seven that we currently know.

Elements of interest in Figure 5 are the sources of infrared radiation and the spectral areas covered by them. We can see here how molecules of certain types have their resonant frequencies limited to certain areas of the spectrum, and we can now understand the physical laws upon which the science of spectroscopy is based: every molecule has its own resonance profile, a string of different frequencies that is unique, since it is generated by the elementary particles of which the molecule is made. It is like a fingerprint, and by spectroscopic analysis these fingerprints are recognized, thus identifying the chemical composition of the substance under examination. A very interesting subject indeed, that has already been covered by many excellent books.

One more item that stands out from Figure 5 is the temperature scale, designated by T. In order to understand the correlation between this scale and the other elements of the chart, let's clarify the meaning of the word "temperature".

WHAT IS TEMPERATURE?

Temperature, everybody knows, is the measure of the level of heat contained in a physical body. What is heat? Pythagoras, holding his lectures at the Temple of the Muses in Crotone, taught his disciples that heat is a manifestation of fire, the fourth of the five elements earth, water, air, fire, and ether, of which the universe is made. Heat, Pythagoras taught, was an invisible fluid that could flow through air, water and earth, bringing everywhere the sacred, primordial energy of fire.

This was 2500 years ago. Today modern science agrees with Pythagoras: heat is energy, kinetic energy of the basic elements of which physical matter is made: molecules, atoms, and subatomic particles; the temperature can be defined as the level of agitation of these particles: the greater the agitation, the higher the energy content and the heat level or temperature.

Also, the greater the agitation, the greater the variations of the electromagnetic field generated by these oscillators. In other words, the higher the temperature, the greater the power emitted by radiation.

Furthermore, the greater the agitation, the higher the frequency of the peak of the emitted radiation band. Figure 5 shows this correlation between T and λ, but to understand the physical law governing this correlation, we must introduce a new concept, a nonexisting item—the *blackbody*.

THE BLACKBODY

We have just seen that every chemical substance has its own radiation spectrum, every compound its own emission band. Now we can imagine a physical body made of an infinite number of different particles, whose resonant frequencies are covering the whole infrared band, without leaving any empty spot.

This imaginary body is called a "blackbody." By definition, it is the ideal emitter, having one oscillator for every frequency. It is perfect—the perfect radiator and the perfect absorber, the perfect material for which the physical laws governing radiation have been formulated and are holding true.

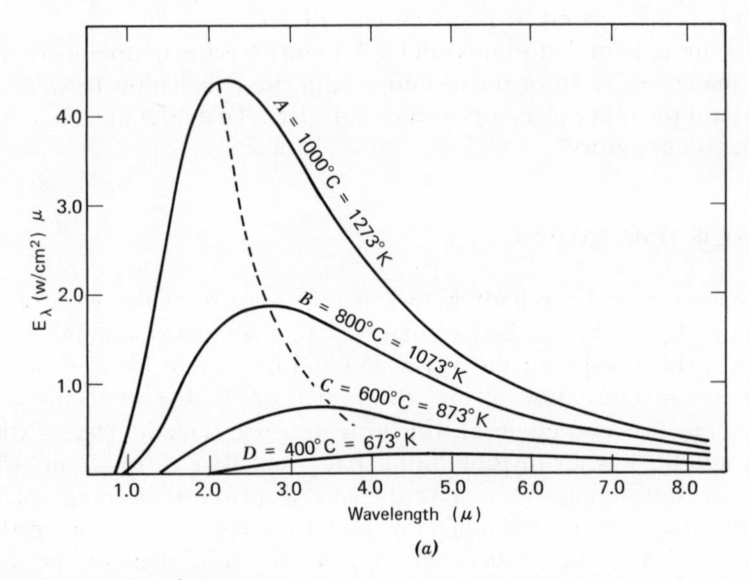

Figure 6. Blackbody radiation.
Stefan-Boltzmann law, $W = \epsilon\sigma T^4$. Wien's displacement law, $\lambda_m = b/T$.

W = radiant flux emitted per unit area (W/cm^2)
ϵ = emissivity (unity for blackbody source)
σ = Stefan-Boltzmann constant [$5.673 \times 10^{-12} W/(cm^2)(°K^4)$]
T = absolute temperature of source (°K)
λ_m = wavelength of maximum radiation (μ)
λ = radiation wavelength (μ)
b = Wien displacement constant (2897 μ°K)

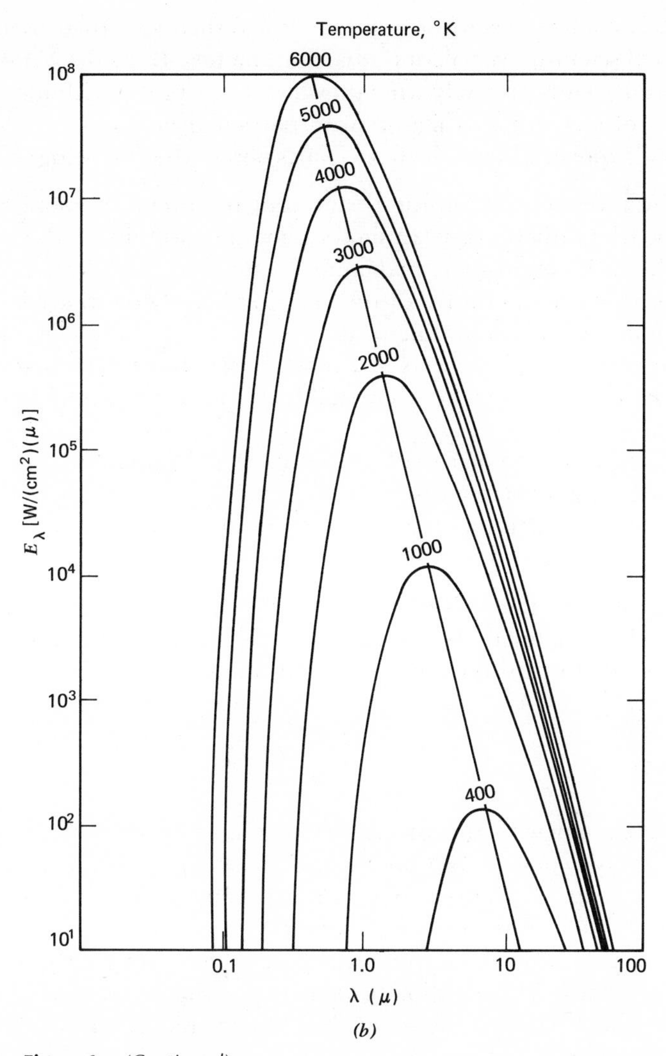

Figure 6. (*Continued*)

Like any other perfect thing, it does not exist on earth. However, its concept is very useful for the applications of the general radiation laws. And very close approximations of it can be made or can be found, as is seen later.

For the time being, it is of interest to see what the blackbody radiation spectrum looks like. Figure 6 does exactly this, *a* with linear ordinates, *b* with logarithmic ordinates. The most obvious feature of these curves is their resemblance to statistical distributions, such as those observed for a variety of scientific, social, or human parameters. Indeed, as Max Planck discovered, that is precisely what they are—statistical distributions of the numbers of elemental oscillators with particular energies.

Even a superficial look at these charts shows the following:

1. The radiation amplitude is different for every frequency and its envelope (the radiation curve) shows smooth continuity, with one single peak where the intensity is maximum.
2. For every temperature of the emitting blackbody, there is a single emission curve corresponding to it.
3. No curve ever intersects any other curve, but for increasing temperatures, every curve runs above all the curves corresponding to lower temperatures.
4. As temperature increases, the peak of the radiation curve moves towards shorter wavelengths.
5. As temperature increases, the amplitude of the emitted radiation increases according to an exponential law.

These then are the laws that are valid for the nonexistent blackbody. How are they related to the radiation by existing physical matter? Simply by a coefficient of proportionality called "emissivity."

EMISSIVITY

A comparison with the properties of visible light might help us to grasp the basic concept. Suppose we have a source of white light, for instance, an electric bulb made of nontransparent opaline glass. Let us call *unity* the measure of the power so radiated. If we now paint the bulb red, or blue, or any other color, the measure of the power radiated will be a fraction of unity. This fraction, usually expressed as a decimal notation, could be called the "emissivity" of the electric bulb, in the visible region of the spectrum.

Similarly, in the infrared region, emissivity could be thought of an "infrared color" and in this term we also want to include the color gray; the only difference is that the "gray" radiation covers the whole frequency spectrum, while the other colors are covering only a fraction of the total spectrum.

Consequently, physical matter emits infrared radiation in accordance to the blackbody radiation laws, except for a factor of proportionality that is called "emissivity" and whose value is always a fraction of unity. Only the blackbody has an emissivity factor of 1.

The emissivity factor of a gray body is constant at all wavelengths; thus its radiation curve is similar (only lower in amplitude) to that of the blackbody. This is not true for those physical bodies that might differ sharply in emissivity at different frequencies, due to the presence or the absence of resonant and antiresonant subatomic particles.

Of course, real objects must possess a "total emissivity," and a "spectral emissivity," the distinction being defined by whether one is concerned with total radiation at any given temperature, or radiation at a particular wavelength. A moment's thought clarifies this distinction—for example, for a gas. Any gas, at a temperature above absolute zero, will radiate as a blackbody in those spectral regions where the gas is totally absorbing. In those spectral regions that are generally called "windows" where the gas is transparent, little radiation is observed. Thus the "total emissivity" of the gas might be quite low, but the "spectral emissivity" in absorbing regions could approximate unity.

It must follow, from such reasoning, that all real objects can have emissivities that vary with wavelength. Also, all real objects can have emissivities that vary with temperature and with physical state (solid, liquid, or gas). Furthermore, it follows that the surface characteristics of the radiating medium should affect emissivity—finish of a solid, molecular layer of a liquid, or envelope of a radiating gas. Indeed, this is observed to the point that no theory exists that can correlate emissivity with basic physical parameters such as conductivity, atomic number, index of refraction, and such. Emissivity is a numeric, always experimentally determined for each object, and for opaque materials it is strictly a surface characteristic, not related to the composition of the material under the surface. Its value ranges from close to zero for mirror-like surfaces, to almost 1.00 for lamp black, zapon black, and such.

BASIC RADIATION LAWS

Kirchhoff and Monsieur de La Palice

Prior to the introduction of the blackbody concept, the experimental plotting of the radiation curves, and the formulation in elegant equations of the basic radiation laws, a german physicist, Kirchhoff, started investigating the thermal behavior of physical matter. He noticed that it would absorb heat when subject to radiation from a warmer body, and would lose heat by radiating into a cooler environment. In his physics laboratory at the University of Heidelberg, he placed a physical body inside a cavity whose walls were kept at constant temperature and measured the heat exchange between the enveloping and the enveloped bodies. No matter which was warmer, eventually the enclosed body, either through radiation or through absorption, would reach the temperature of the enveloping cavity. At this point thermal equilibrium was reached, and Kirchhoff could state his basic law:

When a body is in thermal equilibrium, the amount of energy absorbed equals the amount of energy emitted.

Such an obvious, redundantly self-echoing statement calls to mind Monsieur de la Palice, the French nobleman, after whom every obvious, redundant statement is named a "Lapalicean truth." Remember his story?

Monseigneur Jacques de Chabannes, Seigneur de la Palice, was a gallant, fearless chieftain, who met a hero's death at the battle of Pavia in 1525. In his honor, his soldiers composed a song exalting his valor. Only one verse of the song reached us and does not seem likely to be forgotten. It goes: "Un quart d'heure avant sa mort il était encore en vie." Literally translated, it means: "Fifteen minutes before his death, he was still alive."

However, on second thought, Kirchhoff's law shows a lot of hidden value and leads us to interesting considerations based on the fact that if the amount of energy emitted equals the amount of energy absorbed, the emissivity coefficient ϵ must be equal to the absorptance coefficient α, *no matter what kind of material we are considering.*

Consequently, the following is true:

1. A good absorber is a good radiator.
2. A good reflector or a transparent body is a poor radiator.
3. We can determine emissivity ϵ by measuring absorption α.
4. Since it is impossible for a body to absorb more energy than the total amount radiating onto it, α cannot exceed unity, nor can ϵ. Therefore no body can radiate more than a blackbody at the same temperature.
5. Since all the energy radiating onto an opaque body is either absorbed or reflected away, the expression $1 = \alpha + R = \epsilon + R$ allows us to measure ϵ by measuring R (reflectance).
6. In a hollow enclosure kept at constant temperature, *no matter what material it is made of,* the flux density is the same as it would be if emitted by a blackbody at the same temperature as that of the walls of the cavity, since the infrared energy radiated by every point of the surface is totally absorbed by the other points of the cavity. Therefore, the cavity radiates as a blackbody, all energy within it being continuously absorbed and radiated by its inside walls. This feature provides the basis for the design

of modern cavity radiators that are the closest approximation to the theoretical blackbody.

What is a Cavity?

Any hollow space enclosed within a physical body has the property of exhibiting a level of radiant flux equal to that emitted by a blackbody at the same temperature.

Whenever this radiation must be measured by a detector placed outside the cavity, a hole must be pierced through the cavity's envelope. If the size of this hole is small enough in relation to the size of the cavity, the flux disturbance is so small that it can be disregarded in first approximation.

Figure 7 illustrates schematically the behavior of a ray entering a cav-

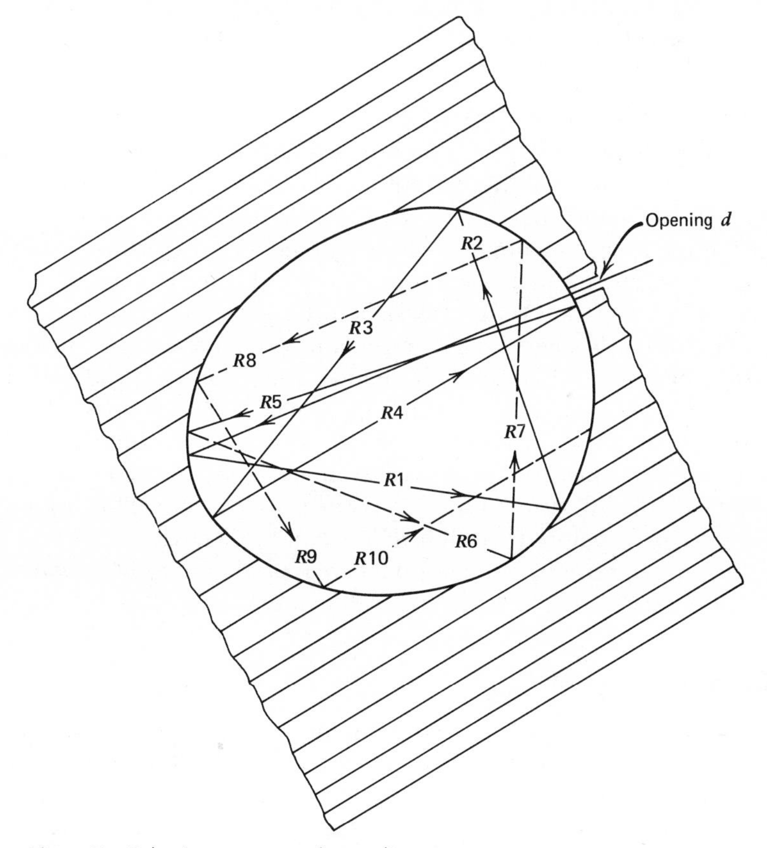

Figure 7. Behavior or ray entering cavity.

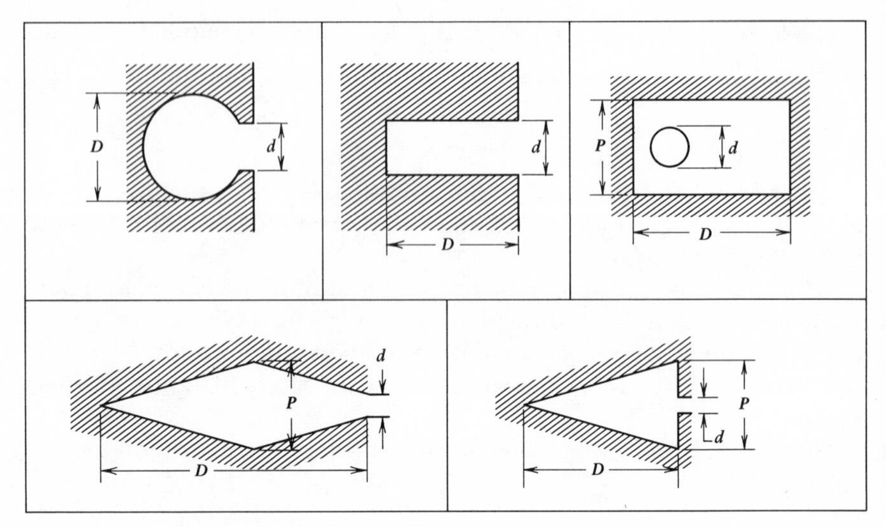

Figure 8. Simple geometrical shapes of black body cavities. (Courtesy Leo J. Neuringer.)

ity: at every point of impact, one fraction of the radiation is absorbed, and the rest is reflected away, until, after n impacts, all the energy has practically been absorbed.

In Figure 7, the opening d behaves almost like a blackbody surface whose temperature equals the temperature of the inside walls of the cavity. It absorbs radiation with an absorptance coefficient almost equal to 1, and per Kirchhoff's law, it emits radiation with an emissivity coefficient almost equal to 1. Blackbody radiation can thus be produced and measured experimentally, even though a real blackbody does not exist.

Figure 8 shows some simple geometrical shapes in which blackbody cavities are manufactured. The most common type is the conical configuration with the aperture located opposite the cone's tip. The inside surfaces are always painted a dull black to avoid reflectivity.

In practice, emissivity values of 0.99 are typical of most blackbody cavities commercially available for laboratory work.

Stefan-Boltzmann's Law

Figure 9 is the graphical representation of the radiation law discovered by Josef Stefan in 1879, and theoretically verified by Ludwig Boltzmann in 1884:

$$W = \epsilon \sigma T^4$$

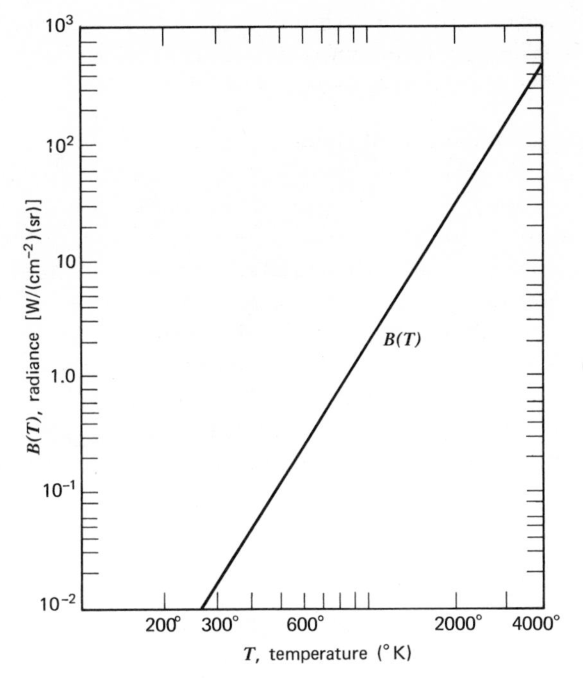

Figure 9. Stefan-Boltzmann law.

where W = radiant power emitted per unit area (W/cm²)
ϵ = emissivity
σ = Stefan-Boltzmann constant [5.673×10^{-12} W/(cm²)(°K⁴)]
T = absolute temperature (°K).

Expressed in words, Stefan-Boltzmann's law states that the power radiated by a physical body is directly proportional to the fourth power of its absolute temperature. The chart of Figure 9 plots $T°$ versus radiance, which is defined as intensity radiated per unit projected area of source per unit solid angle.

Wien Displacement Law

The line connecting the peaks of all radiation curves in Figure 6 is the graphical representation of the displacement law discovered by Wien in 1893:

$$\lambda_m = \frac{b}{T}$$

where λ_m = wavelength of maximum radiation, in microns
b = Wien displacement constant (2897 μ/°K)
T = absolute temperature (°K)

The physical meaning of Wien's law is that, as the temperature increases, the peak of the radiation shifts steadily towards shorter wavelengths. And if we keep in mind the correlation of Figure 3 (the shorter the wavelength, the higher the energy content), we realize that as the temperature increases, the number of molecules having a higher energy content increases in direct proportion.

Wien's law readily explains a well-known physical phenomenon: the change in color of a body whose temperature increases. For instance, when a piece of metal is being heated, first we see a deep red glow. As the thermal level rises, the red becomes brighter, then turns into orange, yellow, and finally white, when the emission spectrum extends to cover also the green and blue bands.

Planck's Equation

Exactly 100 years after the discovery of infrared radiation, Max Planck succeeded in developing a theory that would account for the continuous thermal spectrum, as contrasted with discrete line spectra, emitted by excited materials. This theory presented a departure in physics from purely classical reasoning to the new "quantum" reasoning. Planck's major contribution is best summed up as: Any physical entity whose single "coordinate" executes simple harmonic oscillations can only possess total energies which satisfy the relationship

$$E = nhf$$

where E is total energy, f is the frequency of oscillation, n a "quantum number," and h a universal constant—Planck's constant = 6.625×10^{-34} W sec².

Having constrained the atomic, or molecular, oscillators in an object to such discrete total energies, Planck then developed a mathematical equation for the average energy of a system consisting of such oscillators—in all possible modes—subject to the restriction that the statistical distribution of energy should follow Boltzmann's probability distribution. Thus Planck developed the well-known blackbody distribution law for spectral radiant emittance:

$$W_\lambda = \left[\frac{2\pi hc^2}{\lambda^5}\right]\left[\frac{1}{(hc/e^{k\lambda T}) - 1}\right]$$

where W_λ = spectral emittance [W/(cm^2)(μ)]
e = Naperian base (2.71828)
hc/k = 1.438×10^4 μ °K
K = Boltzmann constant (1.38042×10^{-23} W sec/°K)
c = velocity of light (2.99793×10^{10} cm/sec)

This equation, which fits perfectly well the empirically verified black body radiation curves, represents a turning point in the history of physics, since it introduces the revolutionary concept that energy is emitted and absorbed, not continuously, but in discrete bundles, or "quanta," that are not fractionable.

However, the "quantum" theory is not adequate to explain optical phenomena such as interference (the colors reflected by a soap bubble), diffraction (the spreading of light passing through a tiny hole), and birefrangence in crystals (the splitting in two of a single ray). Instead, the classical wave theory explains these phenomena with ease.

In conclusion, is light a wave or is it made of indivisible corpuscles of energy? For the time being, the answer is: both. More precisely, light behaves as if it were a wave in its propagation, and as if it were of corpuscular nature in its interaction with matter.

Incoherent Radiation and Coherent Radiation

All the radiation laws described thus far are graphically represented by curves encompassing broad sections of the frequency spectrum. This is due to the fact that a blackbody, by definition, contains every oscillator and therefore radiates at every frequency of the spectrum. This type of emission is called "incoherent radiation," as opposed to the "coherent radiation" emitted by oscillators resonant at well-defined frequencies. Common examples of coherent radiation are the electromagnetic waves typical of radio and television transmissions.

In the infrared region, coherent radiation is emitted by lasers, whose frequency is contained in an extremely narrow band. Also semiconductor junctions emit narrow-band radiation, that is called "recombination radiation" because it is the energy liberated as photons by the electrons carrying the current when—at the end of the "carrier lifetime"—they step down from the higher energy level of the conduction band into the lower energy level of the valence band, where the electron-hole pairs recombine (see Figure 10). This recombination radiation usually takes place in the near-infrared (see Figure 11) and its magnitude is directly related (among other things) to the number of current carriers present in the conduction band. It follows that the variations of

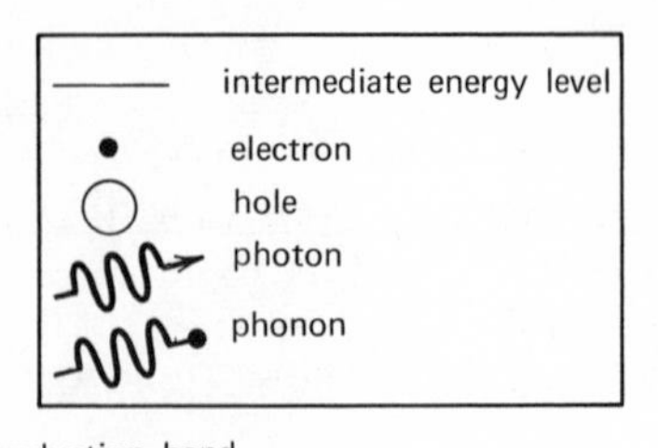

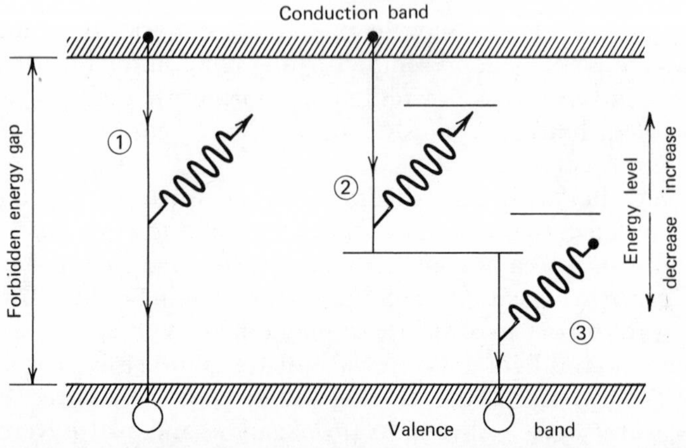

Figure 10. Energy model of recombination radiation in semiconductors.

1. Full jump: electron emits photon in dropping from carrier band to valence band where it recombines with a hole (intrinsic recombination).
2. Partial jump: electron stops at intermediate energy level, emitting a photon of lower energy content than full gap (extrinsic recombination).
3. Partial jump: electron drops another fraction of full gap energy, emitting phonon and finally recombining with a hole.

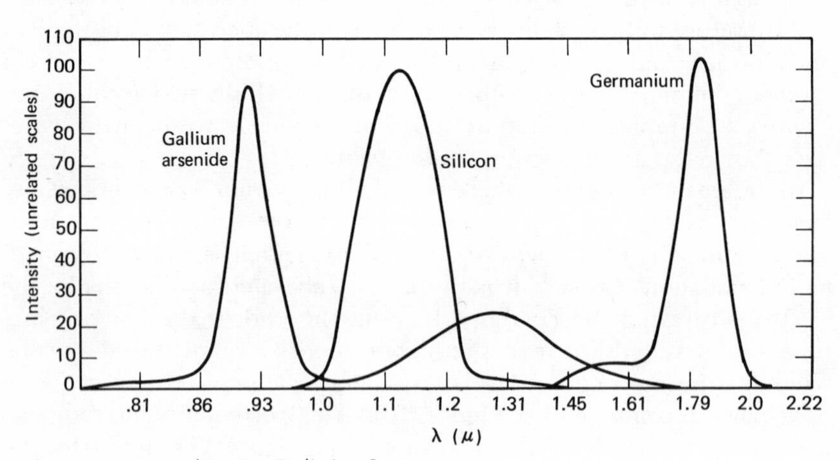

Figure 11. Recombination Radiation Spectra.

the current flow through a semiconductor induce simultaneous variations, of proportional magnitude, of the recombination radiation emitted by the junction through which the current flows. The thought of using a suitable detector to measure these variations to monitor the current's variations through the semiconductor comes immediately to the mind. According to which type of semiconductor is used, the idea might be very easy or very difficult to implement. Recombination radiation has been observed in higher or lower degree in most semiconductor materials, such as, but not limited to, Ge, Si, In Sb, In P, Ga Sb, Ga As, and such. However, what varies within wide limits is their efficiency, that is, the number of photons emitted, divided by the number of electron-hole pairs recombining.

Of the three materials in Figure 11, the efficiency of Ga As is close to 1, while for Si or Ge it drops four or five orders of magnitude, due to the fact that electron transitions are "indirect" instead of "direct."

In other words, one photon is emitted per every electron-hole pair recombining in the "direct gap" regime, while 10,000 or even 100,000 electron – hole pairs recombinations are needed to obtain one photon emitted in the "indirect gap" regime. Where does the rest of the energy lost by the recombining electrons go? Some of it forms phonons, particles of low energy content, that increase the kinetic energy level of the molecular atomic and subatomic particles of which the semiconductor material is made. Some other part of it is radiated at longer wavelengths as "extrinsic" recombination radiation, as it is generated by electron jumps smaller than the full forbidden energy gap.

This process is illustrated in Figure 10. In it, we see that an electron located in the conduction band can descend to the valence band to recombine with a hole in any of the following ways:

1. One full jump; the excess energy is liberated as a photon of a wavelength corresponding to the electron-volt potential difference corresponding to the width of the forbidden gap. The radiation composed of these photons is called "intrinsic recombination radiation."
2. A partial jump, spanning only part of the forbidden gap. The excess energy is emitted as (1) a photon of lower energy content, and the radiation so generated is called "extrinsic recombination radiation"; or (2) a phonon, or an energy particle absorbed by the semiconductor material and whose effect is to increase the kinetic energy content of its molecular and submolecular elements. In this case, no recombination radiation is emitted.

Indirect gap transitions are always present when impurities are present within the forbidden energy gap. "Crystal doping" is the all-

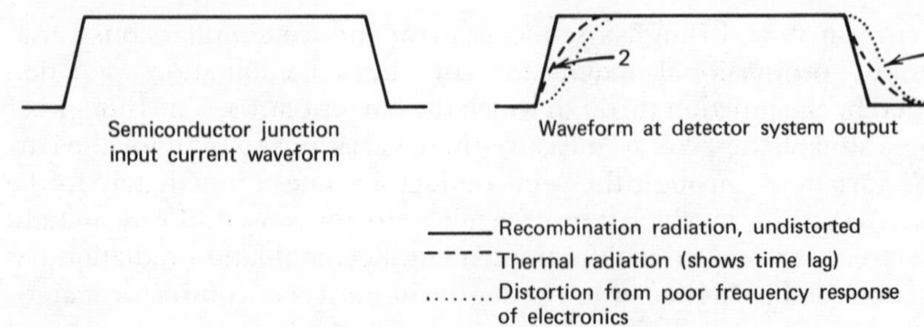

Figure 12. Waveforms, recombination versus thermal radiation.

important process of diffusing these impurities within the lattice structure of the semiconductor crystals.

Another characteristic of recombination radiation is the fact that it takes place without time lag with respect to the flow of current through the junction. As opposed to the generation of incoherent radiation of thermal origin that has a time delay caused by the thermal inertia of the emitter, recombination radiation begins the very instant when current starts flowing, and stops when current is turned off, except for the second order of magnitude delay taken by the electrons to find a hole where to recombine.

Figure 12 shows how to interpret three typical waveforms of the output of an infrared system:

1 Is the waveform of a recombination radiation pulse generated by a current pulse flowing through a semiconductor junction. Basically the two waveforms are identical.

2 Is the waveform representing the infrared radiation of thermal origin emitted by that same junction energized by a similar pulse. We can see the time lag due to the thermal inertia.

3 Is the waveform showing the effect of poor frequency response of the electronics processing the signal.

Description of the work done in this area is included in Part II, Chapter 7, under "Diodes", "Transistors," and "Integrated Circuits."

Other Optical Properties

Whether coherent or incoherent, infrared radiation follows the optical laws that are well known for visible light.

Reflection obeys the $\theta i = \theta r$ relationship, and *reflectivity* is bound to *absorptivity* by the equation: $R + A = 1$ for opaque surfaces.

For transparent bodies, *transmissivity* T is bound to reflectivity and absorptivity by the equation $T + R + A = 1$. *Refractivity* also follows Snell's law, being ruled by the index of *refraction* which in turn for every frequency is dependent on the electrical conductivity of the refracting medium, on its polarization and on its density.

Figure 13 shows the transmission regions of some materials used for infrared optics. These regions are often called "windows" and the transmission coefficient can vary between wide limits in these areas, according to the wavelength. Figure 14 lists the refractive index of some infrared-transparent materials, and also indicates how the index of refraction for a given material changes as a function of wavelength.

Radiation Geometry

The simple geometric relationships that govern the distribution and detection of infrared radiation, in space, are simply understood. As previously stated, a source will emit infrared power in accordance with Planck's law. The total power radiated by a source of finite area is simply

$$P = W \times A$$

where W = radiant emittance
A = geometric radiating area.
P = is expressed in watts

This power, radiated throughout space, provides a spatial density characterized as the radiant intensity, J, of the source—a term analogous to intensity in the visible region. Thus

$$J = \frac{P}{\Omega} \text{ W/sr}$$

where sr = steradian
Ω = total solid angle about the source.

One could also characterize the source by its radiance, N, —a term analogous to brightness in the visible region, such that

$$N = \frac{J}{A} = \frac{W}{\Omega} \text{ W/(cm}^2\text{)(sr)}$$

Now, if the source is an "isotropic" radiator, radiating uniformly throughout all space, $\Omega = 4\pi$. However, if the source is a "Lambertian" radiator, radiating only into a hemisphere of space, $\Omega = \pi$. This latter result (i.e., $\Omega \neq 2\pi$) follows from the fact that any extended source, which radiates only into one hemisphere, cannot be seen end-on.

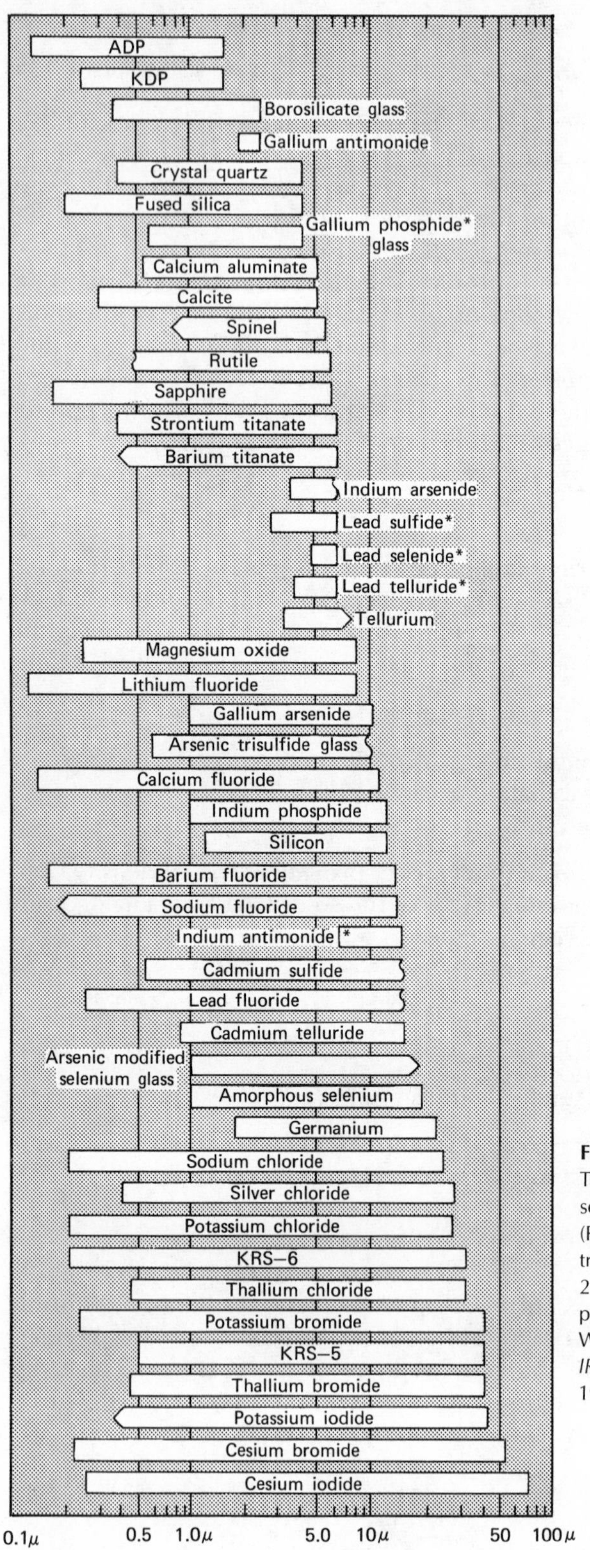

Figure 13. Infrared windows. The transmission regions of selected optical materials. (Regions are for 10% external transmittance or better, for a 2-mm sample at room temperature.) (Courtesy W. L. Wolfe and S. S. Ballard in *IRE Proceedings*, September 1959.)

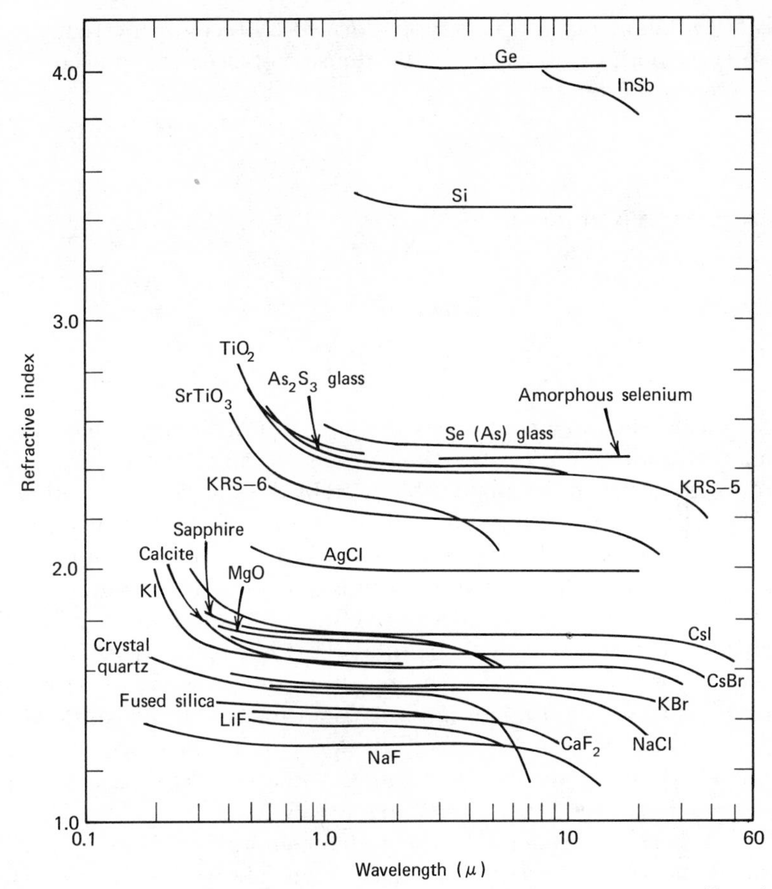

Figure 14. Refractive index of infrared-transparent materials. (Courtesy W. L. Wolfe and S. S. Ballard, *IRE Proceedings,* September 1959.)

The significance of these relationships, for the source, becomes evident when we examine the power received by an infrared system viewing the source. Any such system will possess an angular field of view, ω, defined by the optical system and the detector—or field stop. Being remote from the source, the system can only respond to radiation within its field of view. Therefore, the irradiance, H, at the optical aperture of the system will be

$$H = N\omega \text{ W/cm}^2 \qquad \text{at receiver}$$

or

$$H = \frac{J}{D^2} \qquad D = \text{distance to the source}$$

This last relationship can be simply derived by recognizing that the irradiance is nothing more than the density of received power at the system aperture, or

$$H = \frac{p}{a}$$

where p = received power
a = receiver area

and

$$p = J\left(\frac{a}{D^2}\right)$$

It is quickly seen from these expressions for irradiance that certain primary considerations control the response of an infrared system—the relationship between the angle subtended by the source compared to the angular field of view of the system. If the source is small compared with the system field of view—a "point" source—the received radiation varies with distance, but not with angle about the source. If the source is large compared with the system field of view—an "extended" source—the received radiation varies with neither. This is expressed in Lambert's cosine law, which states that "in any direction propagating from any point of a surface the radiant intensity is a function of the cosine of the angle θ between said direction and the perpendicular to the surface at that point." In other words, the strongest radiation takes place along the normal to the surface, and none tangentially to it.

This pattern distribution of the radiant energy explains why a detector "looking" at an emitting surface will always receive the same amount of energy, no matter what the angle happens to be between the detector's line of sight and the radiating surface. In practice, as the angle θ increases, so does the area viewed by the detector, but this increase exactly matches the reduction of radiated energy as stated in Lambert's law, and as a consequence, the total energy impinging on the detector remains constant. Of course, this holds true only as long as the emitting surface completely fills the field of view of the receiver.

Figure 15 illustrates this condition: as the emitting surface rotates around its axis A, the angle θ between the line of sight and the perpendicular to the surface increases from 0 to 90°. At any angle between these two limits, the area viewed by the detector increases by the same cosine factor which reduces the energy radiated toward the detector. As a result, no change will be apparent in the measurement of the power radiated by the target.

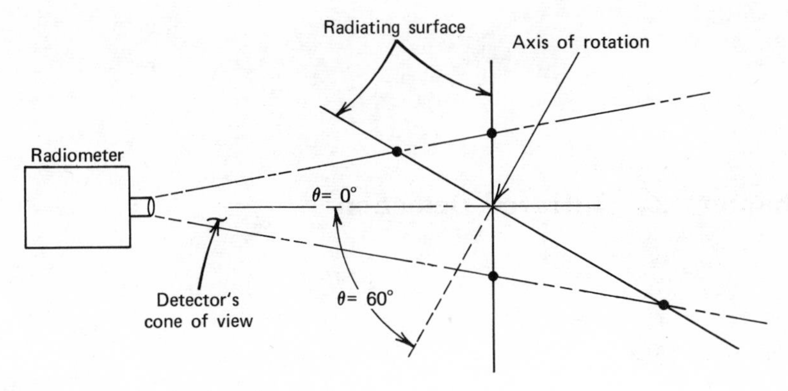

Figure 15. Lambert's cosine law.

Needless to say, the emitting surface does not need to be plane. Any three-dimensional shape will behave in the same manner, since every surface point can be considered a minuscule plane element radiating per Lambert's cosine law and the total power seen by the detector will be the integral of all surface points seen by the detector. This latter result of the geometry of radiation from extended sources materially controls the design of infrared systems by influencing the level of "background" infrared radiation. Thus a simple design rule is to make the source the only object in the field of view by appropriately controlling the field stop of the system. In Chapter 3, we examine those parameters that establish the response of infrared systems.

Chapter 2 Infrared Detection

FUNDAMENTALS

When infrared energy impinges upon physical matter, the portion of it that is absorbed produces one primary effect: it increases the energy content of the atomic and subatomic particles of which physical matter is made, and we commonly detect this phenomenon as an increase in its temperature. This change can be measured by one of the *secondary* effects, such as variation of physical qualities (volume, pressure, refractivity, etc.), of chemical characteristics (as in infrared photography), or of electrical properties (conductivity, dielectric coefficient, secondary emission, etc.). The speed at which these secondary effects manifest themselves covers very wide range, from hours needed by a chemical reaction to a fraction of a nanosecond for a secondary emission effect.

Figure 16 shows the major groups of detectors for the 0.1 to 1000-μ radiation band. We can see how chemical detection through photographic plates is only used in the near infrared and that the quantum detectors are operating in limited spectral areas for each group, while the thermal detectors cover the whole infrared spectrum.

One thing is immediately apparent: for every thermal range, there are available several detectors which have different operating characteristics from which to choose those best suited for any specific application.

In the review that follows, we briefly describe and discuss mainly those detectors that are more closely related to the type of work that is the main subject of this book. However, first a few words on the most common figures of merit are in order.

DETECTIVITY AND NEP

Any detector can only detect signals whose magnitude is greater than the self-generated noise. True or False? Both.

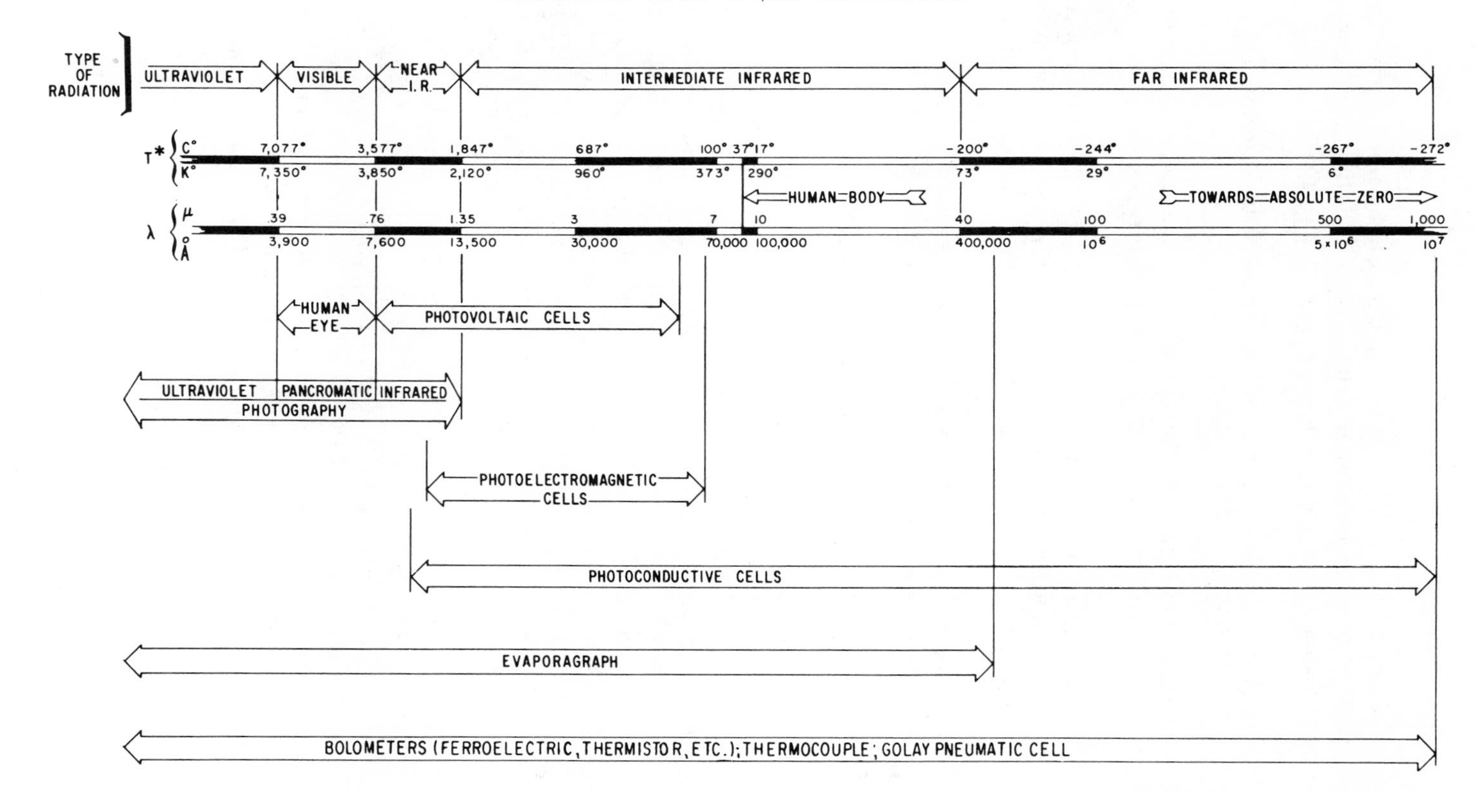

Figure 16. The detectors. T^* is the temperature corresponding to the peak of the spectral radiation band; for every T, radiation occurs (at lower levels) at all other wavelengths.

The statement is true whenever the detector output is processed through conventional electronics which cannot distinguish between noise and signal. Of course, all detector parameters are given in these terms.

The statement is false when special "noise cancellation" techniques are used. These techniques require the following:

1. The signal impinging on the detector must be of repetitive nature, for instance, pulses of constant characteristics repeated at a fixed frequency.
2. Integrating equipment must be used, which adds the output of the detector at the same repetitory rate as the pulses. In this way, each pulse adds its own increment to those of the pulses that already occurred. The noise instead, during this same summing process, builds up at a slower rate, following the r.m.s. rule, since it is made of random "spikes". Thus eventually the pulsed signal emerges above the noise level and can be displayed and measured.

At the time of this writing, only one piece of infrared test equipment, the Semiconductor Junction Analyzer, incorporates noise-canceling electronics – the instrument is described in Chapter 3. All other infrared test equipment does not resort to this uncommon technique. Therefore, we consider true this statement: "a detector can only detect signals whose magnitude is greater than the self-generated noise." Consequently, the noise level becomes the limiting factor on how small a signal can be detected. NEP stands for Noise Equivalent Power and represents the magnitude of a radiation level producing at the detector output a voltage rise equal to the rms voltage generated by the detector's noise. Those elements affecting the NEP value must be specified; for instance, the widely used notation NEP (500°K, 900, 1) indicates that the figure of merit is valid for a 500°K blackbody source, with the radiation chopped at 900 Hz and a reference bandwidth of 1 Hz. The detector area, when not mentioned, is 1 cm^2.

Evidently, the smaller the NEP figure the higher the sensitivity will be. The reciprocal of NEP, or 1/NEP is called D, or *detectivity*. Since most detectors exhibit an NEP figure that is directly proportional to the square root of the detector area, a detectivity figure independent of the detector area is preferably used. It is indicated with D^*, and is equal to D multiplied by the square root of the detector area:

$$D^* = D\sqrt{A_D}$$

The D^*, commonly called D-star, is expressed in $(cm)(Hz)^{1/2}/W$, while NEP is expressed in $W/Hz^{1/2}$.

Responsivity of a detector is defined as the measure of its ability to convert radiation power into output signal voltage. This measure, indicated as $\mathcal{R}$, is expressed as the ratio of detector rms output voltage per watt of the impinging radiation.

Responsivity varies according to the operating conditions of the detector, such as its area, temperature, time constant, and bias voltage. The temperature of the radiating blackbody also affects it, as it is implied by the expression relating it to D^*

$$\mathcal{R} = \frac{D^* V_N}{\sqrt{A_D \Delta f}}$$

where V_N is the rms voltage value of the noise in the Δf bandwidth and A_D is the area of the detector.

THERMAL DETECTORS

Operation of the thermal detectors is based on the measurement of a physical characteristic that varies with temperature. The main groups in this family are the following:

Thermometers (including thermocouples)
Pneumatic cells
Evaporating devices
Bolometers (superconductive, ferroelectric, pyroelectric, thermistor)

Since these devices are capable of absorbing infrared energy of any wavelength, their spectral response covers, theoretically at least, the whole infrared spectrum.

Also, since the secondary effects of the thermal variations of these devices are dependent on their thermal inertia, their time response is limited by their physical size and mass; thus seldom can it attain shorter values than 1/1000 sec.

Thermometers

"Thermometers, in effect, can only take their own temperature," say people, unkindly. Although we do not wish to be overly critical of thermometers, we must add that most of them are slow, sluggish, impractical, and difficult to read. This is why they are very seldom used for measuring infrared radiation, except sometimes to calibrate elements of measuring equipment.

The drawbacks mentioned above are typical of the so-called liquid-in-glass thermometers. They are also partly true for the thermocouples, with the exception that because their physical size can be much smaller, the time response is faster. Also, their output is an electrical signal easier to read and measure.

Contact Measurement

When these thermometers are used to read the temperature of an object by making contact with it, attention must be given to the mass ratio between the measured and the measuring bodies. Figure 17 illustrates the error introduced in the temperature measurement of an electrically heated resistance wire when using a contact thermocouple. In *A* is the visible picture of the wire, attached by one end to a binding post; in *B* is shown the oscilloscope trace representing the temperature of the wire at each point, as measured with a scanning infrared radiometer.

When a small thermocouple is brought to contact with the wire, as shown in *C*, the thermal drain on the wire is such that its temperature at that point drops from above 100°C down to approximately 75°C and this is the measurement given by the thermocouple. The oscilloscope trace shown in *D* points out this large error, which would go undetected when using conventional means of temperature measurement.

This illustration is a strong argument in support of the suggestion that contact measurement might not be the best means. When we care to take realistic thermal measurements, a noncontact technique such as reading the infrared radiation emitted by the target can at least guarantee that the measurement will not upset the thermal condition of the object whose temperature we want to measure.

Noncontact Measurement

Thermometers and thermocouples can be used to measure without physical contact the infrared radiation emitted by a target and to correlate these readings with its temperature. However, in view of the drawbacks mentioned earlier, other thermal detectors have been developed, having more desirable characteristics and whose sensitivity to thermal radiation is much higher; thus temperature readings can be taken with a finer degree of resolution and with more precision.

Among these devices, the best known are pneumatic cells, evaporating devices, and bolometers.

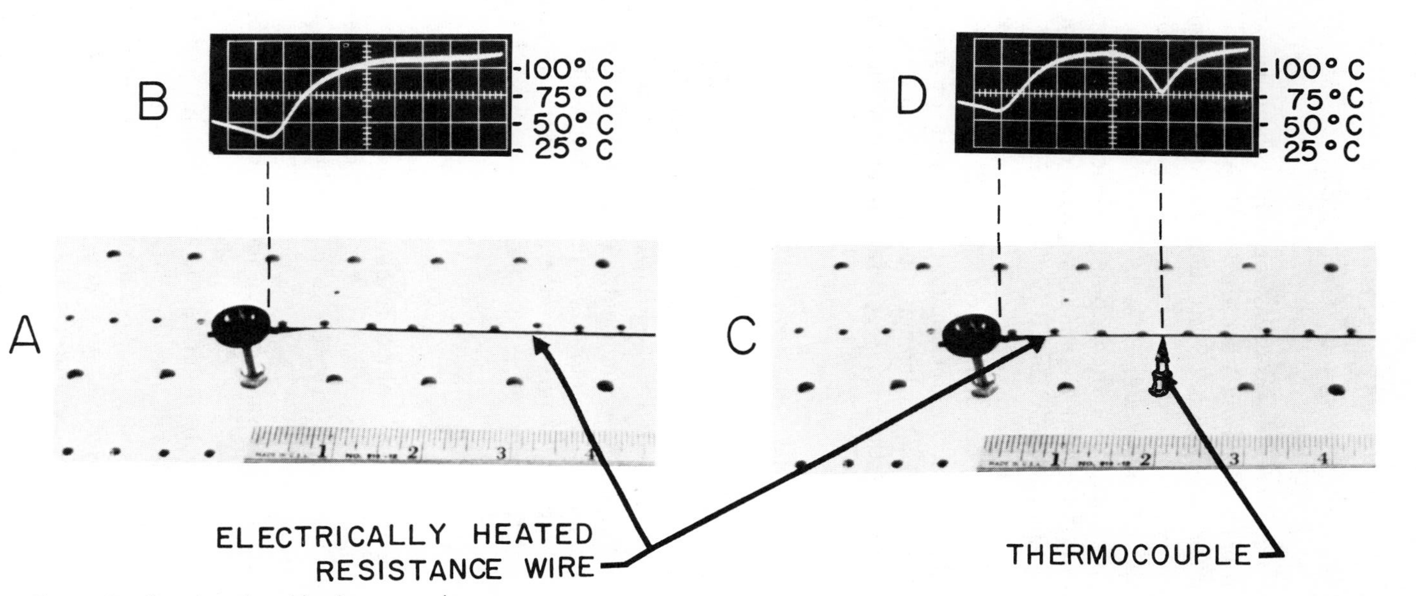

Figure 17. Error introduced by thermocouple.

Pneumatic Cells

The operation of these devices is based on the volume variation of a gas when its temperature changes because of infrared radiation absorption. These volume variations are translated into the physical motion of a mechanical element, whose displacement serves as the measuring element of the impinging radiation.

In the Golay pneumatic cell, the motion is effected by a mirror that is part of a simple optical system. In the "microphone detector" the motion is effected by a capacitor's armature, so that a capacitance variation follows as a consequence.

These pneumatic cells can achieve very high detectivity figures, but mechanical and thermal inertia limit their speed of response to frequencies below 100 Hz. Furthermore, because they are delicate and require skillful handling, their use is generally limited to the laboratory.

Evaporating Devices

The best known instrument in this class is the Evaporograph currently manufactured by Baird-Atomic, which works on the principle that infrared radiation falling on a thin layer of oil evaporates part of it to produce a thickness variation that is then detected according to the colors it reflects under illumination by white light. When coupled with an optical focusing system, and a photographic camera loaded with color film, this system turns into an image converter, as is described further in Chapter 3.

Bolometers

The word bolometer was coined from the greek *βολῶσ*, meaning "power." Their operation is based on the measurement of an electrical-characteristic variation induced by the heat absorbed by a temperature-dependent electrical element.

If the variation affects the value of capacitance, we have the "ferroelectric bolometer."

If the variation affects the value of the DC resistance, we have the metal bolometer, the thermistor (for "thermally sensitive resistor") bolometer, or the superconducting bolometer, according to whether the resistance change is positive, negative, or a step function which usually takes place at very low temperature.

In all cases, the bolometers are nonfrequency selective: at least theoretically they absorb whatever electromagnetic energy impinges upon them, regardless of wavelength. Consequently, their response is represented by a straight horizontal line covering the full spectral range of Figure 16.

The metal bolometers are usually in the form of a thin conductor, whose DC resistance increases with temperature according to a law of this form:

$$R = R_0[1 + \gamma(T - T_0)]$$

The value of γ, for several metals, is around 0.5%/C°, which is the limiting factor for the thermal resolution of the device. The speed of response is inversely proportional to the mass of the bolometer, and in direct proportion to thermal conduction towards the heat sink; it seldom reaches the 100-Hz frequency.

The thermistor bolometers are made of semiconductor material, and their operation is based on the negative resistance characteristic, according to the equation above, where γ is negative, often close to 4%/C°. This makes the thermal resolution of a thermistor bolometer approximately eight times higher than the resolution of a metal bolometer.

Figure 18 shows the current/voltage characteristic of a typical thermistor bolometer. Starting at low voltage and low current levels, the device shows a linear correlation, but as soon as its temperature rises because of the increasing power dissipation, the number of current carriers becoming available increases so that the DC resistance of the device decreases accordingly, and the linearity of the response disappears, until, at point X, the dynamic resistance reverses its sign and becomes negative. From this point on, the reaction is self-accelerating to destruc-

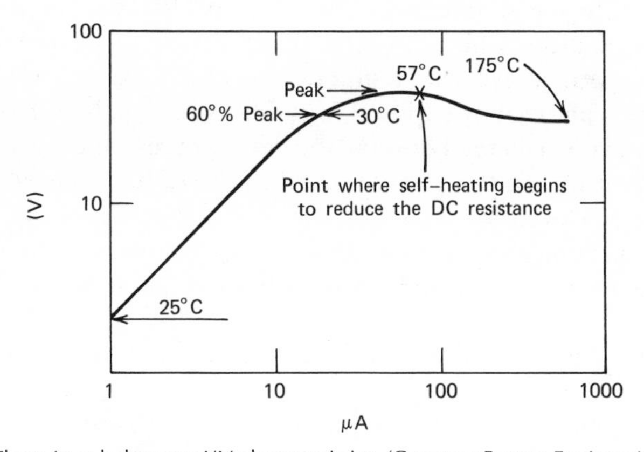

Figure 18. Thermistor bolometer I/V characteristics. (Courtesy Barnes Engineering Company.)

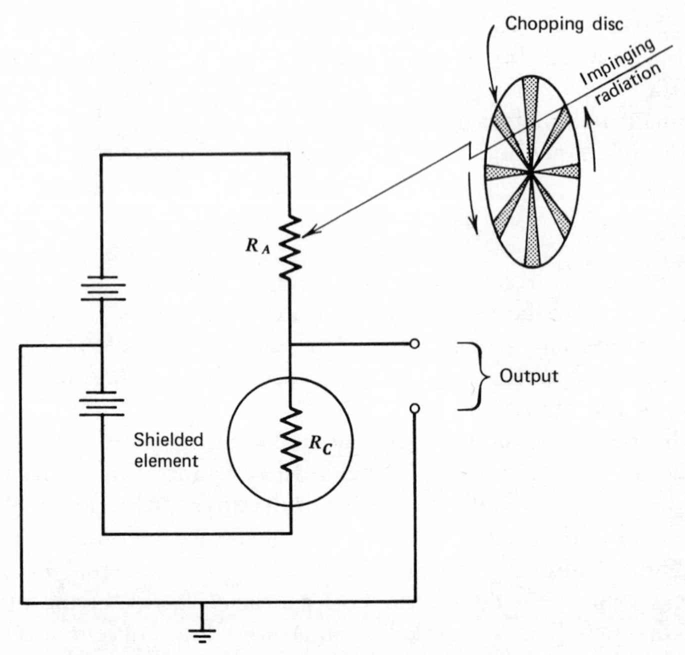

Figure 19. Typical thermistor bolometer circuit. (Courtesy Barnes Engineering Company.)

tion, unless the current value is limited by external means, such as a ballast resistor. Figure 19 shows a thermistor bolometer circuit generally used.

The R_c is a compensating thermistor, and R_A the active thermistor bolometer; both are identical elements, except that R_A is exposed to the impinging radiation, which raises its temperature in proportion to the power absorbed, above the temperature of R_c. When the chopper operates, R_A is alternately exposed to radiation or shielded from it; thus its temperature oscillates between the level corresponding to the ambient temperature, and the level related to the radiant power absorbed. When these two levels are different, they cause variations in resistance of R_A and an AC voltage appears at the output terminals synchronous with the chopping frequency, and amplitude-modulated by the radiation power.

To enhance absorption, the receiving surface of R_A is painted black; to increase its frequency response, its mass is made as small as possible (deposited semiconductor layers are just a few microns thick), while its bonding to the heat sink is made as good as possible. However, this is detrimental to the thermal resolution of the device, because the resistance variation is related to the temperature variation of the bolometer, and a perfect connection to an infinite heat sink would minimize both temperature variations and output signal. A compromise must be reached in the heat-sink design, since response speed and sensitivity are inversely correlated, as shown in Figure 20.

Another compromise must be made for the operating characteristics: in view of the danger of thermal runaway, thermistors must operate below the inversion point X of Figure 18. The most convenient point is approximately at 60% of the peak value, where a relatively small temperature variation induces a large resistance change. The bias voltage value, and the bolometer's temperature determine the operating point on the characteristic curve. A rule-of-thumb generally valid states that bias voltage should be reduced 33% for every 10°C of additional temperature rise.

However, the large difference between the magnitude of the bias voltage and the minute voltage variations due to the bolometer's temperature changes is one of the problems inherent to this type of thermal detectors.

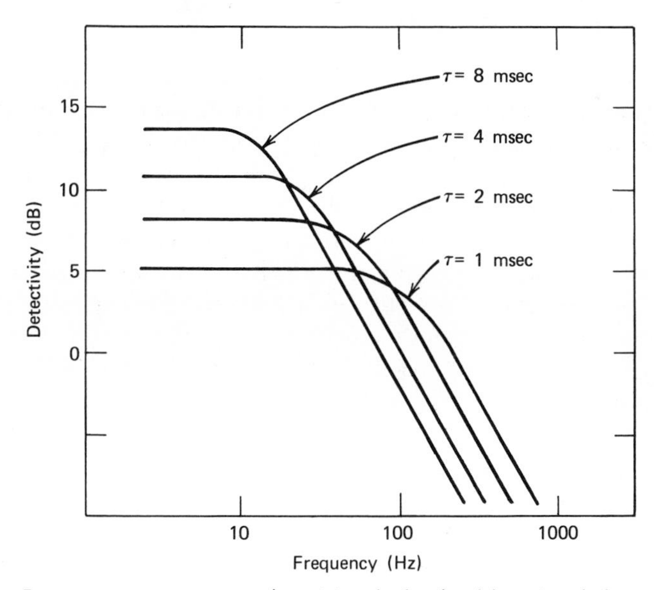

Figure 20. Frequency response versus detectivity of a family of thermistor bolometers. (Courtesy Barnes Engineering Company.)

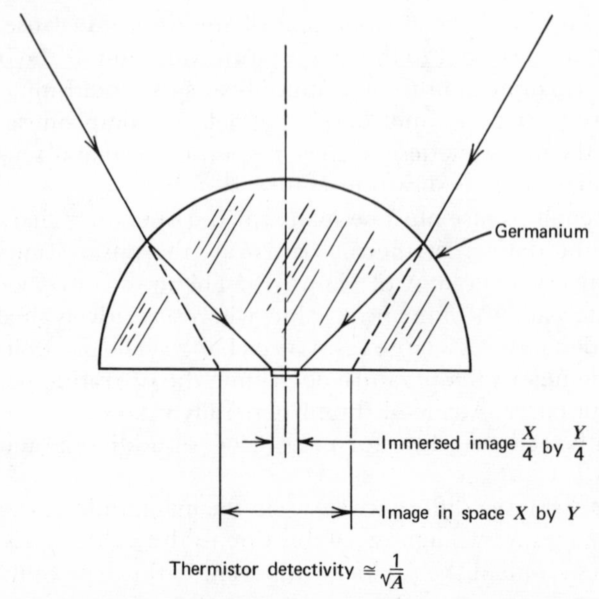

Figure 21. Image reduction with germanium immersion. (Courtesy Barnes Engineering Company.)

The Immersed Thermistor Bolometer

We have seen that the detectivity varies inversely to the square root of the detector area. Thus a reduction of detector area, while maintaining full collection of radiation, will improve performance. This can be achieved by "optical immersion" of the detector, as shown in Figure 21. The principle involves the use of a hemispherical lens with the detector cemented—with a cement of appropriate refractive index to the back of the lens. Under such conditions, the effective reduction of detector area equals n^2, where n is the index of refraction of the immersion lens. Ordinarily, Germanium is the lens material used because of its infrared transmission characteristic and its high index of refraction. Effective area reduction in the order of 16 times is often possible.

The Superconducting Bolometer

The physical principle upon which this class of bolometers is based is the fact that some metals and compounds exhibit very large thermal resistance variations when they are cooled to the border between conductance and superconductance.

At this temperature (that for Niobium nitride is 15°K) the resistance variation becomes sharply nonlinear with relation to the impinging thermal energy; thus the bolometer acquires extremely high sensitivity.

Because today's cryogenic capabilities enable us to attain very low temperatures, the use of superconducting bolometers is possible, although the problem of precisely holding the transition-state temperature makes it difficult to obtain precise, repeatable measurements.

The Ferroelectric Bolometer

Certain dielectric compounds, called ferroelectric materials, exhibit a spontaneous polarization, whose magnitude varies with temperature. When these materials are subject to heating, their crystalline structure undergoes a rapid change at the so-called *Curie temperature.* At this point, their polarization disappears, and their dielectric coefficient varies sharply in value, as shown in Figure 22 for barium titanate ceramic.

It is obvious that a capacitor utilizing a ferroelectric dielectric will undergo a large change in capacitance when its temperature varies around the Curie temperature of its dielectric. A number of ferroelectric compounds are known, whose Curie temperatures cover the range from −260 to 570°C; thus ferroelectric bolometers can be built for a wide range of temperatures.

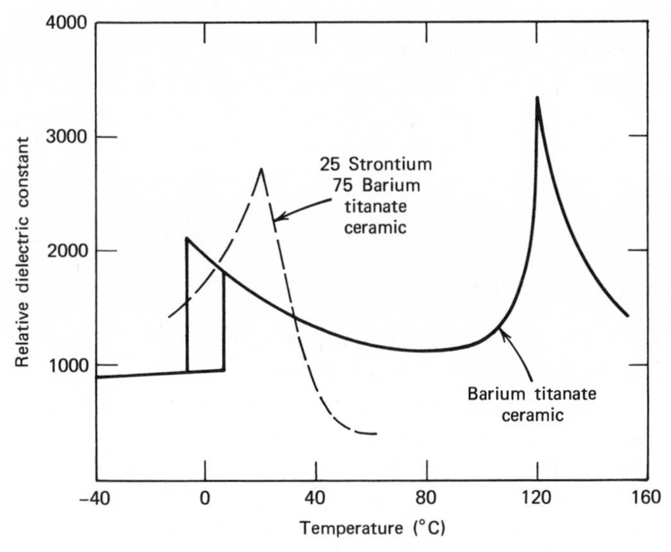

Figure 22. Titanate ferroelectric characteristics. (Courtesy Huggins Laboratories.)

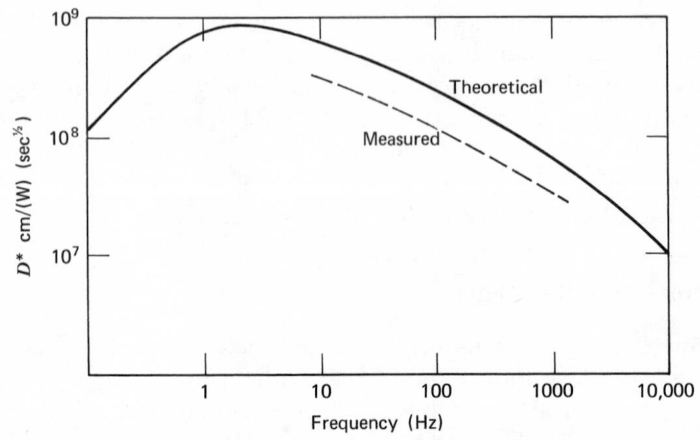

Figure 23. Frequency response of ferroelectric bolometer. (Courtesy Huggins Laboratories.)

Careful choice of the Curie temperature point allows one to obtain ferroelectric bolometers exhibiting up to 5.5% variation of their dielectric constant per degree Celsius as shown in the dashed line of Figure 22, in the temperature region between 25 and 50°C. This is more than the 4% resistance variation offered by the thermistor bolometers. Furthermore, because the ferroelectric bolometer is essentially a capacitor, it is immune from the danger of thermal runaway, which is always present in the operation of thermistor bolometers, and which forces their use at a point located below maximum efficiency.

The frequency response of a ferroelectric bolometer is an inverse function of its thermal mass. Figure 23 shows how D^* varies with frequency for a ferroelectric bolometer Mod. 4001 developed and marketed by Huggins Lab., a company later absorbed by Barnes Engineering Company.

The Pyroelectric Detector

Again, in this case, ferroelectric material is used, specifically triglycine sulfate (TGS) crystals. A spontaneous polarization (electric charge concentration) is exhibited by this material, and this phenomenon is temperature dependent. In practice, a very small capacitor is fabricated, which has the TGS crystal as the dielectric sandwiched between two metal plates. As infrared radiant energy is absorbed by the dielectric, a voltage appears at the two poles of the capacitor, proportional to the magnitude of the impinging radiation.

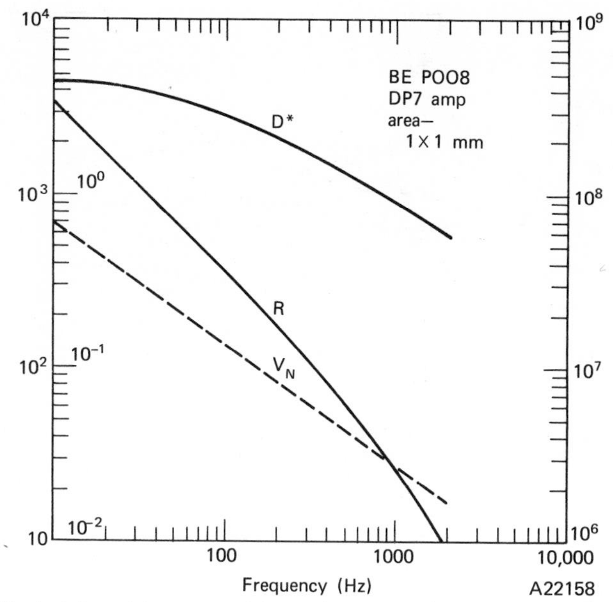

Figure 24. Typical detectivity, responsivity, and noise as a function of frequency for TGS detectors at 23°C. (Courtesy Barnes Engineering Company.)

A major advantage of this detector is that it does not need a bias voltage, with consequent absence of self-generated low-frequency noise which otherwise would be unavoidable.

Typical detectivity, responsivity, and noise characteristics of this detector produced by Barnes Engineering Company are shown in Figure 24. The fact that the curves of D^* and of V_n (noise) run practically parallel through all their length indicates a constant S/N (signal-to-noise) ratio, which is a very desirable characteristic.

CHEMICAL DETECTORS

Photographic plates are the best-known chemical detectors of infrared radiation and Figure 25 shows the spectral response of Kodak type IR 135 film. A Polaroid infrared film is also available and Figure 26 illustrates its spectral response. The ASA sensitivity figures are different, but the frequency coverage of the two films is similar, and outside the visible region their response is strictly limited to the very near infrared.

As a consequence, photographic plates and films are uncapable of de-

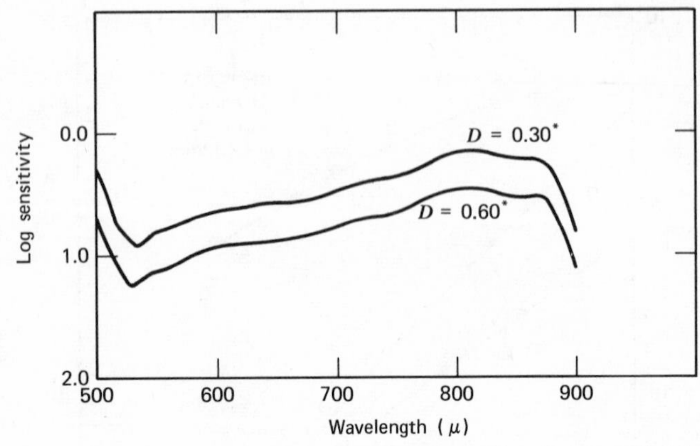

Figure 25. Spectral sensitivity curves of Kodak infrared 135 film. (Courtesy Kodak Company.)

tecting infrared radiation emitted by targets at temperatures in the vicinity of 300°K; thus they are useless for evaluating most of today's hardware and electronics.

Theoretically, it could be possible to produce photographic film for the intermediate and even for the far infrared regions of the spectrum, but their use would be only possible under constant refrigeration, to avoid self-exposure from the infrared radiation ever present at ambient temperature, even if the film should never be taken out of its sealed container.

However, infrared photography makes lovely pictures. Figure 27 is an interesting example of a picture taken in full sunlight; the silvery

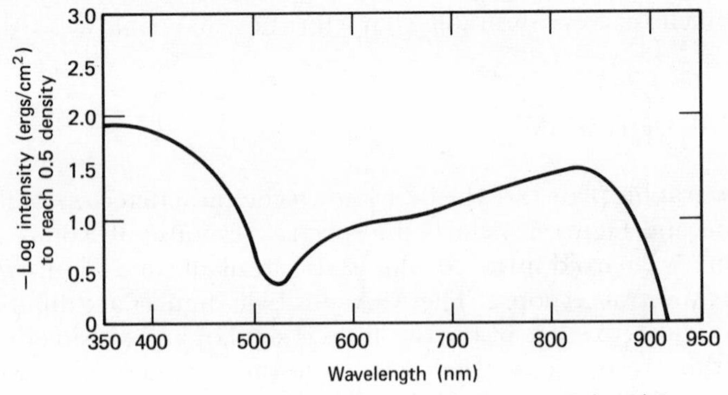

Figure 26. Polaroid film sensitive in the near—infrared. (Courtesy Polaroid Company.)

Figure 27. Infrared photograph of landscape. (Courtesy Eastman Kodak Company.)

appearance of the foliage is due to reflected sunlight, whose infrared content is conspicuous. The dark color of the sky is caused by the very little scattering of the infrared radiation contained by sunlight during its passage through the atmosphere, and contributes to the moonlight effect of the picture. Green leaves also emit infrared radiation as part of the photosynthesis process, but its intensity is so low as to be undetectable in an infrared picture.

QUANTUM DETECTORS

Quantum detectors, or photon detectors, are semiconductor devices whose electrical characteristics are a function of the number of electrical charges made available by the splitting of electron-hole pairs produced by the photons impinging on the semiconductor material of which the detector is made.

To separate an electron from a "hole," the photons must have high enough energy to accelerate the electrons to such a speed that will rip them off their orbits around the atom's nucleus, thus turning them into "free electrons," which are able to move about, following the prevalent electromagnetic field.

This process is shown in Figure 28. When the electron-hole pairs are bound together in the atomic structure, they are said to be located in the valence band. In this configuration their energy content is at a certain level, as indicated in the illustration. When a photon P^1 splits the electron-hole pair, its energy is absorbed by the electron, which raises to the higher energy level called "carrier band" where it remains in a free state until the end of its "carrier lifetime" when it recombines with a hole, losing in the process the excess energy that was keeping it into the carrier band. The energy so liberated is in the form of a photon P^2, and the radiation composed of these photons is called recombination radiation because of the physical process by which it was generated.

The energy region located between the valence band and the carrier band is called the "forbidden gap" of the semiconductor to indicate that neither electrons nor holes can remain in this region as long as the semiconductor material is absolutely pure. This explains why every quantum detector has a long wavelength limit beyond which the incoming radiation produces no effect: the energy content of the longer wavelength

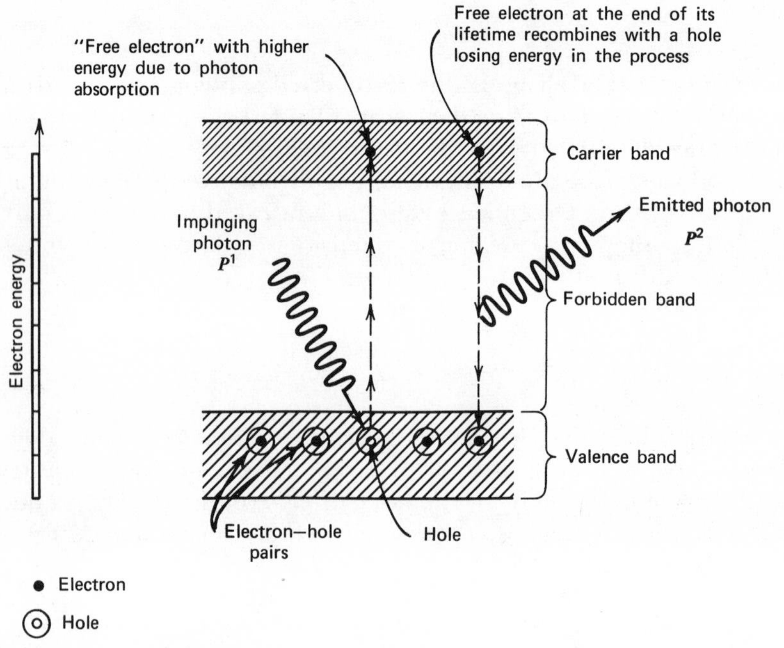

Figure 28. Energy structure in semiconductors.

photons is below the threshold limit set by the forbidden gap. But when the photon energy is above this value, the liberated electrons and holes can do the following:

1. Become available as current carriers, thus decreasing the DC resistance of the semiconductor.
2. Accumulate at the opposite sides of a self-generated potential barrier, thus developing a voltage difference across it.
3. Move in opposite directions because of an external magnetic field, thus again generating a voltage across the semiconductor.

In case 1, we deal with a conductivity effect, and the detectors of this class are called photoconductive. In case 2, we have the photovoltaic detectors, while in case 3, we have the photoelectromagnetic detectors.

Figure 16 shows the spectral region where every one of these detector groups is operative, and Figure 29 shows the detectivity response curves of a number of the most widely used infrared detectors, with the indication of their operating mode.

From this chart, it can also be seen that contrary to the flat response of thermal detectors, the quantum detector response is nonlinear. This nonlinearity, with a peak response, results from two factors. The first factor is simply that for unit watt input, the number of photons per second will increase with increasing wavelength. The second factor is that the quantum efficiency of the detector steadily decreases for photon energies approaching the bandgap energy of the material. Thus with increasing wavelength, these two factors counteract each other with the result that the overall response of the detector passes through a maximum with increasing wavelength of incident radiation, and subsequently drops rapidly toward zero as the reduced energy of the impinging photons falls below the level necessary to "liberate" electrons from their orbits around the nuclei.

The time response of the quantum detectors is on the order of microseconds or less, which is about three orders of magnitude faster than the thermal detectors, and of course, it is independent from their physical mass, since no thermal effect is involved.

Table 1 lists the most important characteristics of the 24 quantum detectors of Figure 29. We can see how for every region of the spectrum one or more detectors exist, capable of performing at peak efficiency in the desired spectral range. Also apparent is the need for cooling at cryogenic temperatures the detectors having their response at the longer wavelengths.

Of special interest for several of the applications described in this book is the Hg-Cd-Te detector listed as number 20 in Table 1 and Figure 29.

Table 1 Characteristics of Available Infrared-Region Photon Detectors[a]

	1	2	3	4	5	6	7	8	9	10	11	12
Detector Material	Si Silicon	Si Silicon	GaAs Gallium Arsenide	Ge Germanium	PbS Lead Sulfide	PbS Lead Sulfide	PbS Lead Sulfide	InAs Indium Arsenide	InAs Indium Arsenide	InAs Indium Arsenide	PbSe Lead Selenide	PbSe Lead Selenide
Operating Mode	(pv)	(pc)	(pv)	(pv)	(pc)	(pc)	(pc)	(pv)	(pv)	(pv)	(pc)	(pc)
Typical peak D^* (cm $Bz^{1/2}$/W) at 1000 Hz modulation frequency	2×10^{12}	5×10^{11}	8×10^{11}	5×10^{10}	8×10^{10}	4×10^{11}	2×10^{11}	6×10^{9}	2×10^{11}	4×10^{11}	2×10^{9}	2×10^{10}
(Wavelength, μ)	0.9	0.9	0.85	1.5	2.5	2.7	3.1	3.5	3.2	3.1	3.4	4.1
(Field of view, degrees)						60	60		60	60		60
(Background temperature, °K)						295	295		295	295		295
Best measured peak D^* (cm $Hz^{1/2}$/W) (conditions as above)	1×10^{13}	1×10^{12}			1.5×10^{11}	7×10^{11}	4×10^{11}	1×10^{10}	3.5×10^{11}	7×10^{11}	2×10^{10}	5×10^{10}
Spectral range exhibiting greater than 50% relative response (μ)	0.6 to 1.0	0.8 to 1.06	0.6 to 0.95	0.9 to 1.7	1.2 to 2.8	1.3 to 3.2	1.4 to 3.8	2.0 to 3.8	2.5 to 3.4	1.8 to 3.3	0.5 to 4.2	2.0 to 5.3
Normal operating temperature (°K)	295	295	295	295	295	195	77	295	195	77	295	195
Operating temperature limits (°K); 50% peak D^* degradation points	–,320	–,350			–,310	130.250	–,160	–,320	–,210	–,180	–,310	–,230
Typical time constant (sec)	5×10^{-7}	5×10^{-6}	1×10^{-6}	1×10^{-7}	3×10^{-4}	5×10^{-3}	3×10^{-3}	$<1 \times 10^{-6}$	$<1 \times 10^{-6}$	5×10^{-7}	2×10^{-6}	3×10^{-5}
Nominal resistance (oΩ)	1×10^{6}	1×10^{6}	1×10^{6}	2×10^{5}	1×10^{6}	1×10^{6}	2×10^{6}	3×10^{1}	5×10^{4}	5×10^{5}	2×10^{6}	5×10^{6}
Area configuration												
Single detectors												
Size range – minimum to maximum (in.)	0.004 to 0.5	0.004 to 0.7	0.004 to 0.060	0.004 to 0.5	0.001 to 1.0	0.001 to 1.0	0.001 to 1.0	0.004 to 0.1	0.004 to 0.1	0.004 to 0.1	0.003 to 0.5	0.003 to 0.5
Shape (round, square, or rectangular)	any	□▭	○□	any	□▭	□▭	□▭	○	○	○	□▭	□▭
Typical package	TO-5/18	Flat Mount	TO 18	TO 5/18, BNC	Flat Mount	Glass Dewar	Glass Dewar	TO 5/18	Glass Dewar	Glass Dewar	Flat Mount	Glass Dewar
Detector arrays												
Minimum size per detector (in.)	0.004	0.004			0.001	0.001	0.001	0.003	0.003	0.003	0.003	0.003
Minimum size per space (in.)	0.002	0.001			0.001	0.001	0.001	0.002	0.002	0.002	0.002	0.002
Dimensions – see code	0.002 in.	0.020 in.			0.001 in.	0.001 in.	0.001 in.	0.002 in.	0.002 in.	0.002 in.	0.001 in.	0.001 in.

[a] pc = Photoconductive mode, pv = photovoltic mode, and pem = photoelectromagnetic mode.

Table 1 (Continued)

	13	14	15	16	17	18	19	20	21	22	23	24
Detector Material	PbSe Lead Selenide	InSb Indium Antimonide	InSb Indium Antimonide	InSb Indium Antimonide	InSb Indium Antimonide	GE:Au Gold-Doped Germanium	Ge:Hg Mercury-Doped Germanium	(Hg-Cd)Te Mercury Cadmium Telluride	Ge:Cd Cadmium-Doped Germanium	Si:Sb Antimony-Doped Silicon	Ge:Cu Copper-Doped Germanium	Ge:Zn Zinc-Doped Germanium
Operating Mode	(pc)	(pem)	(pc)	(pc)	(pv)	(pc)	(pc)	(pv)	(pc)	(pc)	(pc)	(pc)
Typical peak D^* (cm $Bz^{1/2}$/W) at 1000 Hz modulation frequency	3×10^{10}	1×10^{8}	2×10^{8}	8×10^{10}	1×10^{11}	1×10^{10}	2×10^{10}	5×10^{9}	2×10^{10}	1×10^{10}	3×10^{10}	2.5×10^{10}
(Wavelength, μ)	4.8	6.0	6.8	5.3	5.1	5.0	10.5	10.6	16	20	23	36
(Field of view, degrees)	60			60	60	60	60	60	60	60	60	60
(Background temperature, °K)	295			295	295	295	295	295	295	295	295	295
Best measured peak D^* (cm $Hz^{1/2}$/W) (conditions as above)	5×10^{10}	3×10^{8}		1×10^{11}	2×10^{11}	2×10^{10}	5×10^{10}	2×10^{10}	4×10^{10}	2×10^{10}	5×10^{10}	5×10^{10}
Spectral range exhibiting greater than 50% relative response (μ)	2.7 to 6.3	2.0 to 7.0	3.6 to 7.3	3.0 to 5.4	2.0 to 5.4	3.0 to 7.5	6 to 14	9 to 13	11 to 20	12 to 23	15 to 27	20 to 40
Normal operating temperature (°K)	77	295	295	77	77	60	27	77	4.2	4.2	4.2	4.2
Operating temperature limits (°K); 50% peak D^* degradation points	–,160			–,95	–,105	–,80	–,40	–,100	–,26	–,10	–,20	–,6
Typical time constant (sec)	4×10^{-5}	2×10^{-7}	1×10^{-6}	6×10^{-6}	$<1 \times 10^{-6}$	1×10^{-7}	2×10^{-7}	$<1 \times 10^{-8}$	1×10^{-7}	1×10^{-7}	5×10^{-7}	2×10^{-8}
Nominal resistance (oΩ)	5×10^{6}	1×10^{1}	2×10^{1}	1×10^{4}	1×10^{5}	1×10^{5}	1×10^{5}	2.5×10^{1}	1×10^{5}	7×10^{6}	1×10^{5}	2.5×10^{5}
Area configuration												
Single detectors												
Size range – minimum to maximum (in.)	0.003 to 0.5	0.015 to 0.040	0.040 to 0.1	0.003 to 0.1	0.003 to 0.1	0.003 to 0.1	0.003 to 0.1	0.020 to 0.080	0.003 to 0.1	0.004 to 0.1	0.003 to 0.1	0.003 to 0.1
Shape (round, square, or rectangular)	□▭	□	□▭	□▭	□○	□▭	□▭	□▭		□▭	□▭	□▭
Typical package	Glass Dewar	Metal Cont.	Flat Mount	Glass Dewar	Glass Dewar	Glass Dewar	Metal Dewar	Glass or Metal Dewar	Metal Dewar	Metal Dewar	Metal Dewar	Metal Dewar
Detector arrays								Developmental				
Minimum size per detector (in.)	0.003			0.003	0.003	0.003	0.003		0.003	0.004	0.003	0.003
Minimum size per space (in.)	0.002			0.002	0.002	0.002	0.002		0.002	0.004	0.002	0.002
Dimensions – see code	0.001 in.			0.001 in.	0.002 in.	0.002 in.	0.002 in.		0.002 in.	0.004 in.	0.002 in.	0.002 in.

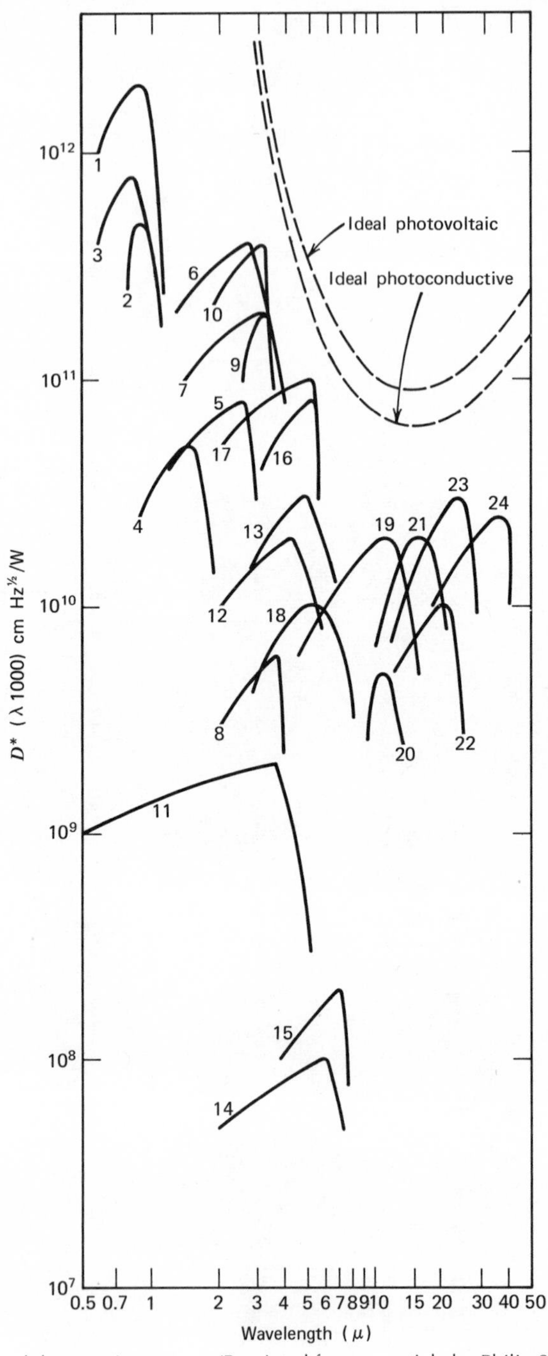

Figure 29. Infrared detectors' response. (Reprinted from an article by Philip Shapiro, in *Electronics*, Jan. 20, 1969. Copyright McGraw-Hill, Inc., 1969.)

Although detectors of this material can be produced with a spectral response peak at virtually any desired wavelength, according to the chemical mix of Hg-Te and Cd-Te, the unit shown in the chart of Figure 29 has the following characteristics:

1. Response speed: as low as 3 nsec
2. Spectral response: peak in the 10-μ region
3. Cooling requirements: 77°K

The device is made of a HgTe-CdTe single crystal, and can be used either in the photoconductive or in the photovoltaic mode of operation. In the United States, it is made by Honeywell Radiation Center, Boston, Mass., and by Santa Barbara Research Center, Goleta, Calif. Its performance specs are classified, since it is used in aircraft and satellite reconnaissance systems.

Unclassified detectors of the same material are produced in England

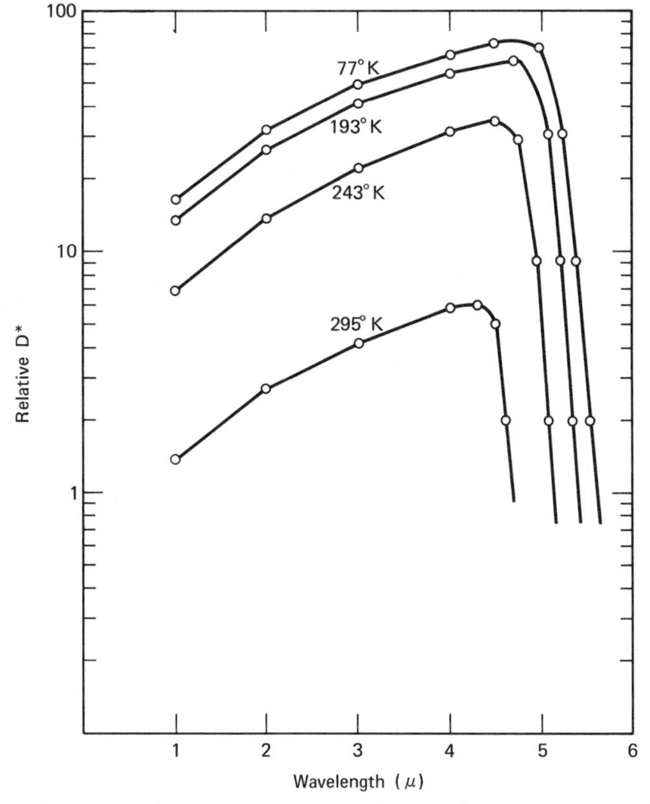

Figure 30. Relative spectral response of mercury cadmium telluride 3–5 μ infrared detectors. (Courtesy Mullard Ltd.)

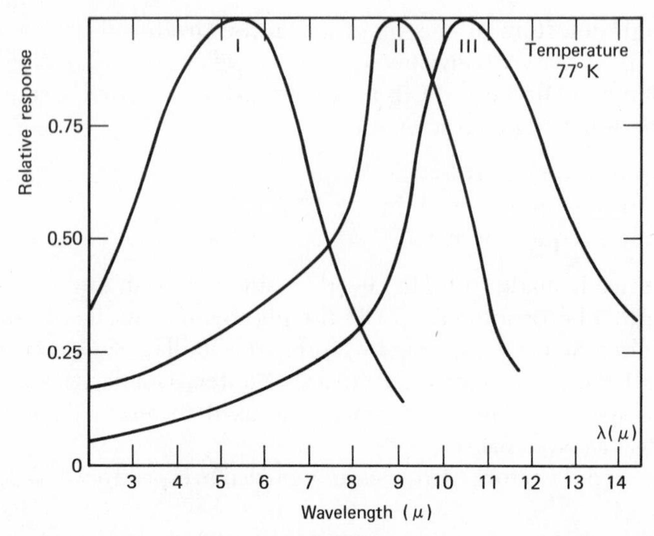

Figure 31. Relative response curves of Hg-Cd-Te detectors. Curves I, II, and III are functions of different Cd/Hg ratios. (Courtesy S.A.T., Paris, France.)

by Mullard, Ltd., and they are made commercially available in two groups. One group includes detectors whose peak spectral response is in the 3 to 5 μ, while the other group has its peak sensitivity in the 8 to 10 μ region. The detectors of group one can be used at different levels of cooling, and even without cooling, as shown in Figure 30. However, the latter condition carries a D^* loss of one magnitude when compared with operation at liquid nitrogen temperature. A compromise solution is the use of a multistage thermoelectric cooling device. Under these conditions, the Hg-Cd-Te detector exhibits a higher D^* than the popular In-Sb detector cooled at 77°K.

Similar detectors made of the same material are also produced in Paris, France by S.A.T., and Figure 31 shows the spectral response of three detectors of this type, corresponding to three different ratios of Cadmium and Mercury in the composition of the crystals.

From the data above, it appears that the HgCdTe detector is exceptionally well suited to the measurement of infrared radiation emitted by targets whose temperature is in the vicinity of 25°C, which is generally known as the "ambient temperature."

Photodiodes

A special group of photodetectors consists of the photodiodes. These units, as their name implies, are composed of two regions of opposite polarity separated by a junction. They operate in the photovoltaic mode and their spectral response matches that of the photoconductive detectors made of the same intrinsic semiconductor material at the corresponding temperature.

For instance, in Figure 29 the Germanium detector spectrum marked as number 4, peaks at 1.5 μ: this is the spectral response of intrinsic Ge, and Ge photodiodes exhibit exactly the same response.

In conventional photodiodes, one electron-hole pair is separated by an impinging photon having at least the energy of the forbidden gap, and the electrical charges so liberated are made available at the poles of the diode as a voltage potential.

Figure 32 shows the spectral response of Silicon and Germanium photo diodes.[2] The first has its peak in the visible region, while the latter peaks in the near infrared.

Recently, "avalanche" photodiodes have been produced.[3] These units could be described as photodiodes with a built-in photomultiplier. This is achieved by introducing a high electrical field in the area of the *p-n* junction. This electrical field accelerates the electrons liberated by impinging photons when they initially split electron-hole pairs. The electrons so accelerated acquire higher energy, and in turn can split more electron-hole pairs into free carriers. These in turn are accelerated by the electrical field and in moving along, they split further electron-hole pairs, and so on, until they reach the collection regions at the ends of the junction.

The result is that for every single photon impinging onto the diode, as

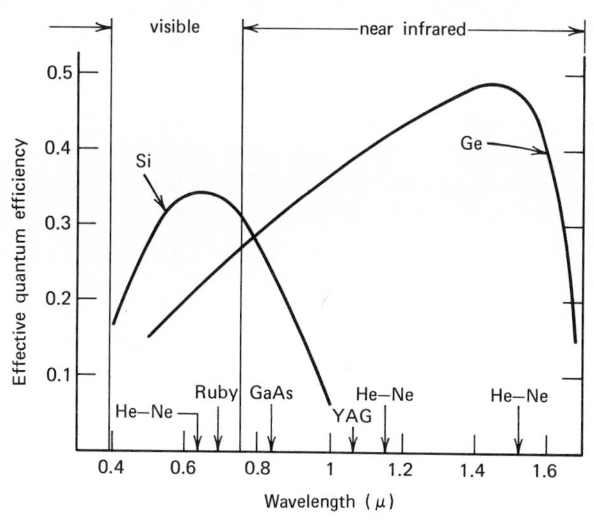

Figure 32. Photodiode response curves. (Courtesy Melchior and Lynch, Bell Labs.)

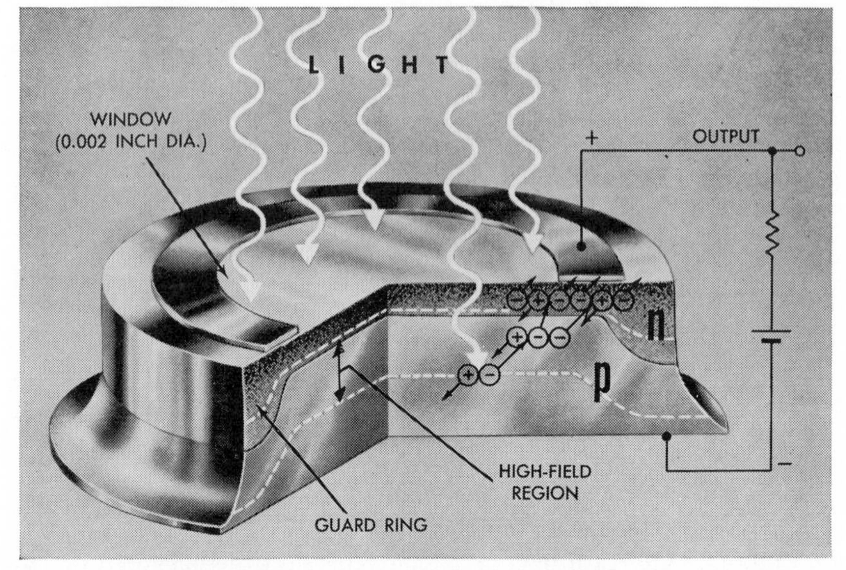

Figure 33. Germanium avalanche photodiode. (Courtesy Melchior and Lynch, Bell Labs.)

many as 250 electron-hole pairs can be made available instead of the single electron-hole pair generated in conventional photodiodes.

The basic structure of an avalanche photodiode is shown in Figure 33,[2] along with a schematic indication of its electrical operation.

COOLING DEVICES

Many of the applications described in this book require detector sensitivity in the spectral region comprised between 5 and 15 μ.

Quantum detectors able to meet this requirement must be cooled to temperatures approaching absolute zero. This is due to the fact that the width of the forbidden energy gap must be small enough to allow the impinging low-energy photons to cause electron transitions from the valence to the carrier band. Therefore, the overall thermal energy content of the detector must be brought low enough to avoid self-generation of these transitions because of the detector's own temperature. Otherwise, self-saturation would occur, and sensitivity to low-energy radiation impinging from outside would be lost.

For instance, the Ge:Au detector shown in No. 18 of Fig. 29 needs cooling to 77°K for a D^* peak at 5 μ while Ge:Cu (No. 23 of Fig. 29) has to be cooled to 4°K for a D^* peak at 22 μ.

To reach these low temperatures, extensive use is made of various liquified gases, utilized at their boiling point. The most used among these gases are nitrogen, neon, hydrogen, and helium, although several others cover the same thermal area, as shown by Table 2.

To cool the detectors with liquid gases, Dewar flasks are used. These are similar to thermos bottles, with the difference that they are often made of metal, and that they can be of a two-stage design, that is one thermos bottle is contained inside a larger thermos bottle. The two-step thermal gradient thus obtained when the inside bottle is filled with a colder liquid while the enveloping bottle holds a less cold liquid ensures better insulation of the inner area and reduces its thermal losses. Longer operation is thus assured since the gas evaporation takes place at a slower rate. Figure 34*a* shows a double Dewar flask that allows an uninterrupted working period of about 8 hours when the inside container is filled with liquid helium while the outside chamber holds liquid nitrogen. The infrared detector, as shown in the crossection sketch of *b*, is mounted on the so-called "cold finger" protruding from the inner container, which is practically at liquid Helium temperature, and operates in a vacuum. The impinging radiation enters through a hermetically sealed window located just in front of the detector.

A variation of the double-Dewar concept is the gas-shielded unit, shown in *c*. In this case only one liquid gas is used, and the gas developing from its boiling condition is made to circulate through the enveloping chamber of the double Dewar, before being discharged to the outside atmosphere. Thermal insulation of the liquid contained in the inner chamber is thus achieved by its own gas, eliminating the complications connected with the use of two different liquefied gases.

Another method of cooling detectors to temperatures as low as 30°K is to use one of the many cryogenic refrigerators designed for infrared

Table 2 Boiling Points of Some Gases Used in Cryogenic Systems

GAS	°K
Helium	4.4
Hydrogen	20.5
Neon	27.1
Xenon	65.9
Nitrogen	77.2
Argon	87.3
Oxygen	90.0
Krypton	120.7

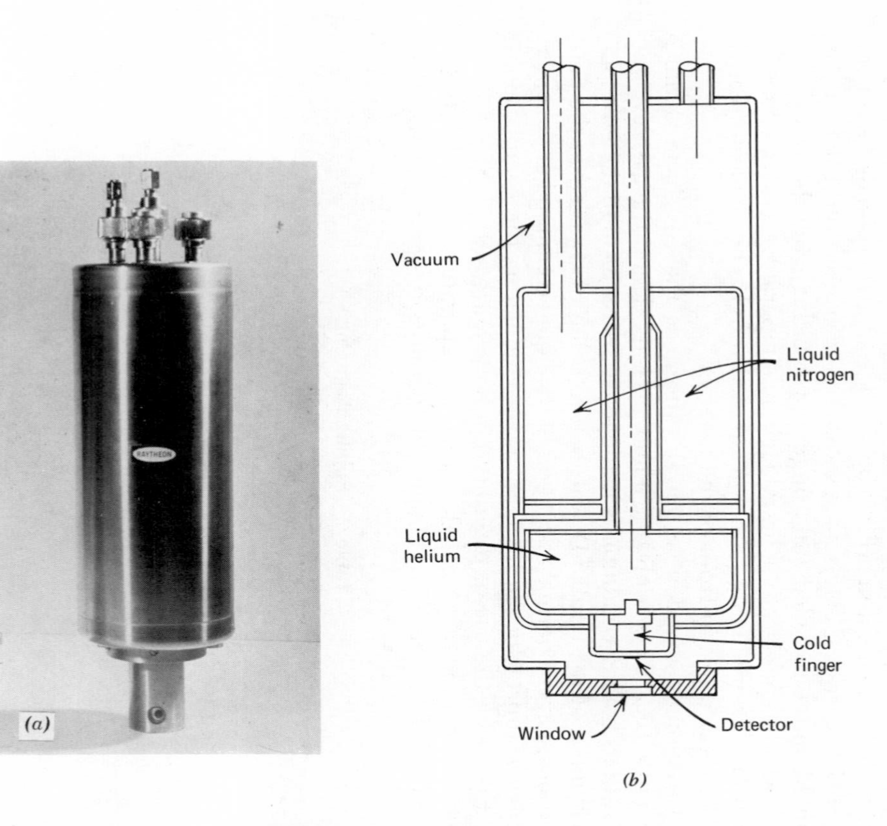

Figure 34. Double Dewar Flasks. (Courtesy Raytheon Company.)

cooling. There are several closed-cycle coolers using Joule-Thomson expansion valves, or expansion engines, to reach low temperatures; but one recent development that has done much to make detector refrigeration more practical is the availability of coolers based on the Stirling cycle. Figure 35 shows the four phases of a Stirling cycle refrigerator: in phase I the compressor piston moves up, while the expander piston remains stationary; during this phase isothermal compression of the gas takes place, since the heat developed by the compression is eliminated by the cooling fins of the cylinder. In phase II, the compressor piston moves up while the expander piston moves down at the same speed. In this way, the compressed gas is transferred from one cylinder to the other, at constant pressure and volume. In phase III, the compressor piston remains stationary at the top of its chamber, while the expander piston moves down, decreasing the pressure of the gas and drawing heat from the head of the compressor. Phase IV is the reciprocal of phase II: the gas is transferred back to the compressor cylinder at constant volume and pressure.

Cryogenic generators are produced by several companies in the United States. Figure 36 shows the Mark 7 model, made by Malaker Corporation. It is a closed-cycle cooler, not requiring additional liquified gases or make-up gas, that can reach temperatures as low as 30°K, which is the optimum temperature for Ge:Hg detectors. This unit is very com-

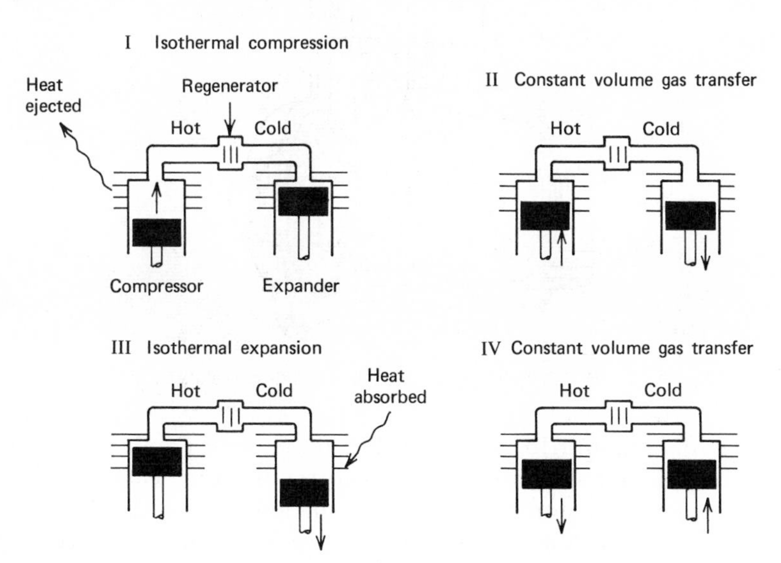

Figure 35. Modified Stirling cycle—basic system used in unit to obtain cryogenic temperatures. (Courtesy Malaker Corp.)

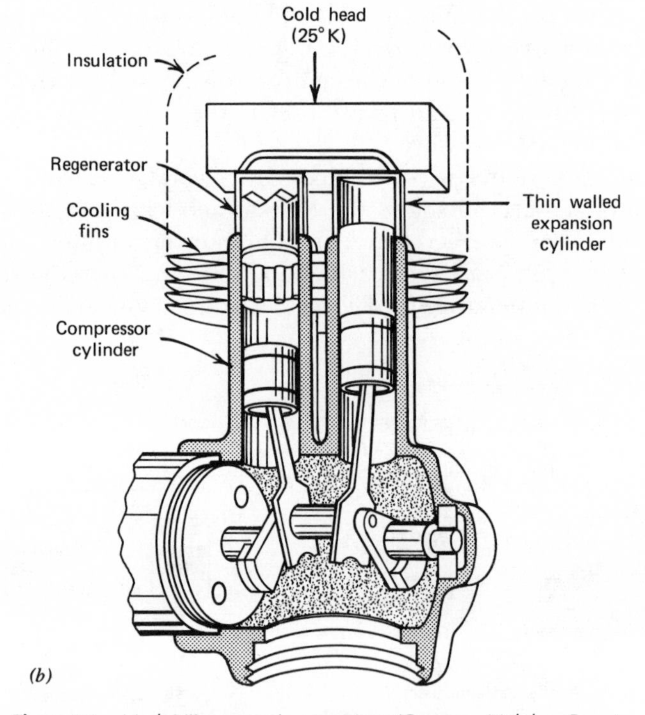

Figure 36. Mark VII cryogenic generator. (Courtesy Malaker Corp.)

pact (12 in. long and 15 lb in weight) can operate in any attitude, and either forced air or water can be used to remove the heat of compression. The gas used for its operation is helium. Figure 37 shows two curves, correlating time versus temperature, and time versus refrigerative power for this model. Besides single-stage units such as this, two-stage coolers are also available for reaching lower temperatures or for removal of larger quantities of heat.

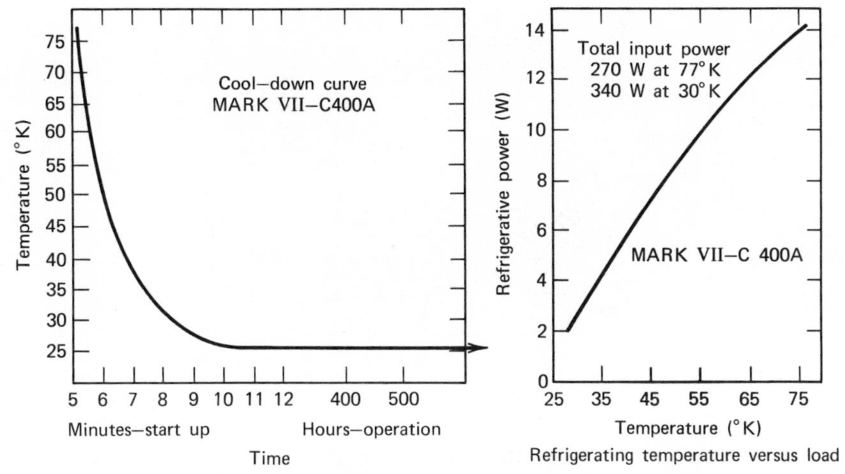

Figure 37. The Mark VII characteristic curves. (Courtesy Malaker Company.)

FREQUENCY CONVERSION

Another way to detect infrared radiation is to turn it into a visible signal: this frequency conversion can be achieved in several ways, by taking advantage of different physical phenomena, such as:

1. Luminescence of phosphor compounds, further subdivided in (1) photoluminescence, and: (2) electroluminescence.
2. Secondary emission of photons from primary impinging radiation.
3. Refractivity changes in cholesterol compounds.

Conversion of infrared radiation into visible light can be achieved in many different ways, but in view of the fact that the light-detecting element is usually the human eye, the frequency converters most used are of the imaging type, where the different levels of infrared radiation emitted by the target are turned into corresponding different levels of visible light, which have the same geometrical distribution, so that the image structure is preserved.

Phosphor Luminescence

Basically, the light emission by phosphor compounds is a recombination radiation phenomenon. "Phosphor" is the general designation for certain dielectrics made of a crystalline structure, doped by impurities whose energy levels are located within the forbidden energy gap. Figure 38 is an example of this condition: the energy levels just above the valence band are called activator levels, while those located just under the conduction band are called trapping levels.

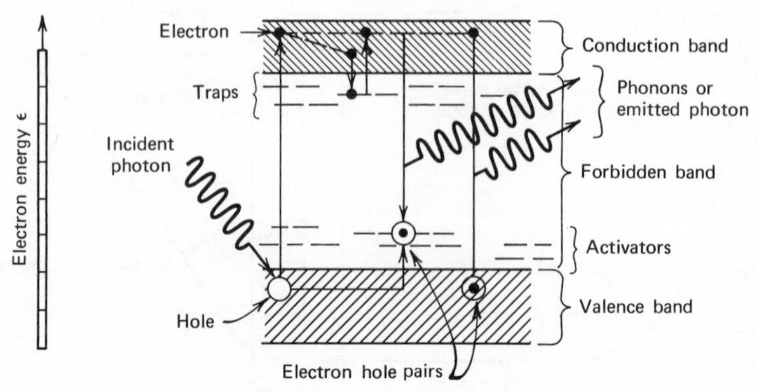

Figure 38. Energy structure of phosphors.

When radiation of adequate energy content (usually blue or ultraviolet light) impinges upon a phosphor compound, electron-hole pairs located in the valence band are separated and the electrons are brought to the higher energy level of the conduction band. At the end of their lifetime, they recombine into the valence band, after having lost their excess energy either by transfer to another particle or by emission of a photon, in which case electromagnetic radiation of wavelength corresponding to the photon's energy is emitted. When the wavelength falls within the visible light range, the phenomenon takes the name photoluminescence.

The impurities located at energy levels within the forbidden gap can affect the photoluminescence phenomenon in two ways: those at activator levels contribute free electrons when struck by lower energy impinging photons, while those at trapping levels act as intermediate energy steps for the electrons that are degrading from the conduction to the valence band, thus reducing the energy level of the photons emitted at each step. In this way the wavelength of the emitted radiation is shifted out of the visible spectrum. This is called "quenching effect" and for a certain group of phosphors is produced by impinging infrared radiation.

The opposite effect is called "phosphor stimulation." It takes place for certain phosphors when external excitation is removed. At this point, a number of electrons remain trapped at the trapping levels, and impinging infrared radiation raises their energy enough to liberate them, so that they can decay to the valence band, releasing photons in the visible spectrum.

By careful choice of the doping elements, phosphors can be produced that will exhibit quenching or stimulation by infrared within well-defined thermal limits.

Table 3 is a list of several thermographic phosphors commercially

Table 3 Radelin Thermographic Phosphors

Radelin Number	Fluorescent Color	Effective Temperature Ranges (°C)	Approximate[a] Sensitivity (%)	Chemical Base	Average Particle Size (μ)
1807	Orange	Room Temp. to 80	25	ZnCdS: Ag, Ni	9
3251	Orange	35 to 130	13	ZnCdS: Ag, Ni	9
2090	Green	120 to 280	7	ZnS: Cu, Ni	9
3003	Green	250 to 400	7	ZnS: Cu	9

[a] Change in brightness per degree Celsius change in temperature.

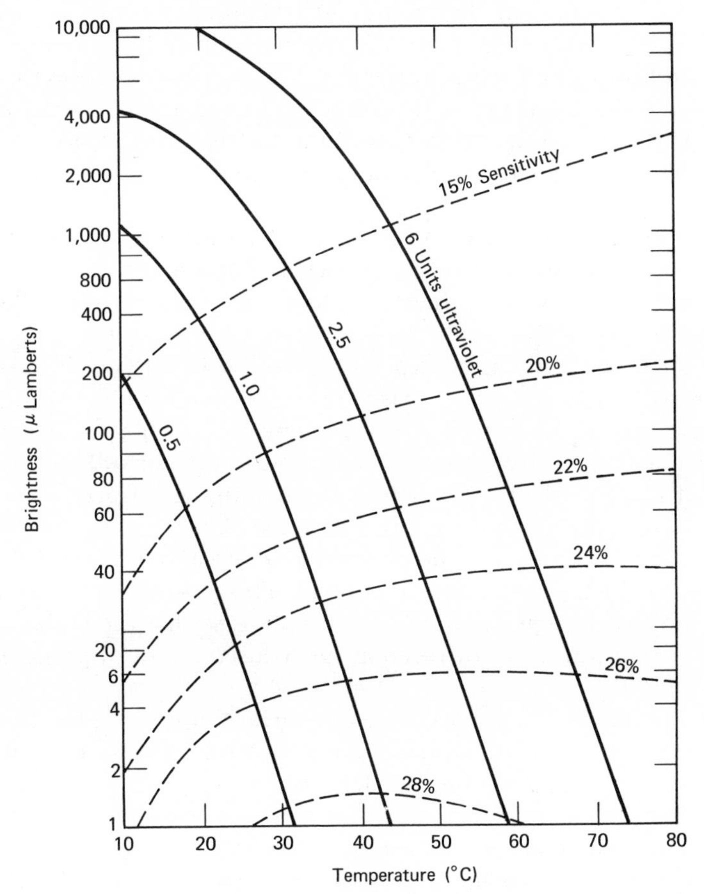

Figure 39. Performance of radelin thermographic phosphor 1807. Solid curves show brightness vs temperature at ultraviolet excitation. Dashed curves correlate sensitivity (percentage brightness change per degree Celsius change of temperature) with brightness and temperature. (Courtesy U.S. Radium Corporation.)

available and their principal characteristics. All are of the quenching type.

Figure 39 is a diagram showing the characteristic curves of Radelin thermographic phosphor 1807. The solid lines indicate brightness versus temperature at four levels of ultraviolet excitation. Dotted lines indicate the sensitivity, that is, the brightness variation per every degree Celsius in the temperature of the compound.

Electroluminescence

Phosphor excitation that will produce visible luminescence can also be achieved by means of an alternating AC field imposed across a phosphor layer. In this instance, doping is not required, since the presence of activator and trapping levels is unwanted. The energy needed to raise the electrons from the valence to the conduction band is supplied by the external AC field, and the brightness of the luminescence is proportional to the voltage of the applied field and to its frequency, as shown in Figure 40.

Electroluminescent panels can be used as infrared-to-visible converters when an infrared image is focused onto a surface capable of modulating point-by-point the strength of the AC signal applied across the phosphor layer.

One such converter is shown in Figure 41. It was produced commercially by Sylvania, and it makes use of a photoconductive layer to modulate the AC excitation at every point of the phosphor's surface. Modulation takes place because the photoconductor's DC resistance varies inversely with the strength of the impinging infrared radiation: thus for every point of the surface where the infrared image is focused, the resistance variations of the photoconductor cause inverse variations of the AC field applied to that same point. This in turn causes variations of the induced luminescence of the phosphor, so that the visible image faithfully reproduces the infrared image focused on the opposite side of the converter.

A photographic film applied onto this surface can record the image in a permanent way, and can also store the low-level light emitted during a long exposure if its instantaneous level is too low.

The thermal range of operation for this type of converter is dependent on the characteristics of the photoconductor used: in view of the extreme difficulty of cooling to cryogenic temperature, only photoconductors operating at ambient temperature are used, and therefore the converter is only sensitive to the near infrared.

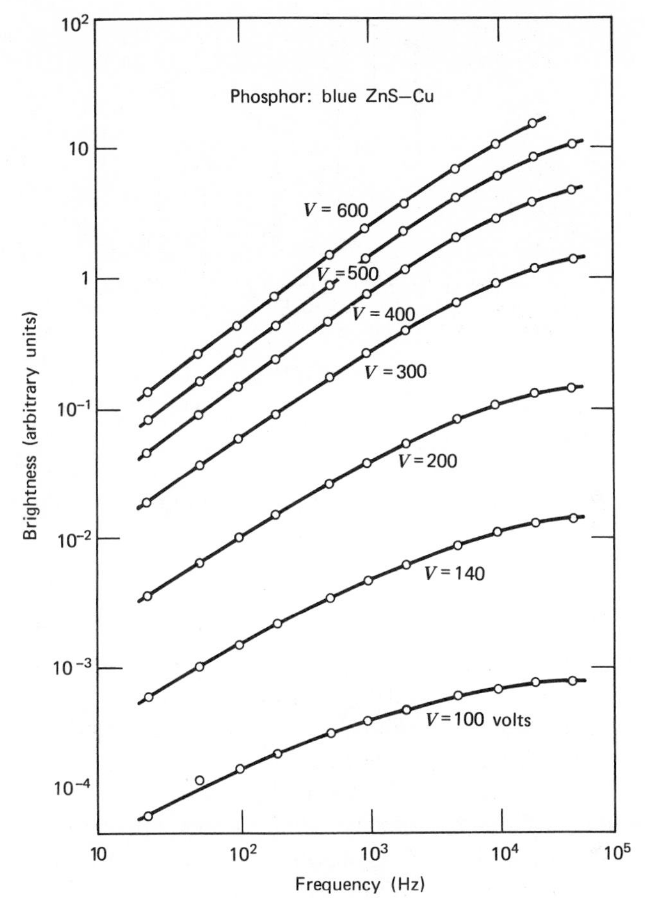

Figure 40. Frequency dependence of the brightness of a blue-emitting electroluminescent ZnS: Cu phosphor for various applied voltages. (Courtesy Lehmann.)

Ferroelectric Conversion

To operate in the far-infrared spectrum, modulation of the AC voltage can be achieved by replacing the photoconductor with a ferroelectric layer. In this case, the modulation takes place by capacitance variation, instead of resistance variation, but the result is similar. The ferroelectric layer is a compound of barium titanate and strontium titanate, (as described in the section on Ferroelectric Bolometers) in such proportions as to make its capacitance variations highest in the desired temper-

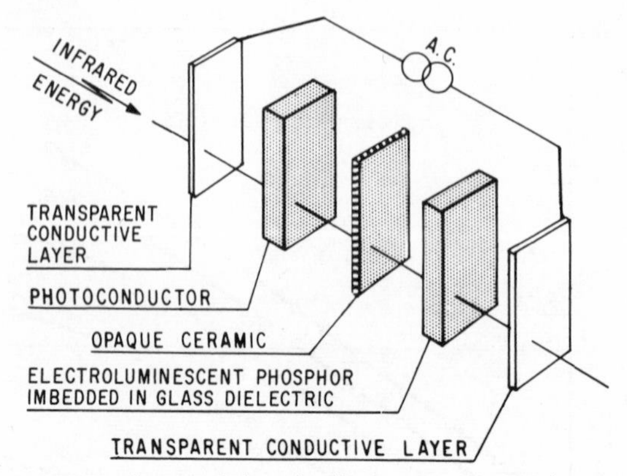

Figure 41. Image converter for near-infrared radiation. (Courtesy Sylvania.)

ature range. Figure 42 illustrates the arrangement that solves the problem of long-wave image conversion.

Infrared Vidicon

The signal-storage capability upon which is based the operation of the vidicon tubes is especially useful in the infrared range, where the energy content of the radiation emitted by the target becomes progressively lower as the wavelength increases.

Figure 43 is a schematic diagram of the infrared vidicon tube type Z-7808 made by General Electric. The operating principle can be summarized as follows: the video information is generated by the current variations occurring in the scanning electron beam when it recharges the areas of the retina that have been partially discharged by leakage through the photoconductive material of which the retina is made. Focusing an infrared image on the front surface of the retina creates different DC resistance values for every point, in inverse proportion to the intensity of the impinging radiation.

As shown in the illustration, an optical system of reflective or refractive elements is located in front of the vidicon tube. The incoming radiation passes through the infrared transparent window and is focused on the front surface of the retina. This surface is kept at a positive potential with respect to the cathode and as long as no infrared radiation falls onto it, the photoconductive material of which the retina is made constitutes a barrier between this positive-charged surface, and the back surface, which is negatively charged by the electron-scanning beam emitted by the cathode.

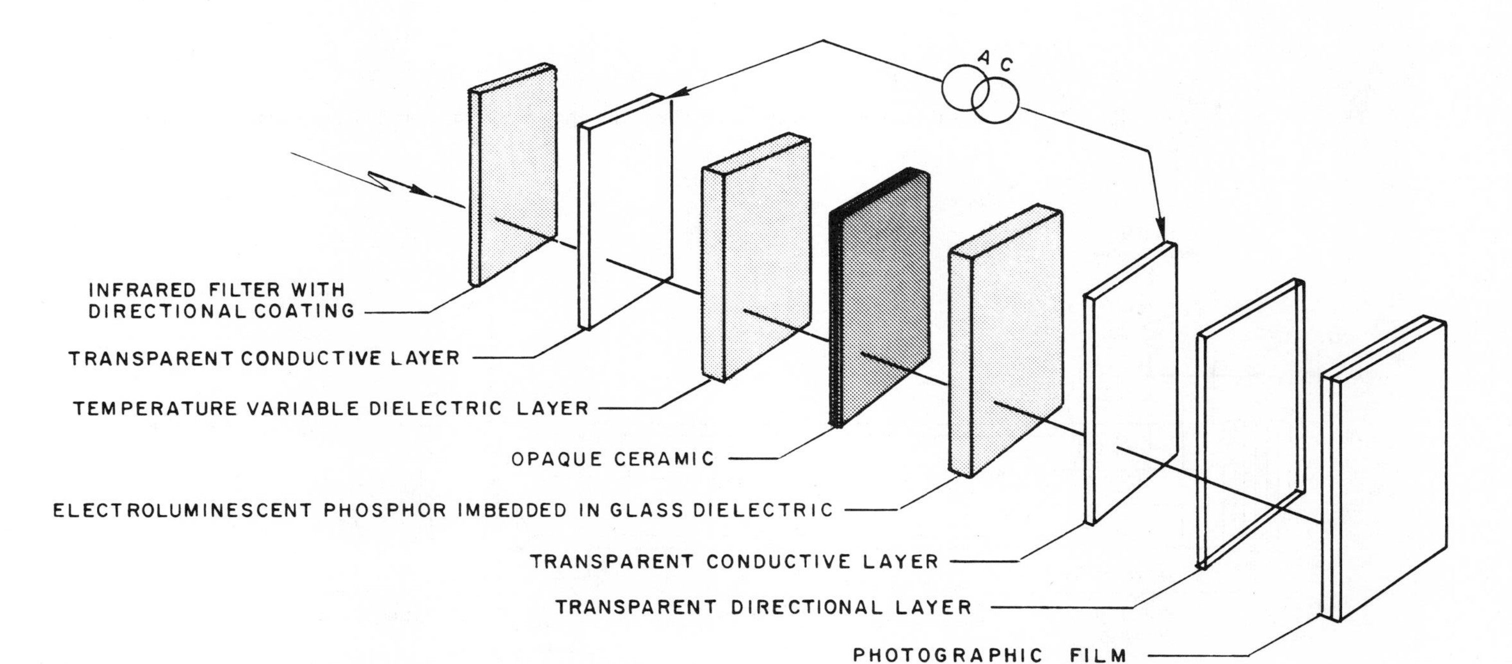

Figure 42. Ferroelectric image converter for far-infrared radiation.

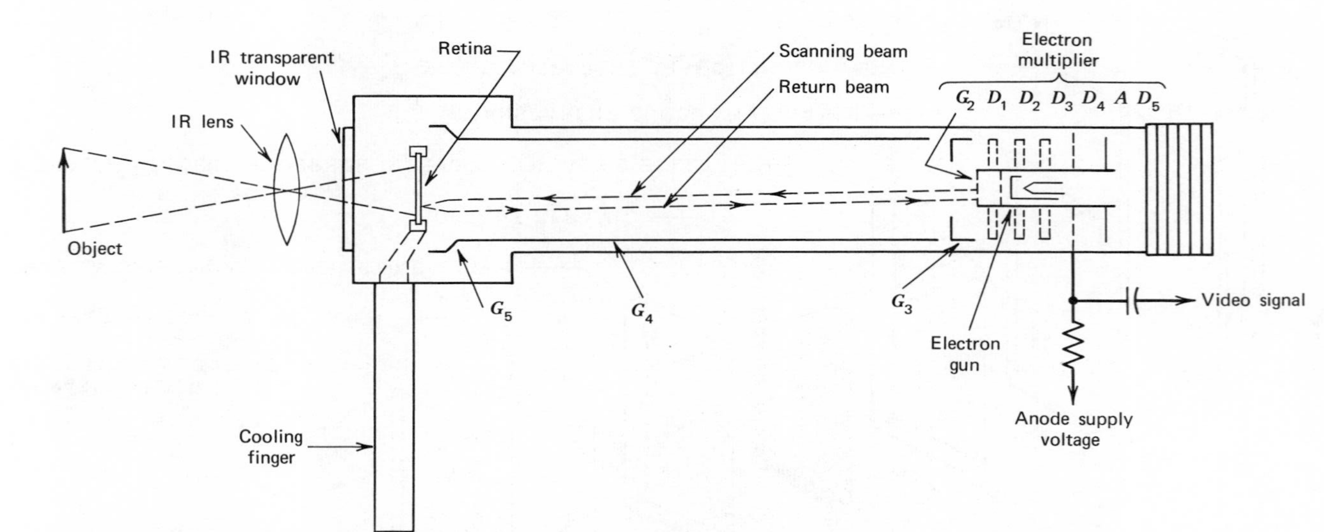

Figure 43. Schematic diagram G.E. type Z-7808 infrared vidicon. (Courtesy General Electric Co.)

However, as soon as infrared radiation falls on the retina, electrical leakage between front and back of the retina takes place, in direct proportion to the amount of infrared energy impinging on every elementary area of the retina. Consequently, the electron scanning beam on its next passage will deposit on the depleted areas just enough negative charges to reestablish the cathode potential. As a result, the strength of the electron beam that returns to the anode G_2D_1 surrounding the electron gun will vary in inverse proportion to the infrared energy that illuminates every point of the retina.

The path of the electron beam from the cathode to the retina and back is shown in dotted lines with arrows pointing the direction of travel. Focusing, alignment, and deflection coils are wound around the tube's neck, but are not shown in the illustration. Grid G_5 slows down the incoming electrons for better landing characteristics, while G_3 serves to control the path of electrons emitted by D_1 and approaching D_2. The remaining dynodes are D_3, D_4, and D_5, which together make the electron multiplier of the return signal. At every dynode stage, several electrons are ejected for every impinging electron, for the well-known principle of secondary emission, until finally at the last dynode stage, the amplified electron beam current leaves the tube, via the anode. Video signals, or variations in beam current corresponding to infrared radiation on the vidicon target, are also amplified and are capacity coupled into a preamplifier.

The amplification provided by the electron multiplier maintains a high signal-to-noise ratio and permits the use of a lower gain preamplifier. Optimum performance is obtained when the output noise of the tube, composed of the random noise of the electron beam amplified by the electron multiplier, exceeds the preamplifier input noise.

Figure 44 is a picture of the above-described infrared vidicon tube. The cooling of the photoconductive material of the retina is obtained by using liquid nitrogen in the cryogenic container located at the front end of the tube.

This tube is sensitive in the intermediate infrared spectral region and its performance specifications are classified.

Liquid Crystals

This apparently self-contradicting term is currently used to indicate those organic compounds that, while in liquid form, exhibit optical anisotropy typical of crystalline solids. This behavior is characterized by a difference in the speed of light propagating along two different direc-

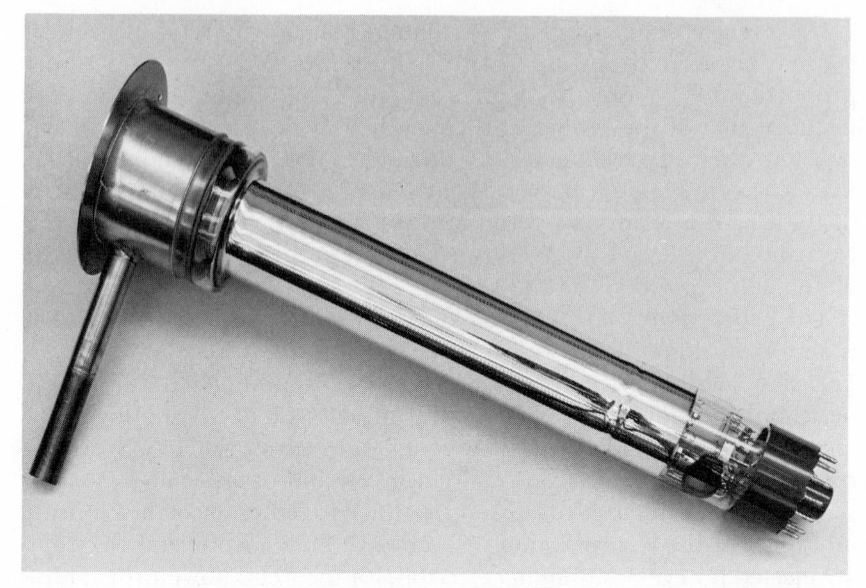

Figure 44. The G.E. infrared vidicon. (Courtesy General Electric Co.)

tions, so that refraction and reflection take place according to the laws of birefringence. The phenomenon is due to the fact that the molecules are aggregated in orderly configurations of groups or layers, where they are all stacked parallel to each other.

Liquid crystal properties are exhibited by a relatively large number of organic compounds, but only those producing dichromism have been used to convert temperature into color information. The physical phenomenon used for this conversion is the unequal absorption coefficient for the two polarized light beams into which the impinging rays are split. The nonabsorbed polarized beam is either transmitted or reflected, after having rotated a certain angle, whose magnitude depends on the temperature of the compound. This produces various colors in the visible spectrum, which are indicative of the temperature of the surface on which the compound is adhering.

The derivatives of cholesterol, called esters, exhibit the above-mentioned phenomenon to a high degree. Figure 45 shows how the molecular layers are stacked together, while the molecules' axes (see arrows) are progressively rotated always the same angle from one layer to the next. This configuration in turn produces a rotation of the polarization plane of the incident light, to an extent that is several orders of magnitude larger than the corresponding phenomenon in solid crystals. For instance,[4] at 20°C, a 1-mm section of quartz will rotate light of a wavelength of 656 mμ (red) 17.25 degrees, while the same sample will rotate light of a wavelength of 448 mμ (blue) 39.24 degrees. A 1-mm section of an optically active substance that shows the cholesteric structure (cyanobenzylideneaminocinnamate) at 75°C rotates light of wavelength

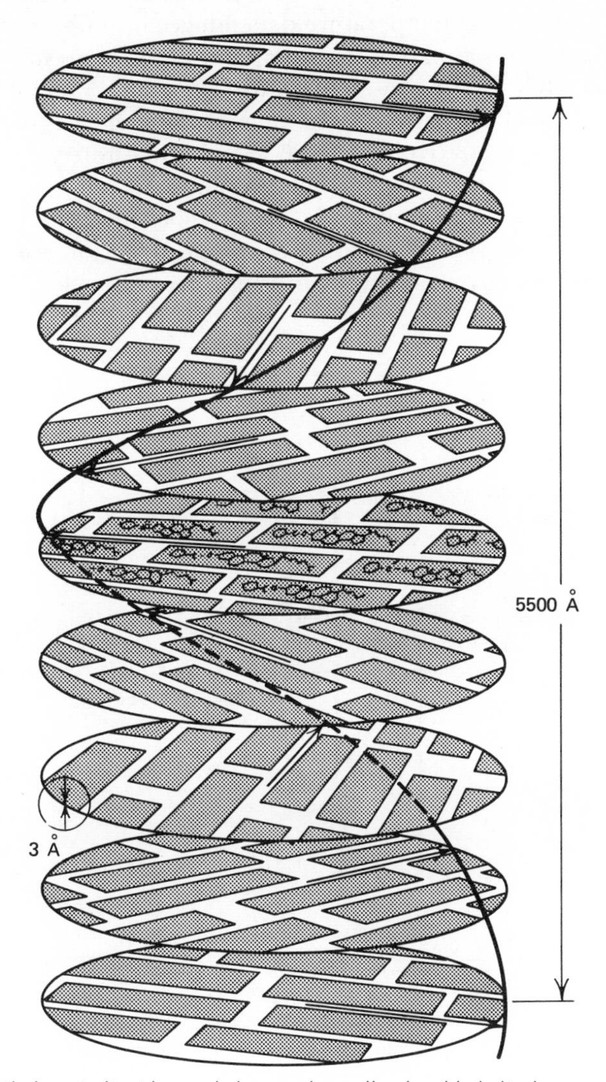

Figure 45. Cholesteric liquid crystals have a thermally alterable helical structure. (Courtesy Lauriente and Ferguson.)

of 452 mμ (blue) 27,650 degrees. No other substances are known which compare to liquid crystals in their capacity to rotate polarized light.

Figure 46 shows how the color range of the reflected light varies with the number of carbon atoms contained in the acid group of the cholesteric molecule.[5] It is interesting to note that at a given temperature, a given compound will always exhibit the same color. Also invariable is the rate of change from color to color for every given material.

Figure 47 shows the temperature dependence of different cholesterol esters mixtures. It is apparent that compound *D* changes smoothly from red to aqua while varying in temperature between 25 and 45°C. Compound *A* instead changes from red to blue in one single degree Celsius of temperature excursion (from 25 to 26°C), and therefore is best used for display of fine thermal variations in said range.

The time response of compound *A* is on the order of 1 sec for changes from red to blue or vice versa. Other compounds instead are not reversible, but exhibit color change only during a cooling cycle.

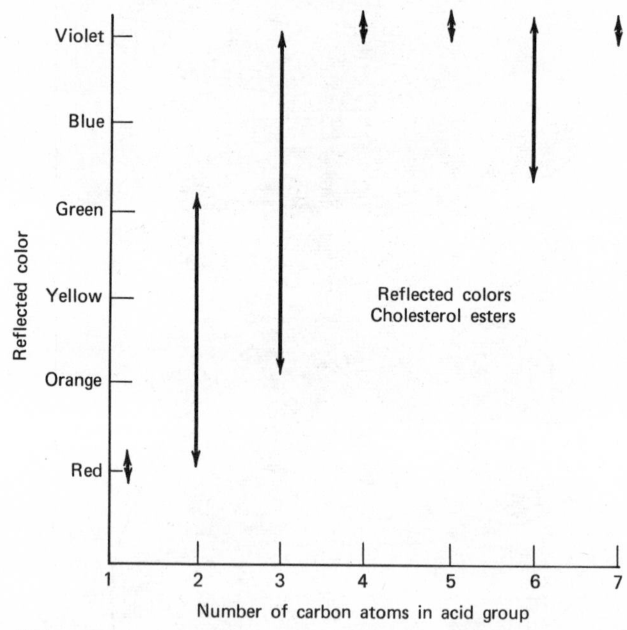

Figure 46. Light scattering as a function of cholesterol ester structure. (Courtesy Woodmansee.)

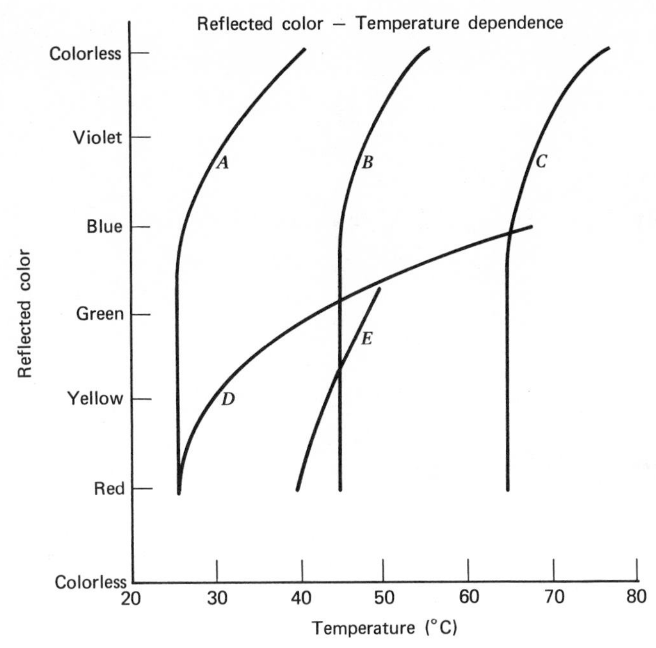

Figure 47. Temperature dependence of light scattering for mixtures of cholesterol esters. (Courtesy Woodmansee.)

Due to the fact that the colors are produced by reflection and not by transmission, these compounds should be deposited onto black surfaces. Since a layer 10-μ thick is sufficient to produce accurate color display, a single gram of mixture dissolved in adequate volume of solvent can cover 1000 cm^2 of surface.

Chapter 3 Infrared Measuring Equipment

BASIC STRUCTURE OF INFRARED TEST EQUIPMENT

Remote Temperature Sensing

This chapter is restricted to a discussion of that class of infrared measuring equipment that is capable of determining the temperature of remote sources. While the range of applications of infrared techniques to spectroscopy, missile guidance, reconnaissance, flow process control, and such is quite broad, our primary interest is the acquisition of information as determinable from the thermal behavior of materials.

Infrared System Considerations

Figure 48 is a block diagram of a typical infrared system configuration. Without entering into the specific details of each element, we can, at this point, examine the overall response of such a system through analysis of the parameters which define the elements.

The infrared detector, characterized by a noise equivalent power (NEP), already discussed at the beginning of chapter 2. collects infrared radiation incident upon the system aperture, A, so as to provide a system having a noise equivalent irradiance, NEH, equal to

$$\mathrm{NEH} = \frac{\mathrm{NEP}}{\eta A}$$

where:

η = effective reduction of geometric area by obstruction and/or filtering.

The source emitting infrared radiation provides an irradiance, H, at the aperture of the infrared system (see Chapter 1) equal to

$$H = \frac{J}{D^2} \qquad \text{for a simple "point" source}$$

or

$$H = N\omega \qquad \text{for an "extended" source}$$

where ω = field of view of system for which the transmission of the intervening atmospheric path must be accounted. Thus the simplest expression for signal to noise, in the infrared system, is

$$\frac{S}{N} = \frac{H}{\text{NEH}}$$

From this simple statement, with considerable manipulation of expressions for each of the terms, we can derive a general expression for the noise equivalent temperature difference (NETD) of a typical system. Inasmuch as such a derivation is of instructive value, it will be carried out.

First, in considering the source, the emittance of the source is given by the Planck equation

$$W_\lambda = \frac{c_1}{\lambda^5}\left(\frac{1}{e^{c_2/\lambda T} - 1}\right) \qquad \text{W/(cm}^2)(\mu)$$

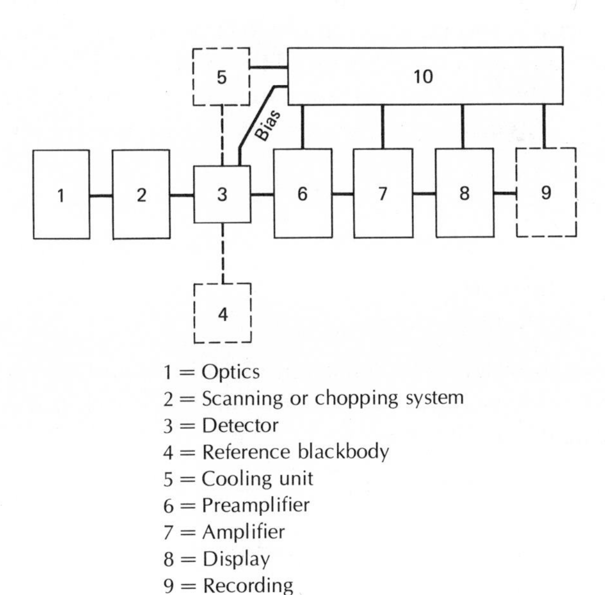

Figure 48. Basic structure of infrared test equipment. Note that the dotted line elements are not always present.

by differentation,

$$\frac{dw_\lambda}{dT} = \frac{-c_1c_2}{\lambda^6T^2}\left(\frac{e^{c_2/\lambda T}}{(e^{c_2/\lambda T}-1)^2}\right)$$

In incremental form, for a response bandwidth, $\Delta\lambda$,

$$\Delta w = \frac{-c_1c_2}{\lambda^6T^2}\left(\frac{e^{c_2/\lambda T}}{(e^{c_2/\lambda T}-1)^2}\right)\Delta\lambda\ \Delta T$$

or, approximately

$$\Delta w = \frac{-5.4\times 10^8}{\lambda^6T^2}\ \frac{\Delta\lambda\ \Delta T}{e^{1.44\times 10^4/\lambda T}} \qquad \mathrm{W/cm^2} \qquad (\lambda,\ \mu;\ \mathrm{T},\ ^\circ\mathrm{K})$$

This last expression, for the incremental change in source emittance resulting from an incremental change in absolute temperature—as detected in a spectral bandpass—is approximately correct (10%).

For an extended real source of a given emissivity ϵ, the corresponding incremental change in radiance is

$$\Delta N = \frac{\epsilon\Delta w}{\pi} = \frac{-1.72\times 10^8}{\lambda^6T^2}\ \frac{\epsilon\ \Delta\lambda\ \Delta T}{e^{1.44\times 10^4/\lambda T}} \qquad \mathrm{W/(cm^2)|sr)}$$

When this change is sensed by the infrared system as equal to the electrical noise arising from the infrared detector, we can obtain a measure of the noise equivalent temperature difference, ΔT.

We saw that the irradiance at the system was determined by the field of view of the system. In a simple radiometer, this field of view is simply:

$$\omega = \frac{A_D}{f^2}$$

where
A_D = area of the detector
f = focal length of the optical system

The noise equivalent irradiance, on the other hand, was determined by the effective collecting area of the system. In a simple radiometer, the noise equivalent irradiance is simply

$$\mathrm{NEH} = \frac{\mathrm{NEP}}{\eta\dfrac{\pi}{4}d^2}$$

where d = diameter of optics

$$\mathrm{NEP} = \frac{(\Delta f)^{1/2}(A_D)^{1/2}}{D^*}$$

where Δf = noise electrical bandwidth,
D^* = detector figure of merit.

In the expression above D^*—the figure of merit of the infrared detector—is a quantity that has already been discussed in Chapter 2. Given all the expressions above for detection performance, we can develop an equation for NETD as when

$$\frac{\Delta H}{\text{NEH}} = \frac{-1.35 \times 10^8}{\lambda^6 T^2} \frac{\epsilon t_\lambda \, \Delta\lambda \, \Delta T}{e^{1.44 \times 10^4/\lambda T}} \eta D^* \left(\frac{A_D}{\Delta f}\right)^{1/2} \left(\frac{d}{f}\right)^2 = 1$$

$$\text{NETD} = \Delta T = \frac{-\lambda^6 T^2 e^{1.44 \times 10^4/\lambda T}}{1.35 \times 10^8 \, \epsilon t_\lambda \, \Delta\lambda} \left(\frac{\Delta f}{A_D}\right)^{1/2} \left(\frac{1}{\eta D^*}\right) \left(\frac{f}{d}\right)^2$$

where t_λ = transmission of path

This general system performance equation provides helpful design insights for remote temperature sensing. To achieve very high sensitivity and small NETD, one should:

1. Utilize long wavelength infrared detection—the exponential factor of more significance is λ^6.
2. Increase the emissivity of the source.
3. Minimize the length of atmospheric path or reduce path absorption by evacuation or flushing with nonpolar gases.
4. Utilize as large a spectral bandwidth consistent with optical efficiency.
5. Maximize the overall efficiency of the optical system by increased transmission and reduced obstructions.
6. Minimize the electronic bandwidth consistent with fidelity of signal transmission.
7. Utilize detectors with highest attainable figure of merit (D^*) and largest area which matches the focused image of the source.
8. Utilize "fast" optical systems—low *F/no.*

Example

As an example of a typical infrared system performance which can be estimated from this equation, consider:

Source: $T = 300°\text{K}$
$\epsilon = 0.5$

Path $t_\lambda = 0.7$
System: $D^* = 10^{10}$

$$\lambda = 10\ \mu$$
$$\Delta\lambda = 8\text{-}14\ \mu$$
$$\eta = 0.25$$
$$\frac{f}{d} = 1.0$$
$$\Delta f = 10\ \text{kHz}$$
$$A_D = 1\ \text{mm} \times 1\ \text{mm}$$

Applying these values in the general system performance equation, we obtain the value of the thermal resolution of the system:

$$\text{NETD} = 0.015°\text{K}$$

The Optical System

In simple terms, the function of the optics in an infrared system is that of collecting as much radiant energy while minimizing noise. In a sense, this establishes a requirement for "optical gain": magnify the receiving area of a bare detector yet maintain the noise level equivalent to that from a detector of small receiving area. Thus the effective "optical gain" is the ratio of the optical receiving area to the detector area.

In the preceding section, we saw how this "optical gain" affects the sensitivity of an infrared system, and surmised that the geometrical area, of the optical system, was effectively degraded by obscurations, transmission of elements, and spectral bandpass—the factor η. It is not the author's intention to enter into a discussion of many optical system designs, but rather to categorize those which are of practical utility. At the outset, therefore, we limit discussions to reflecting optical systems with the knowledge that such achromatic systems can be made more effective—for infrared radiation—than any lossy refracting systems.

Infrared optical systems can be categorized into telescopes and microscopes, schematically represented in Figures 49 and 50, for which corresponding angular magnifications are shown. The lens representations, in the figure, are only symbolic. Of course, "optical gain" equals the square of linear—or angular—magnification. For each case, it is seen that a generalized "objective" forms an image of the source which is then viewed by some collecting element—an element which transfers radiant energy to the infrared detector. This collecting element, in every case, is actually a collimating element which has the objective-focused source at its focal point. The goal of optical design is to fill the detector with

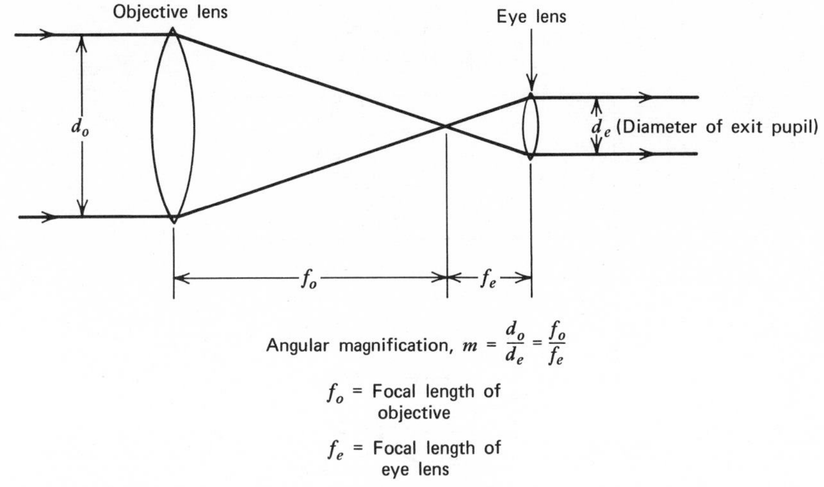

Figure 49. Generalized telescope. (Courtesy R. Barbera.)

collimated radiation which has been increased in angular density by the objective.

Typical reflecting telescope objectives are shown in Figure 51. For perfect imagery, all utilize a primary element which is a parabola. While the parabola suffers from astigmatism off-axis, for collimated radiation it forms the theoretically perfect image. No other simple mathematical surface has a single focal point for all rays, which are parallel to the axis. Examination of the previous system considerations shows that overall

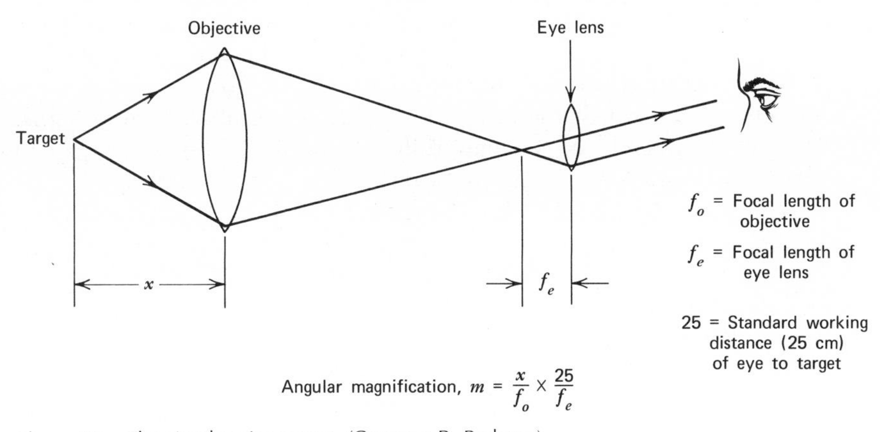

Figure 50. The simple microscope. (Courtesy R. Barbera.)

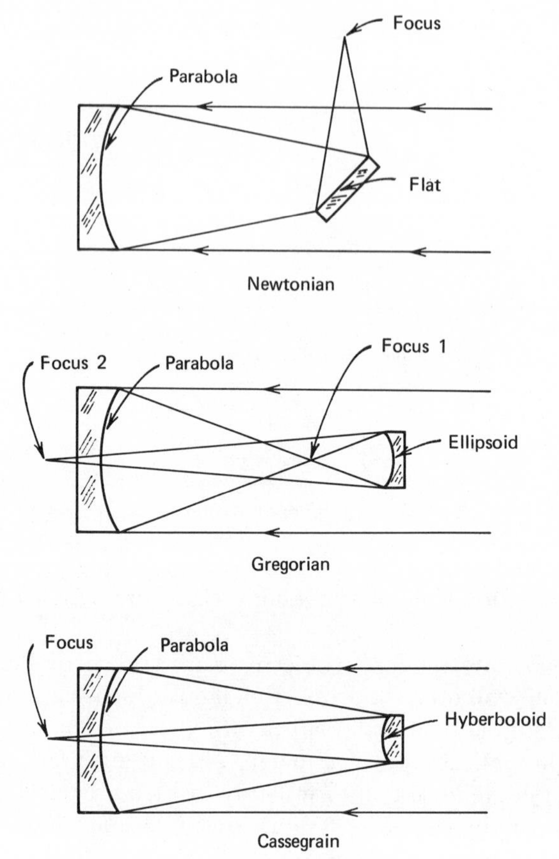

Figure 51. Typical reflecting telescope objectives. (Courtesy R. Barbera.)

system performance depends upon focal length. Thus the variations of Figure 51 demonstrate techniques for modifying focal length—while maintaining element diameter constant—for practical package designs. The first objective, the Newtonian, utilizes a simple flat mirror to deviate the focused rays. Several flats could be used to achieve folding.

In the second example, the Gregorian, long focal lengths can be achieved with theoretically perfect performance for axial rays by utilizing a folding ellipsoid. The ellipsoid, a surface with two focal points, is situated so that one focal point coincides with the focal point of the primary parabola. A modification of this technique is that of the cassegrain, which utilizes a hyperboloid instead of an ellipsoid. Again, theoretically perfect performance is achieved for axial rays, all mathematical surfaces being so defined.

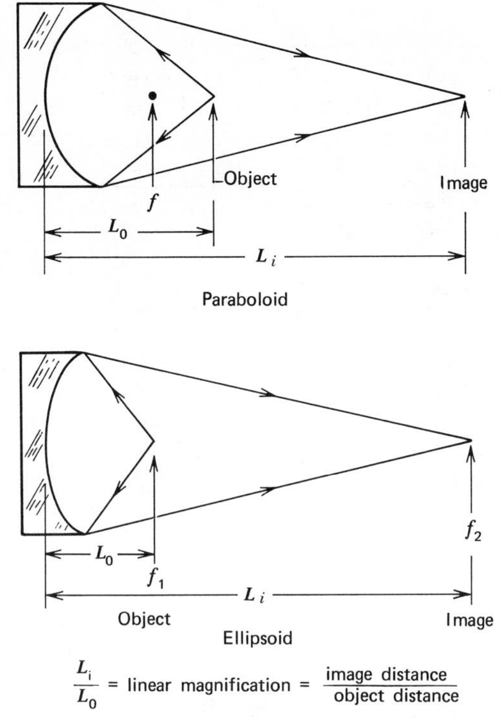

Figure 52. Typical reflecting microscope objectives.

Figure 52 illustrates typical reflecting microscope designs. Inasmuch as linear magnification is the ratio of image to object distance, surfaces which allow ready attainment of such real distances while permitting theoretically perfect performance, must be used. In the first example, a paraboloid is used with the object just outside the focus. Unfortunately, astigmatism results when the parabola is so used because of the incident angles. However, the astigmatic aberration can be minimized by utilizing "slow" optical systems, systems with large *f/no.* This conclusion follows from the fact that the magnitude of astigmatic aberration is proportional to sin θ tan θ, where θ is the angle of the rays.

In the second example, the problem of astigmatism is overcome by utilizing an ellipsoid with source and image at conjugate foci. Again, a mathematical surface which inherently possesses desired optical properties is the choice. Transfer of the focused rays to the detector can be accomplished by a collimator of appropriate design, or the detector can be placed at the focus directly.

From the foregoing, it should be apparent that optical systems with rather simple configurations can be adequate for most infrared applications. Reflecting elements are preferred by virtue of their achromatic properties and their relative ease of fabrication. Reasonable image quality—if not theoretically perfect—can be achieved with minimal design effort. Advantages and disadvantages of optical systems made of reflective and of refractive elements are summarized in Table 4.

Table 4 Comparison of Optical Systems

	Optical System	
Image Degradation	Surface Reflective	Refractive
Chromatic aberration	None	Unavoidable. Requires compensation
Transmission losses	None	Wavelength—dependent. Thickness—dependent. Losses as high as 50% are not uncommon for thicknesses of just a few millimeters.
Reflection losses	2 to 5%. Coating helps to reduce these losses	10 to 30% for two surfaces. Coating helps to reduce these losses
Obscuration	Up to 10% due to presence of holes for the passage of the optical field	None
Coma	Present. Can be completely compensated. in special systems (Cassegrain et al.)	Present. Can be in part compensated.
Spherical aberration	Present. Can be completely eliminated.	Present. Can be in part compensated

Refractive systems can be used for narrow spectral bands, or when chromatic aberration can be disregarded and transmission losses can be afforded. Many materials having good transmission to infrared radiation are available for the manufacture of lenses (see Figure 13). Among the most widely used are KRS-5, germanium, silicon, sapphire, barium fluoride, and arsenic trisulphide. Design of infrared refractive optical systems is governed by the laws of classic optics. No matter which type of optical system is used, its field of view is generally given in angular units (degrees, milliradians, steradians). It is interesting to note that the distance of the target does not affect the detector reading, as long as the field of view is contained within the limits of the target. Figure 53 illustrates this fact, the reason for it being the compensating effect of area increase versus radiating power decrease, both quantities varying equally, but in opposite direction, with the square of the distance.

Furthermore, detector readings are also independent of angle to the target or even surface curvature, as justified by Lambert's cosine law (see Chapter 1, Optical Properties) as also shown in Figure 53.

A Special Optical System

When the target is rotating at uniform speed, it is generally desirable to present to the detector a stationary optical field. This is achieved by the Spinvision System, developed by Comstock & Wescott, Inc., for use in conjunction with an AGA Thermovision fast scan system. This is done by the introduction along the axis of the optical field, of a corner mirror system rotating in synchronism with the tire, but at half the speed. This effectively arrests the image of the tire so that the infrared camera observes a derotated heat picture through the surface reflective mirrors of the system. Since the rotational speed of the SPINVISION is variable, it can be synchronized with any tire speed.

Figure 54 shows schematically how this is achieved. The system found its first application in the infrared testing of rubber tires, searching for hidden defects such as layer de-lamination and voids.

Limitations

However, all optical systems of the type described so far have limitations of physical nature, due to their material size, weight, alignment, optical finish of the surfaces, and such. And they all must comply with the requirement that the infrared detector have a direct, unimpeded view of the target—no opaque obstacles in between, no shrouding envelopes, no hiding around corners. Unless the detector can view the object, it cannot measure its radiation. Light, and the term includes infrared radiation, travels only in a straight line, and the detector cannot look around corners or through solid obstacles.

Or can it? Well, perhaps with some help it could. Leaving alone the Einstein-postulated bending of light by a gravity field, we could try some simpler means of looking around a corner. A simple mirror might do the trick, although in a very rigid way. Much more flexibility (literally) is

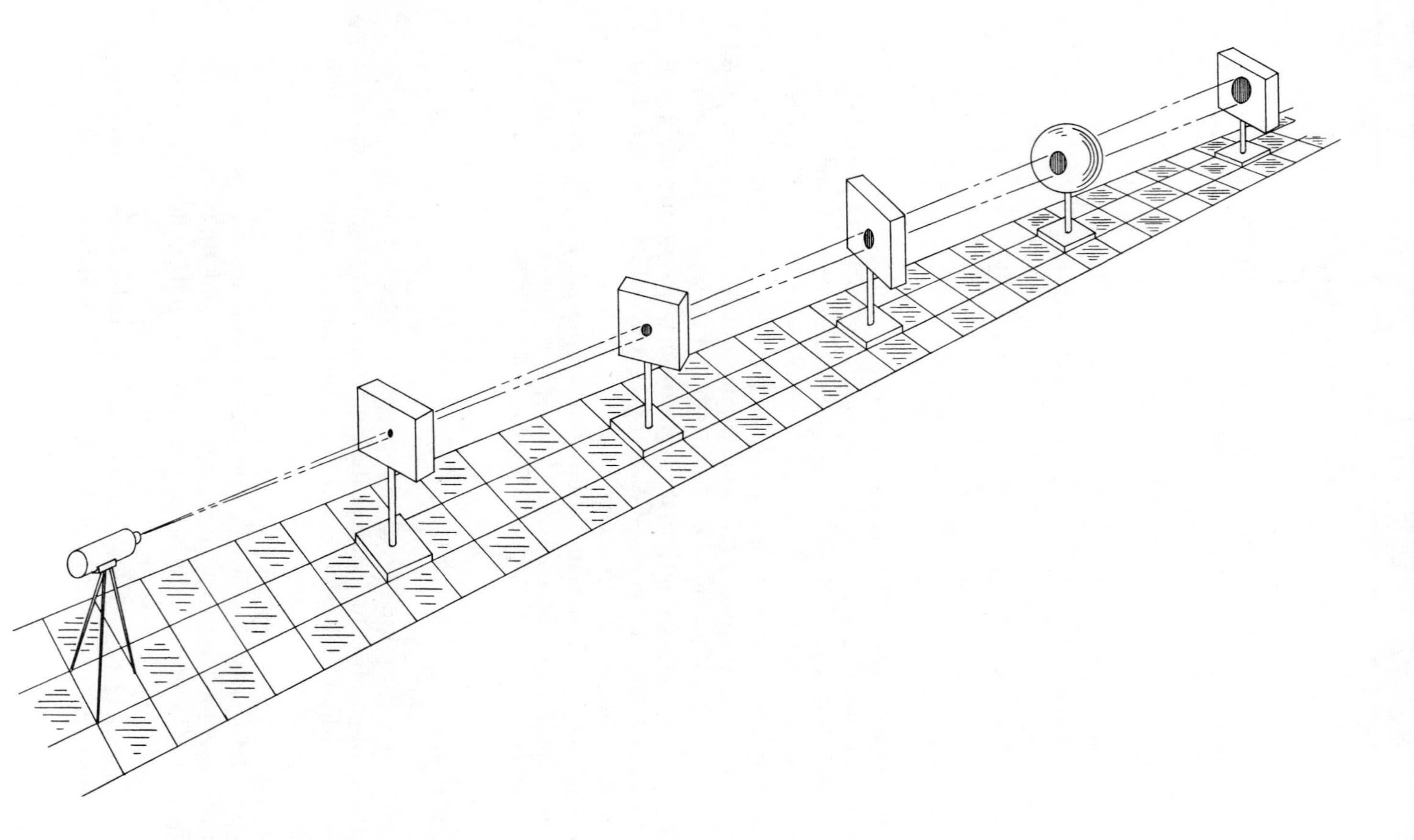

Figure 53. Field of view of a detector.

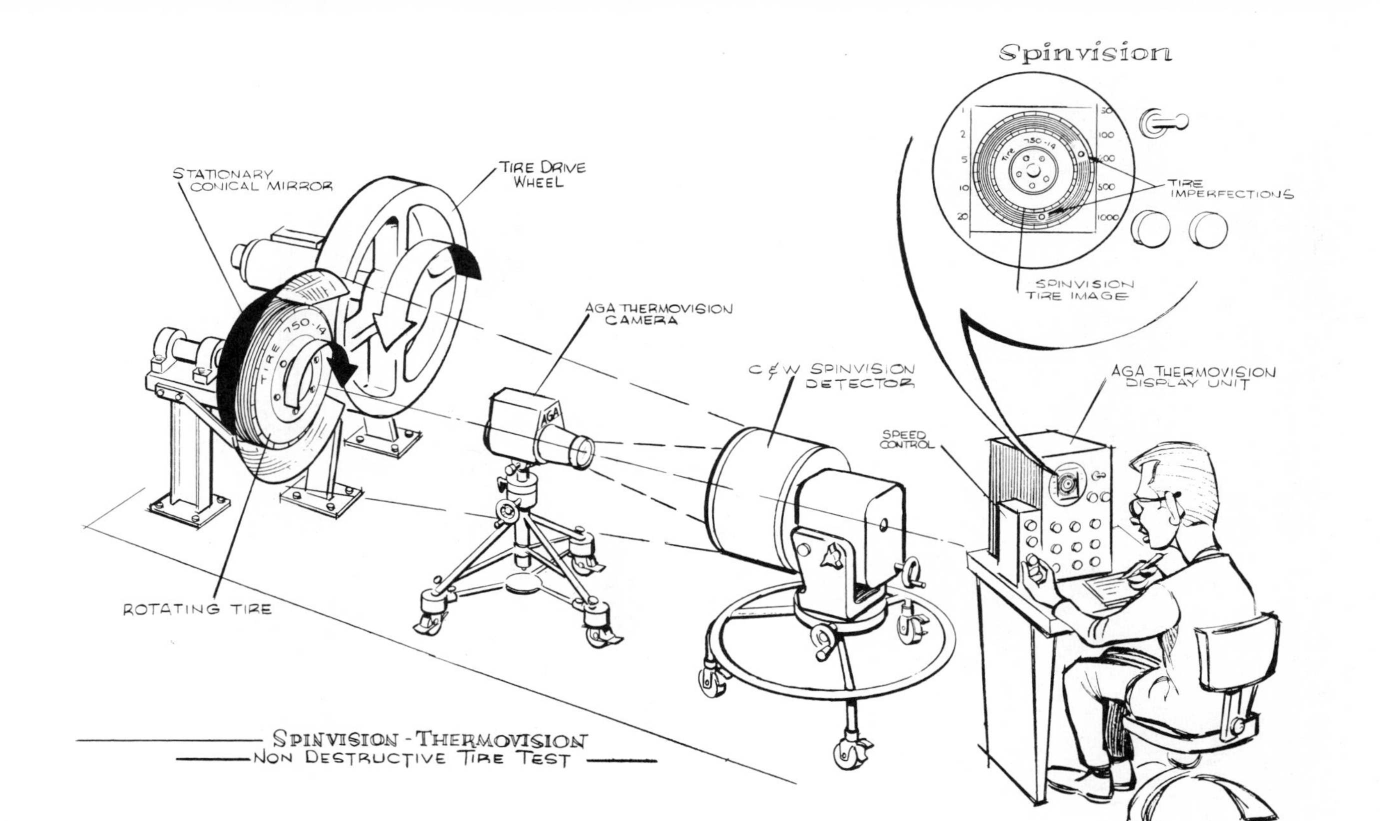

Figure 54. Spinvision system. (Courtesy Comstock and Wescott.)

offered by optical fibers. Before we proceed any further, it seems in order to spend a few words on this subject.

Optical Fibers

Optical fibers are transparent linear elements inside which light propagates by total internal reflection. Figure 55 shows the physical principle on which the fiber operation is based. Most of the fibers presently available are made of glass.

From the illustration, we can see that all light rays that, after entering the front surface, acquire an inclination smaller than the critical angle are totally reflected inside the fiber, and keep traveling in this fashion until they reach the opposite end or are totally absorbed, whichever comes first. For a fiber having a critical angle of 67°, Figure 55 shows an acceptance angle of 70°, which means that all rays incident onto the fiber's front surface at a 35° angle or less with its axis will be trapped inside the fiber by total internal reflection.

On the other hand, all incident rays entering the fiber with an inclination larger than the 35° angle will leave the fiber at the first contact with its surface. This behavior is commonly called "spilling."

The value of the critical angle is a function of the ratio between the refractive indexes n_1 and n_2 of the glass of which the fiber is made and of the medium surrounding it.

Thus by controlling the ratio n_1/n_2, we have a means of increasing or decreasing the acceptance angle of fiber optics, a feature that, as we see later, can be used to obtain special performance characteristics.

Cladding is the name of the technique used to establish a permanent n_1/n_2 ratio. It consists of coating the fiber with a layer of solid material

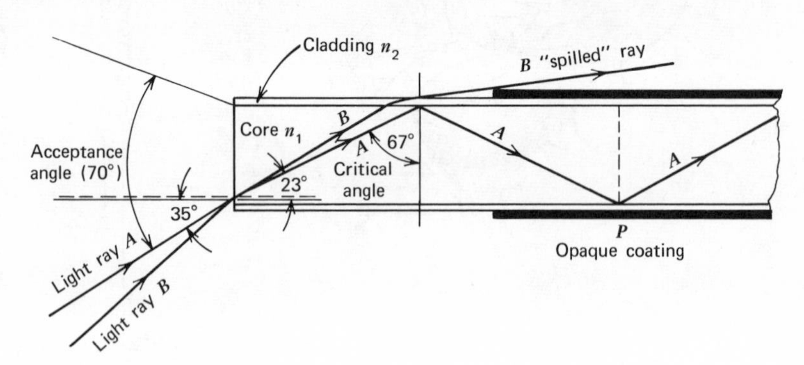

Figure 55. Ray propagation in optical fiber.

(usually a different type of glass) having the desired index of refraction. Without this coating, two fibers touching each other would "spill" light into each other, since they have the same refractive index. The presence of the cladding prevents this from happening. A rather common combination is lead glass for the core and borosilicate glass for the cladding.

Figure 55 shows a fiber enveloped by a layer of cladding, whose outside surface, to the right of point *P*, is in turn coated with an absorbent layer of material. This will prevent stray rays, such as those surpassing the critical angle (for instance, where the fiber makes a curve) from spilling out: the outer coat will merely absorb them. Also, at a curve, light cannot get into the fiber from outside, and every chance of interference with the rays traveling along the fiber is eliminated.

How tight a curve can a fiber make without spilling away most of the trapped light? As long as the ratio of the bend radius to the fiber diameter is above 40, the losses are negligible. This means that a fiber with a 25 μ core diameter can be wound around a 1-mm mandrel with no significant transmission loss.

What is the average number of total internal reflections that a light ray undergoes while traveling along a fiber? It depends on the average inclination of the rays entering its front end, and we have seen that this inclination is a function of the n_1/n_2 ratio. The closer to 1 is the value of this ratio, the smaller will be the angle made by the entering rays with the fiber's axis and the lower the number of internal reflections per unit length.

The core size of the fiber also has direct bearing on the number of internal reflections: for conventional acceptance angles around 30° from the fiber axis, about 500 reflections per inch of length will be the average for thick fibers (0.010 in. or 250 μ in diameter), such as the ones recently made available of plastic material. For thin glass fibers (about 20 μ in diameter) as high as 4000 reflections per inch is probably a realistic average.

Besides their ability to carry light around corners and through airtight walls, optical fibers can provide quite large acceptance angles for incident radiation; thereby making them comparable to "fast" conventional optical systems: that is, systems having a large numerical aperture (NA) number. This NA is the numerical aperture number of an optical system, and it is a measure of its ability to accept incident light rays. This, of course, is a function of the limit angle of acceptance: the larger this is, the larger the cone of radiation entering the optical fiber, transmitted along its length and out of its output end.

For example, arsenic-trisulphide fibers can have a half-angle acceptance of approximately 45° at the input. This corresponds to a "numerical aperture" of 0.707. Converting this "numerical aperture" to conventional *f/no* we obtain a value of 0.714. Fibers made of selenium and tellurium compounds can achieve a half-angle acceptance of approximately 30° at the input, corresponding to a "numerical aperture" of 0.5, and a conventional *f/no* of 1.0.

When compared with typical radiometers, which have optical *f/no* ranging from 4 to 6, we can see that optical fibers can capture appreciably higher radiating power. On the basis of comparison of equivalent *f/no*, such optical fibers are theoretically more effective by factors ranging from 16 to 70, or the square of the relative *f/no*. Figure 56 graphically illustrates a comparison between the cone of radiating energy collected by a conventional telescope and an optical fiber.

On the other hand, we have the losses. Plastic fibers made of Crofon have most of their transmission spectrum in the visible range, with two areas around 1 μ of wavelength where some near-infrared transmission takes place. See Figure 57.

Lanthanate glass fibers have a poorer response in the visible spectrum, but transmit further into the infrared. Figure 58 shows their transmissivity spectrum. Of special interest is the very large NA of these fibers: it reaches the full value of 1.0. Transparent even further down in the

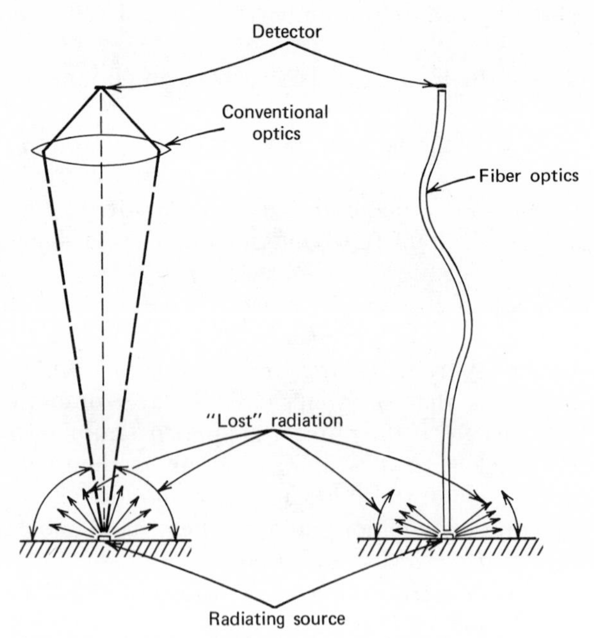

Figure 56. Capture of radiant energy. Telescope vs fiber optics.

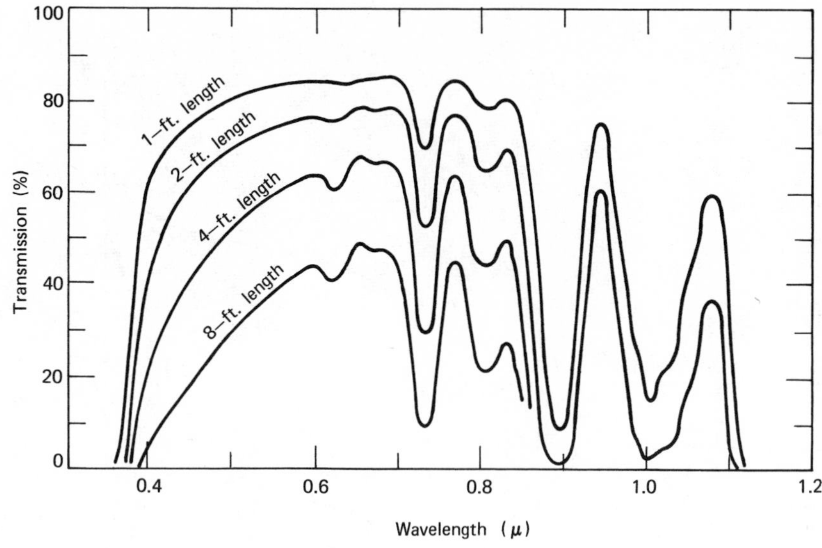

Figure 57. Spectral transmission of crofon fibers. (Courtesy DuPont.)

intermediate infrared are arsenic sulphur glass fibers. See Figure 59. These fibers too have a large NA: 0.8, corresponding to an *f/no* of 0.63, as a result of their large acceptance angle.

Arsenic-trisulphide glass is also used to transmit infrared radiation. Its spectral transmission is shown in Figure 60. These fibers have a NA = 0.7, are available with AsS cladding, and have orange-red color. In view of their chemical composition, they must be kept at a safe distance from coffee pots and lunch bags.

A new breed of fibers is made of As-Se-Te or Ge-Se-Te compounds. The transmissivity spectrum of one glass of this family is shown in Figure 61; it was developed by Servo Corporation of America, and its composition is 30% As, 30% Si, and 40% Te.

The principal characteristics of fibers made of this glass is the ability of transmitting infrared radiation all the way to 12 μ which includes the spectral peak of the radiation emitted by surfaces at conventional ambient temperature (approximately 300°K). Their NA is 0.5, corresponding to a 30° half-angle of acceptance, and they are opaque in the visible spectrum.

Fiber Bundles

Thus far we have discussed single fibers. For practical purposes, however, fiber bundles are most commonly used. A fiber bundle is an as-

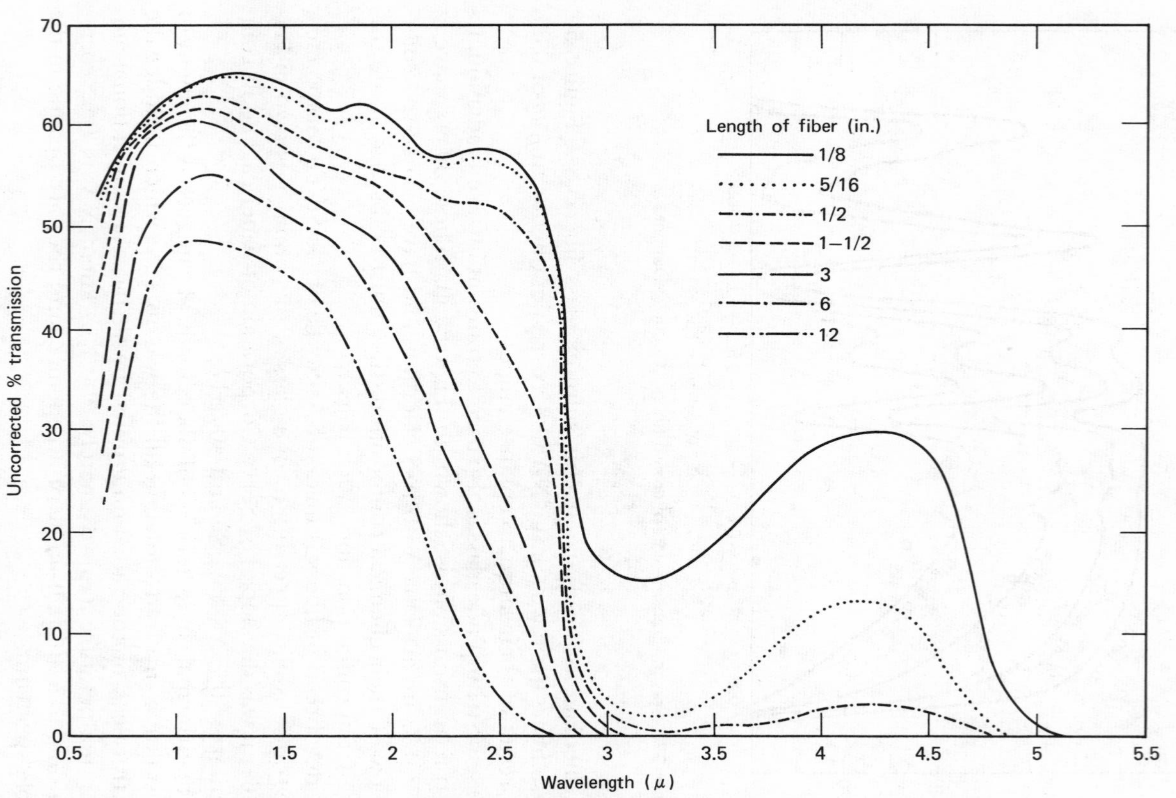

Figure 58. Spectral transmission of lanthamate fibers. (Courtesy R. J. Simms.)

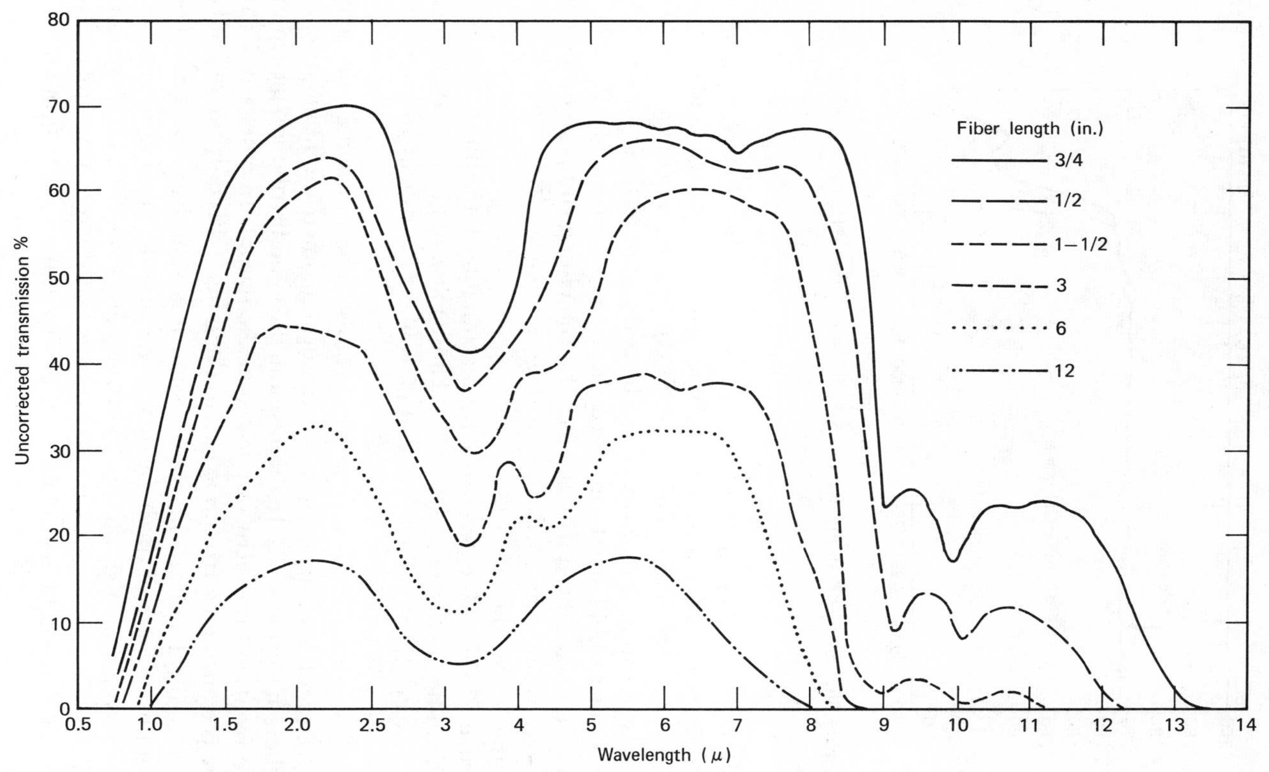

Figure 59. Spectral transmission of As-S glass fibers. (Courtesy R. J. Simms.)

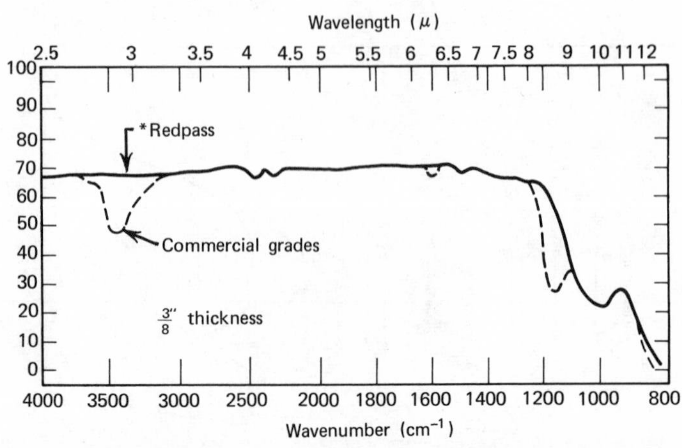

Figure 60. Spectral transmission of AS_2S_3 fibers. (Courtesy TEEG Research, Inc.)

sembly of a number of fibers, anywhere from just a few to several hundreds, lined inside a containing pipe that can be either rigid or flexible. In Figure 62 fiber bundles of different length and composition are shown. As it can be seen from the picture, the ends of these bundles are made rigid to hold firmly in place the terminations of all the fibers. Usually this is achieved with the use of a cementing compound, such as an epoxy resin, that hardens to a degree adequate to permit optical finishing of the end surfaces.

Figure 63 shows the geometric representation of an end surface of a small bundle of Crofon plastic fibers. Those made of glass are very much like it, except for the difference that glass fibers are usually thinner, and thus more numerous.

Fiber bundles are divided in two groups: coherent and incoherent. The coherent ones have the same identical geometrical distribution of the fibers at the two ends. In this fashion, the light distribution picked up at the front end is exactly duplicated at the output. In other words, an image is transferred from one end to the other, with only the degradation resulting from (a) the transmission losses, and (b) the resolution allowed by the size of the individual fibers and the spacing in between.

The incoherent fiber bundles, instead, have a random distribution of the fibers at the two ends, and no image is transferred from one end to the other. Only light is available at the output, in quantity proportional to the total light input, minus the transmission losses.

In the transmission of images through coherent fiber bundles, the

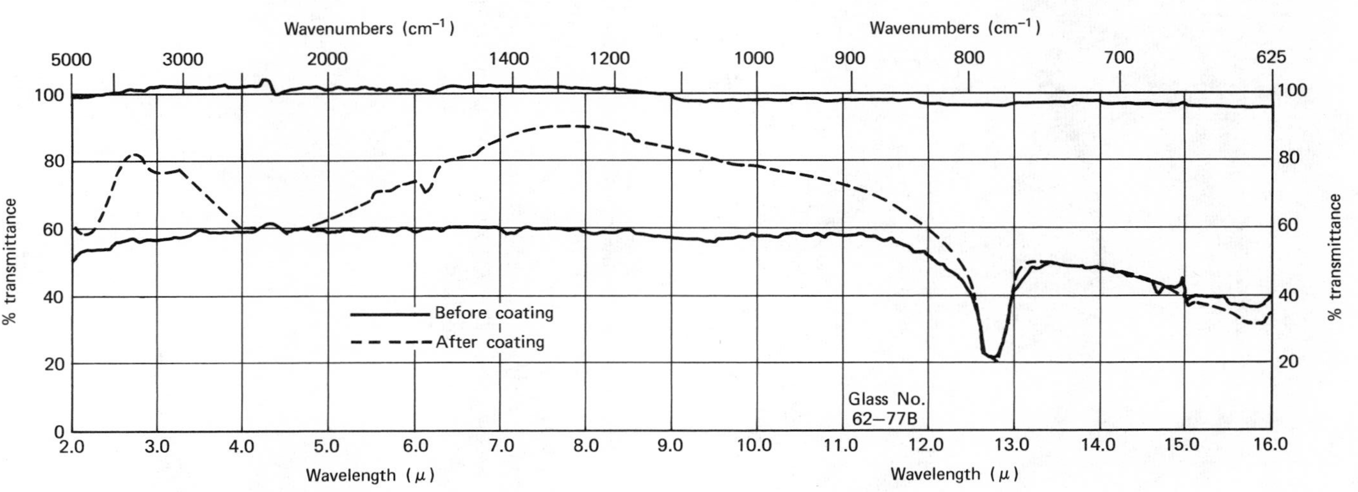

Figure 61. Spectral transmission of As-Se-Te glass. (Courtesy Servo Corporation of America.)

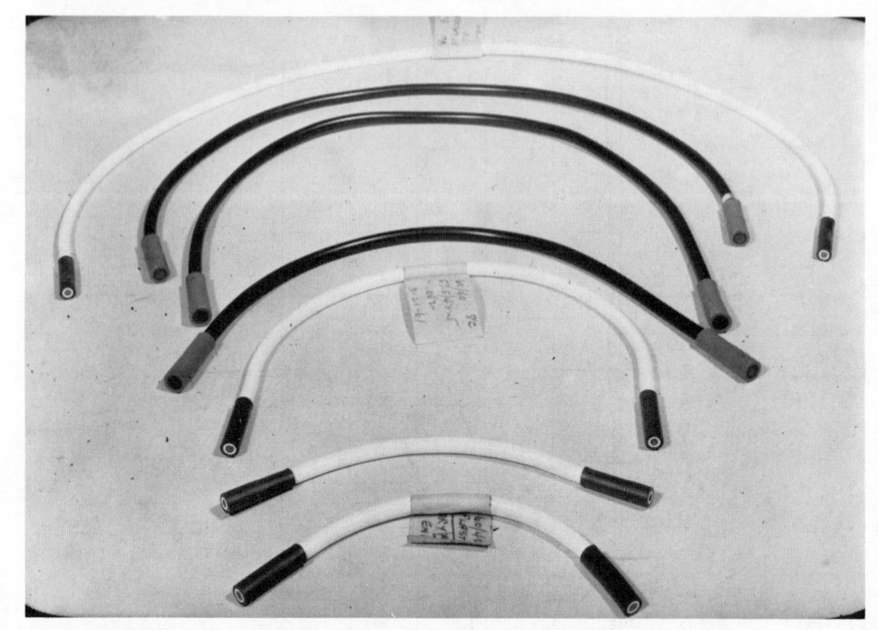

Figure 62. Optical fiber bundles. (Courtesy American Optical Company.)

capability of controlling the acceptance angle can be used to enhance resolution. We have already seen how by reducing the difference between the refractive indexes of the core and of the cladding it is possible to proportionally reduce the acceptance angle at the fiber front end. In this way, every fiber "sees" only what is in front of it, with consequent finer resolution capability.

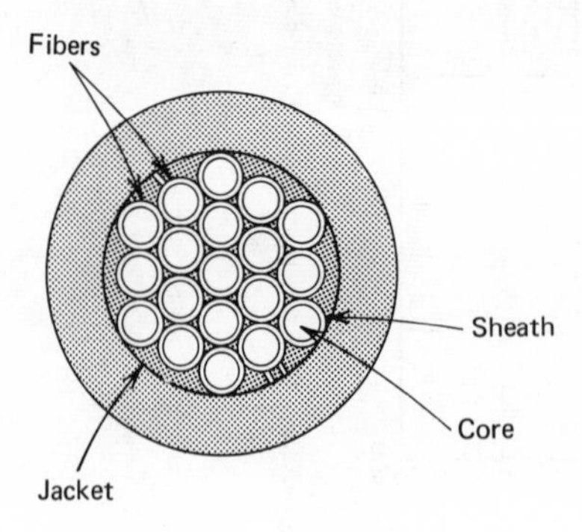

Figure 63. End view of Crofon fiber bundle. (Courtesy Du Pont.)

The Chopping System

A large number of radiometric systems and actually all those mentioned in this book incorporate a "chopper" that is a moving device that periodically interrupts the flow of infrared radiation from the target to the detector. In this way, the detector's output is an AC signal, easier to handle than a DC signal. In the better designed systems, during the time that the shutter is closed, the detector is made to "see" the radiation of a reference body, that is kept at known temperature, to establish a base for the measurement of the radiation from the target.

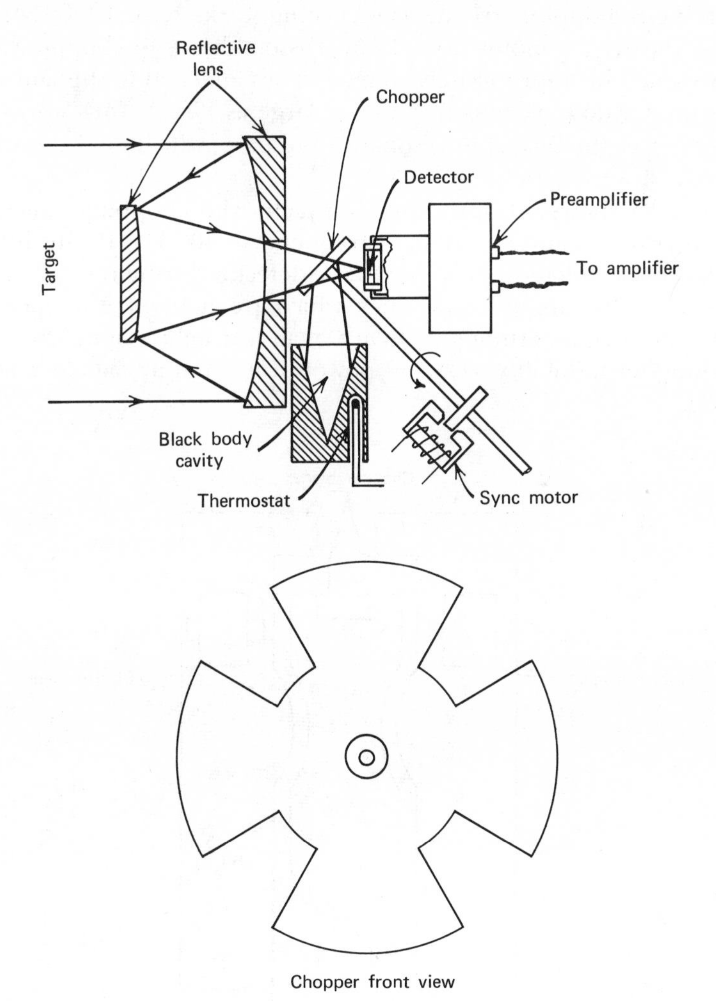

Figure 64. Chopper in reflective system. (Courtesy Barnes Engineering Company.)

The detector's characteristics dictate the best chopping frequency, that is, the frequency that allows the highest sensitivity of the system. For thermistor bolometers of current design, the chopping frequency can be as low as 10 Hz or as high as 100 Hz. For photodetectors, the frequency is at least one order of magnitude higher, due to their much faster time response.

Figure 64 shows how a chopper is incorporated in a reflective optical system: the chopping element is a rotating mirror that, when cutting off the radiation from the target, folds the reference blackbody's radiation onto the detector.

For high chopping frequencies, tuning forks have lately been replacing the driving motor-slotted wheel assemblies. The chopping action is performed by appropriately shaped vanes attached to the end of the fork's tines, whose excursion can be as large as $\frac{3}{8}$ in. A transistorized oscillator drives the fork at its resonant frequency, which can be anywhere between 30 and 25,000 Hz.

Figure 65 shows a tuning fork complete with chopping vanes. It is made by Bulova, and its resonant frequency is 400 Hz. In this instance the blackbody radiation provided to the detector during the "off" time is supplied by the tuning fork's vanes, which are at ambient temperature.

The advantages of tuning forks are small size, light weight, great accuracy, long-term stability, very low power drain, and negligible heat generation.

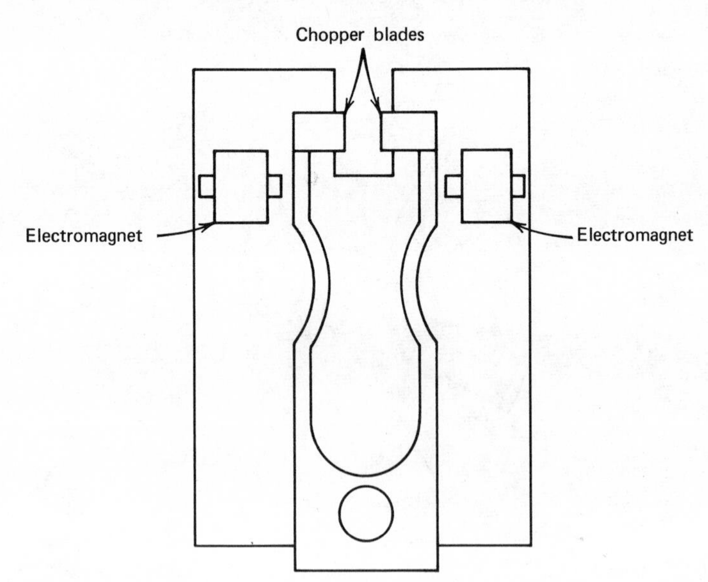

Figure 65. Bulova tuning fork. (Courtesy Bulova Company.)

The Scanning System

When discrete measurements of radiation emitted by different areas of a two-dimensional target must be made, some way of scanning the target must be implemented. This can be achieved in one of the following ways:

1. Linear raster scan similar to the TV picture system.
2. Spiral scan similar to the industrial TV imaging system.
3. Point-to-point scan following a predetermined pattern that can be of any configuration.

The scanning can be implemented by:

1. Moving the detector.
2. Moving the target.
3. Moving the optical field.
4. A combination of the systems above.

The first can be used only for slow-speed operation and has the drawback of subjecting the detector to vibration that can increase the noise content of the output signal.

The second can only be implemented for very small targets: here too, the scanning speed is limited.

The third is usually achieved with the use of oscillating or rotating mirrors that are located along the axis of the optical field. High speed can best be attained with rotating mirrors, although their efficiency is low, since during most of the rotation cycle the mirror reflects into the detector the surrounding background, as shown in Figure 66.

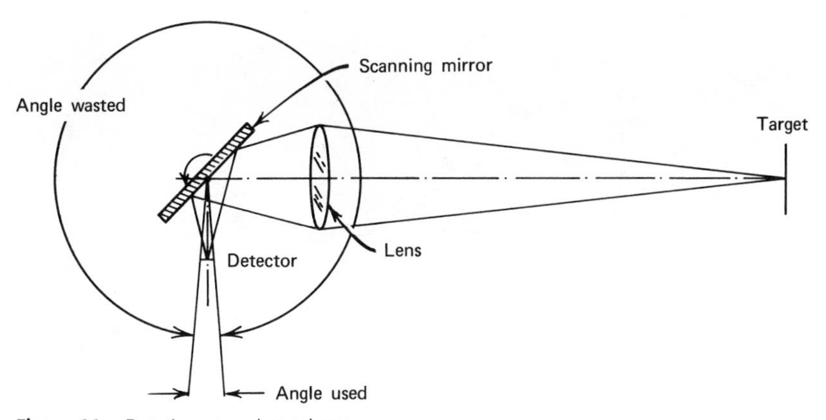

Figure 66. Rotating scanning mirror.

For linear, TV-like rasters, one single mirror, rotating and tilting at the same time, has been used with satisfactory results.

The optical field can also be displaced by moving transparent prisms or "wedges," as shown in Figure 67 if the chromatic aberration, the reflection, and the variable transmission losses can be either tolerated or compensated for.

Figure 68 shows a wedge system capable of scanning a surface in the x and y direction: if the speed ratio of the two wheels can be varied, the system will have the capability of varying the number of lines per frame. Prisms can be translated, as in the illustrations, or rotated in a full circle or in oscillating fashion. The last two modes are also typical of mirrors: the full rotation has the drawback of low efficiency, as already seen, while the oscillating motion is affected by uneven speed of operation, since it consists of accelerating, linear, and decelerating motions, separated by two reversals of direction. Since only the linear portion can perform scanning at even rate, acceleration, deceleration, stop, and return must be blanked out, thus greatly reducing the efficiency of the oscillating mode.

Modified Corner Reflector System

The problem of achieving a scanning motion having linear speed and almost no losses of the kind described above was solved by the author by

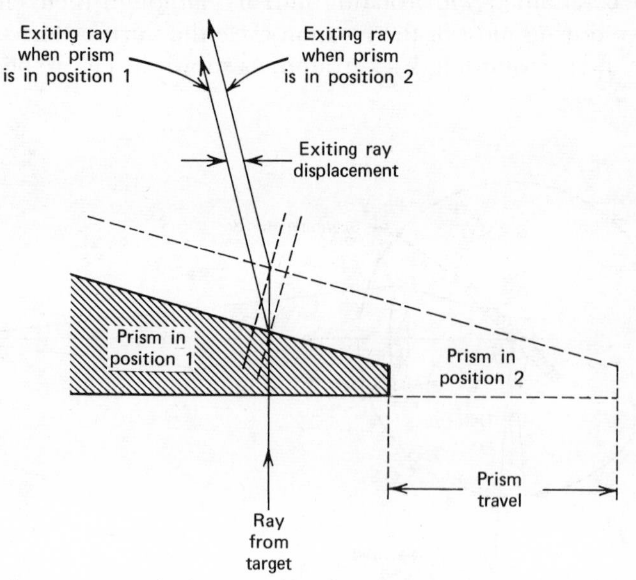

Figure 67. Ray displacement through prism movement.

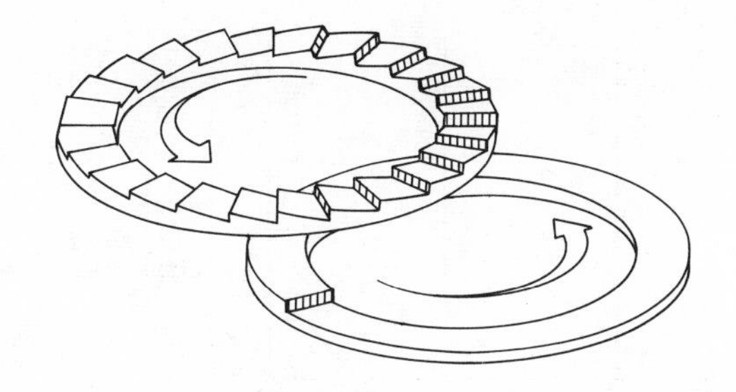

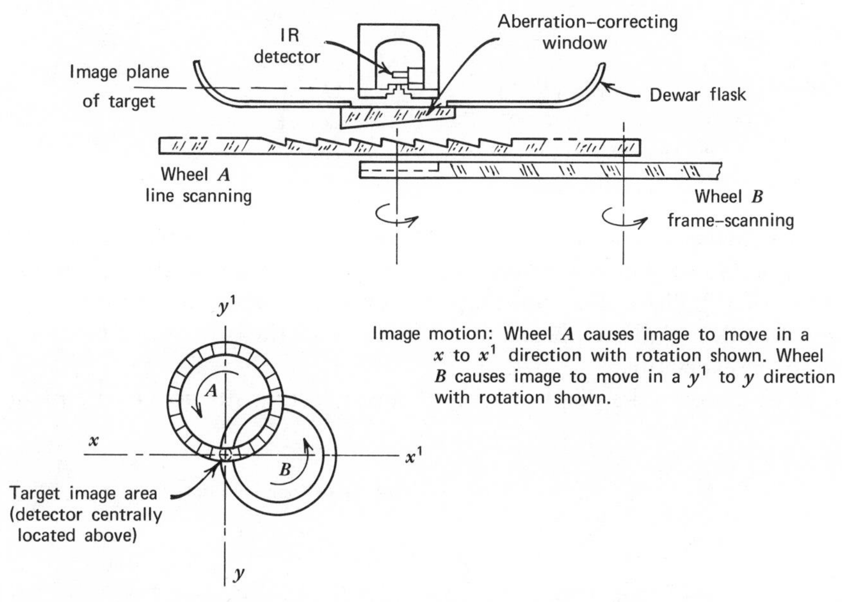

Figure 68. Prismatic scan system.

modifying the "corner reflector" system. In its conventional configuration, this system is composed of two flat mirrors rigidly tied together at 90° angle. It can be seen from Figure 69 that a horizontal motion of this pair of mirrors by one unit in length produces a two unit displacement of the outgoing ray, without any change in optical path length (i.e., without shift of the focal plane). Such an optical arrangement could be used in a scanning system, but of course, in this configuration, it would have all the efficiency losses typical of the alternating motion. However, all these losses can be eliminated with the solution shown in Figure 70, where the corner reflector is formed by the shaded areas of two heli-

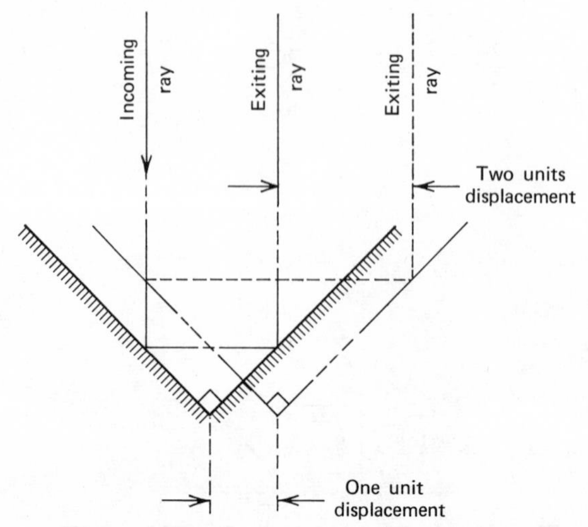

Figure 69. Conventional corner reflector.

coidal surfaces. The pitch of these helices equals the displacement that the mirrors of Figure 69 must travel to complete the scanning motion. The two helices are rotating at uniform speed and in perfect synchronism, and are phased in such a way that the two "steps" face each other at every turn.

If we observe the displacement of the reflecting surfaces in the plane

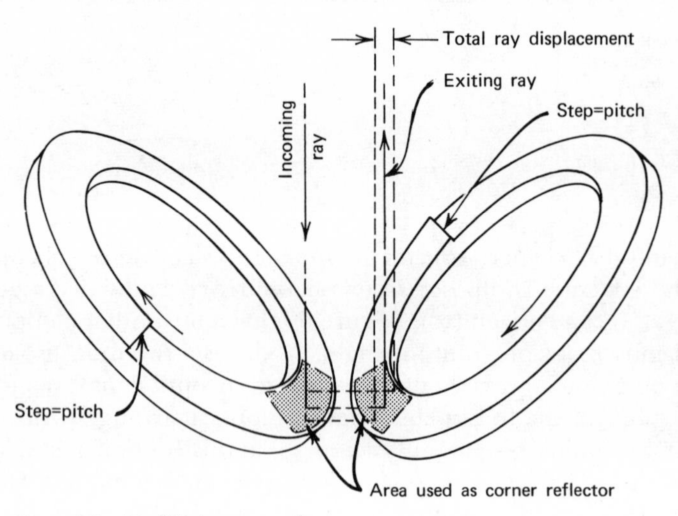

Figure 70. Modified corner reflector.

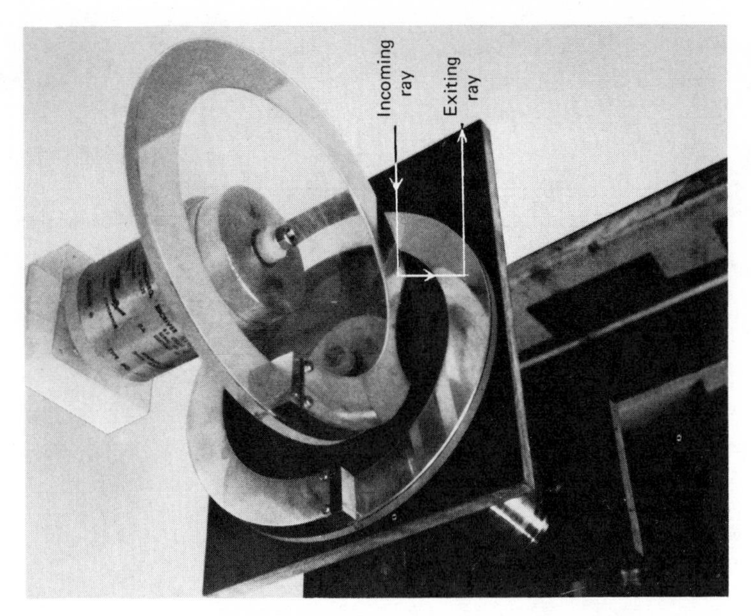

Figure 71. Distortion compensating arrangement of modified corner reflection.

passing through the axes of rotation of both helices, we notice that for every turn (starting from the position where the two steps are facing each other) the two reflecting surfaces move in the X direction at uniform speed and at the end of a 360° revolution of the helices they fly back in no time to the starting position.

This solution achieves a perfect sawtooth displacement of the incoming optical ray, with linear speed and zero return time. If the incoming optical field is composed of many rays instead of a single one, the time lost for the return is equal to that fraction that it takes for the step in the helical surface to cross the whole optical field.

For a single ray there is no distortion. For an optical field deflected by a finite area of the mirrors, there is distortion, due to the fact that the helicoidal surfaces are not true planes. The amount of distortion can be largely compensated by locating the two helices in such a way that the outside rim of one faces the inside rim of the other, and vice versa, as shown in Figure 71. Extensive calculations of the residual amount of distortion left in the system under this configuration have been made and they show that this quantity is controlled by the ratio between the pitch of the helices, the diameter of the wheels and the area intercepted by the optical beam.

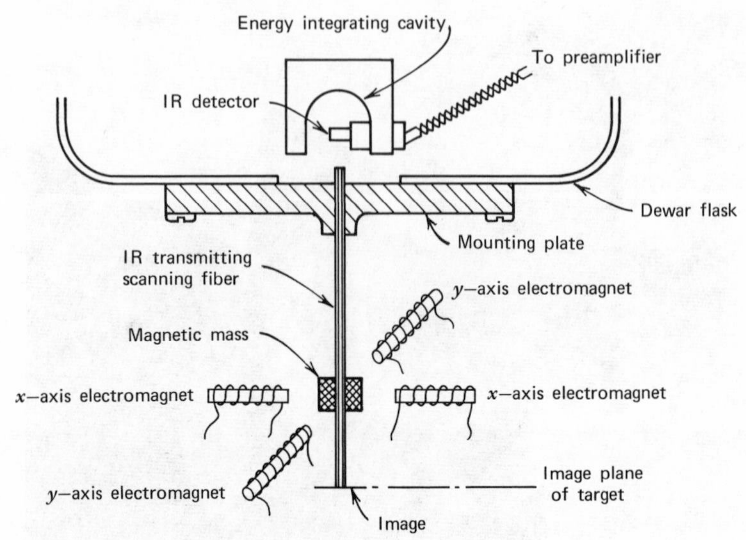

Figure 72. Infrared transmitting fiber scan system.

The Scanning Fiber

Another scanning system devised by the author, for very small targets, is composed by a single optical fiber, or by a small fiber bundle, made of infrared-transparent glass. The front end of this fiber, that as we have already seen is, really, a light-guide, scans the target, or, better yet, the image of the target in its focal plane. The infrared energy entering the front end travels through the length of the fiber and exits at the opposite end towards the detector. Due to the possibility of choosing any convenient size for the target image, the area resolution that can be achieved can vary within very wide limits.

Figure 72 describes the fiber scan system. In the illustration, the fiber motion is achieved by means of two electromagnetic fields at 90° to each other, acting on a magnetic mass located near the tip of the fiber.

Computer-Controlled Scanner

Perhaps the most exacting requirement for a scanning system designed to yield perfectly repeatable infrared signatures is the need for constant scan speed. Should this requirement not be met, the raster would be composed of lines of varying length, spaced at varying intervals from each other, and the location of the points of the target would correspondingly change in any type of output display.

For imaging devices, this would result in a loss of detail, a defect that might be tolerated. Much more serious instead is the condition for those systems where the analog signal from the detector is turned into digital information. In these systems the data are taken at preaddressed points, called "windows," located along the scan line, in correspondence of those elements of the target that are of interest. This approach greatly simplifies the signal processing, by drastically reducing the number of points to be monitored. In the case of electronic printed board assemblies, for instance, the "windows" will be located in correspondence of the center of every component part of the assembly. In this way, only the information about these component parts is processed, while all nonessential data related to the mounting board, wires, and such are ignored. Of course, to obtain repeatable results with this approach, it is necessary to assure perfect coincidence between the point to be monitored and the corresponding window. Whenever this was not assured (such as in the *Compare* System described on p. 296), serious difficulties were encountered due to the fact that the speed of the scanning elements was not perfectly constant.

The problem was solved in the *Inspect* System where the computer that opens the "windows" also controls the scanning elements by means of a shaft encoder mounted on the mirror's rotating axis. In this way, perfect coincidence between points to be monitored and "windows" is assured, and signature repeatability is consistent.

The Detector

The heart of any infrared system is the detector. In Chapter 2 we have already seen the major families of detectors and their operating characteristics. Most infrared test equipment presently existing uses the following detectors:

1. *Thermistor bolometers,* often immersed or hyperimmersed in a germanium lens, uncooled.
2. *Pyroelectric bolometer,* uncooled.
3. *Indium arsenide* cell, uncooled, photovoltaic mode of operation, long wavelength cutoff $\approx 4\ \mu$.
4. *Lead sulphide* cell, cooled or uncooled, photovoltaic mode of operation, long wavelength cutoffs: 2.8 μ at ambient, 3.8 μ at 77°K.
5. *Indium antimonide* cell, cooled at 77°K, photovoltaic mode of operation, long wavelength cutoff 6 μ.
6. *Indium antimonide* cell, cooled at 77°K, photoconductive mode; long wavelength cutoff $\approx 6\ \mu$.

7. *Mercury-Cadmium-telluride* cell, cooled at 77°K, photovoltaic mode, long wavelength cutoff 14 μ.

8. *Gold-doped Germanium* cell, cooled at 77°K, photoconductive mode, long wavelength cutoff 9 μ.

9. *Mercury-doped Germanium* cell, cooled at 30°K, photoconductive mode, long wavelength cutoff 13 μ.

10. *Copper-doped Germanium* cell, cooled at 4°K (liquid helium boiling point), photoconductive mode, long wavelength cutoff 30 μ.

11. *Germanium avalanche* photodiode, cooled at 77°k, for detecting recombination radiation emitted by Silicon or Germanium semiconductor devices.

Detectors can be used as single units, in pairs, or as multiple units, such as linear arrays or mosaics for the simultaneous coverage of two-dimensional images. When used as single units, they are made to receive, alternatively, radiation from the target and radiation from a reference body: a controlled-temperature blackbody in the better designed systems, or an ambient-temperature chopping element in the simpler units. The magnitude of the resulting AC signal is proportional to the difference in radiation level between target and reference body.

When used in pairs, the two detectors (this configuration is widely used for thermistor bolometers) are mounted in a bridge setup with opposite identical bias, as already illustrated in Figure 19. The operation of

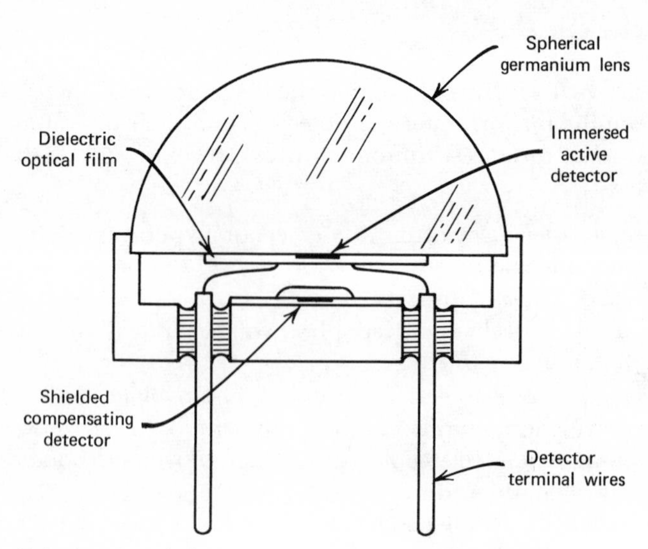

Figure 73. Twin thermistor bolometer arrangement for ambient temperature compensation. (Courtesy Barnes Engineering Company.)

Figure 74. Detector mosaic. (Courtesy Barnes Engineering Company.)

the system is based on the fact that the "active" detector receives the radiation from the target, while the "compensating" detector is carefully shielded from any radiation except that related to ambient temperature. The generated signals being of opposite polarity, they cancel out the effect of ambient temperature, and the output of the system is only related to the value of the impinging radiation. Figure 73 illustrates the physical implementation of a system of this type, utilizing thermistor bolometers.

Multiple detector units are used when faster scanning or special performance are desired. According whether the individual detectors are located in line or cover a surface, the devices are called detector arrays or detector mosaics.

Figure 74 shows one such assembly composed of 100 separate detectors that have been deposited, by evaporation in vacuum, on the same substrate. Each of these elementary units reaches an outside connector through individual conductors, and their output can be processed to reproduce in a crude visible presentation the infrared image focused onto the area covered by the detector mosaic.

Figure 48 showed two elements connected by dotted lines to the detector; a cooling unit and a blackbody reference element. The cooling unit is not always necessary, while the reference element is always present, although sometimes disguised as a chopper or supplied by the surrounding background.

The Control Electronics

The raw signal yielded by the detector must be amplified and appropriately processed for adequate display of the desired information. To this effect, the following elements must be considered:

1. Detector electrical configuration.
2. Noise characteristics and filtering methods.
3. Signal bandwith.
4. Amplifier characteristics.

General Configurations

Figure 75 shows the common connections for a photodiode detector. Figure 75*a* shows the *voltage mode* type of operation. A DC bias supply is connected across the detector and a load resistor, $R1$. The impedance of the detector is much higher than that of $R1$. The signal is AC coupled off the load resistor and connected into an amplifying system. Radiation impinging on the detector changes its impedance and a current is produced through $R1$. Signal current flowing through $R1$ produces the signal voltage. The input impedance to the amplifying system must be much higher than $R1$. The choice of $R1$ is dictated by considerations of desired frequency response, tolerable noise level, and desired gain. The simplified equivalent circuit of the detector and load resistance is shown in Figure 76. In a given application this circuit must be used to calculate expected frequency response and expected noise voltage.

Figure 75*b* shows a *current mode* configuration. Signal current produced by impinging radiation is coupled directly to a transimpedance amplifier. The input impedance is $Z_i = R_f/A$. The Z_i can typically be as low as 10 Ω. At such a low impedance level the effective signal bandwidth can be larger than in the voltage mode and noise levels at the amplifier input can be significantly lower. Also for some detectors the current mode produces better stability in the detector operating point and produces a signal much lower in harmonic content.

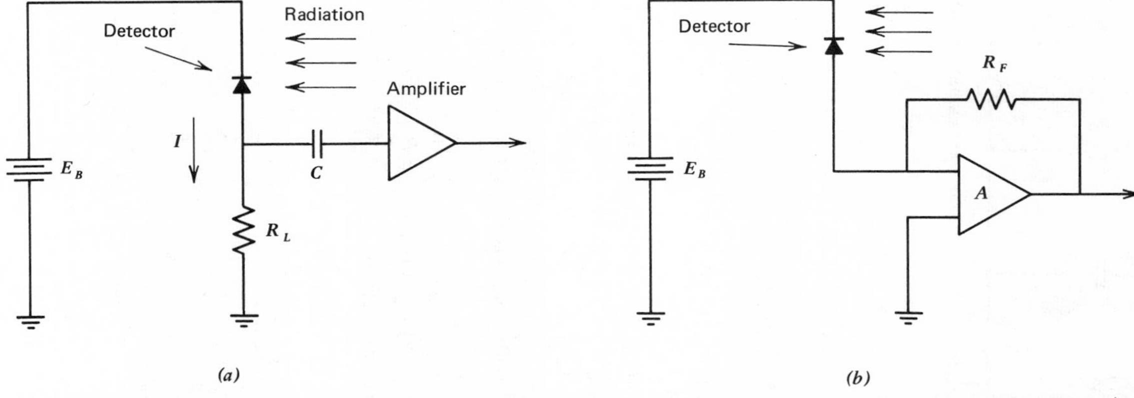

Figure 75. (a) Voltage mode. (b) Current mode. Electrical configurations for photodiode detectors. (Courtesy J. P. Ward.)

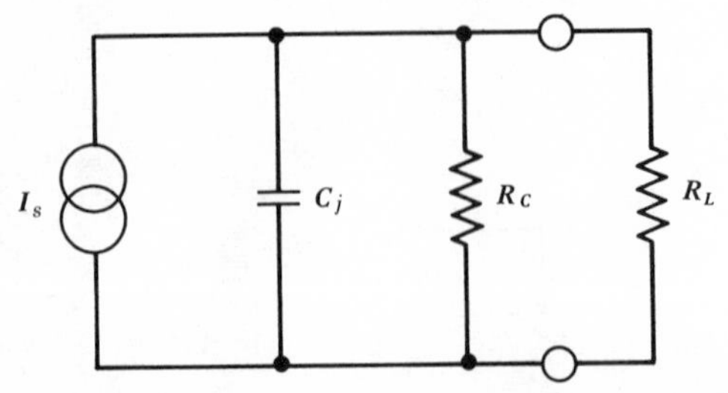

Figure 76. Simplified equivalent circuit voltage mode configuration. (Courtesy J.P. Ward.)

Noise Considerations

The major noise contributors in the system are as follows:

1. Detector noise (see Figure 77).
2. Bias supply noise.
3. Amplifier equivalent input noise.
4. Resistor thermal noise.

Detector noise is principally shot noise above about 100 Hz. Below 100 Hz, $1/f$ noise, also called flicker noise, predominates.

Bias supply noise is supply rectifier ripple, and variation with line and load. Obviously, in low-signal applications, the bias supply contributions to noise must be controlled consistent with an acceptable signal to noise ratio.

Amplifier noise is shown by a typical curve in Figure 78. The region from DC to 10 Hz for the voltage curve is the "flicker region." The region above 10 Hz is "shot noise" dominated. The i_n curve in Figure 78

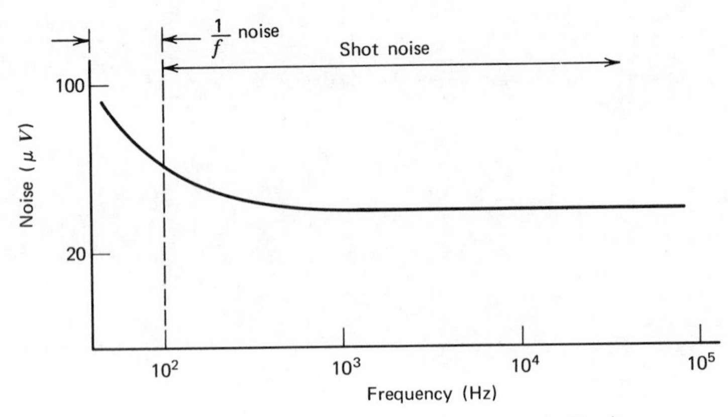

Figure 77. Typical detector noise spectrum. (Courtesy J. P. Ward.)

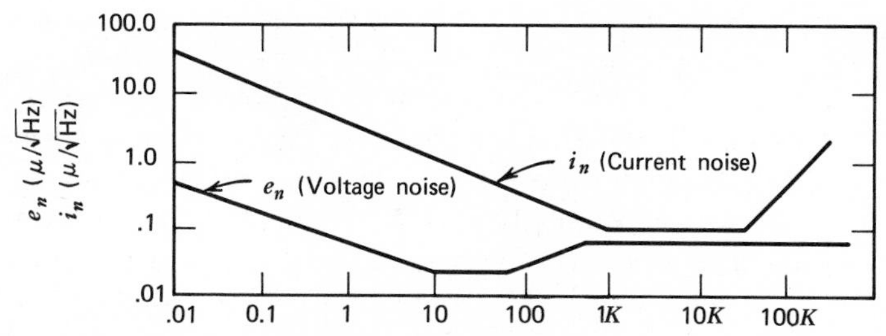

Figure 78. Asymptotes of typical amplifier voltage and current noise per root hertz bandwidth. (Courtesy J. P. Ward.)

is dominated by flicker noise up to about 800 Hz. The level section after the flicker region is shot noise. The increasing region above 10*K* Hz is noise current due to the noise voltage divided by the capacitive reactance due to the amplifier input capacitance. Of course, a particular amplifier will have different noise characteristics from that shown in Figure 78 and the selection of a proper amplifier must be determined by the constraints of the problem at hand.

For white noise, the spectral density, e_n is constant. The rms white noise in a given bandwidth, E_n, can be calculated as $e_n \sqrt{f2 - f1}$. Shot noise and Johnson (thermal noise) are both white noise. Flicker noise can be calculated as $E_n = K\sqrt{\ln\,(f2/f1)}$, where K is the value of e_n at 1 Hz.

Resistor thermal noise can be calculated using the expression Erms = $\sqrt{4KTRB}$. Here

K = Boltzman's constant = 1.374×10^{-23} J/°K
T = Absolute temperature (°K)
R = Resistance (Ω)
B = Bandwidth (Hz)

Resistor noise contributions are generated in detector load resistors and in first stage amplifier feedback resistors.

All the above-described noise sources must be rms summed and used to determine effects on system performance. Typical design techniques to improve signal to noise ratios are as follows:

1. Chop the infrared radiation at a fixed frequency and pass the signal through a narrowband filter centered at the chopping frequency.
2. Design a gated integrator that samples the signal during the gate time and integrates samples to enhance the signal and average out the noise.
3. Select a signal frequency that utilizes a low noise region of the detector and amplifier characteristics.

Bandwidth Considerations

The application dictates the bandwidth required. For instance, in a scanning application, the spatial resolution required coupled with the scanning speed will determine the upper limit of frequency response required. This limit is usually below 1*K* Hz for thermal detectors since their response does not exceed 1 msec. For quantum detectors used in fast scanning systems, the high frequency response might exceed 100*K* Hz. Low frequency cutoff is usually determined by the repetition frequency of the signal. System bandwidth should be kept as narrow as is feasible.

Amplifier Characteristics

In addition to low noise, amplifiers for infrared systems must provide:

1. Gain.
2. Linearity.
3. Wide dynamic range.

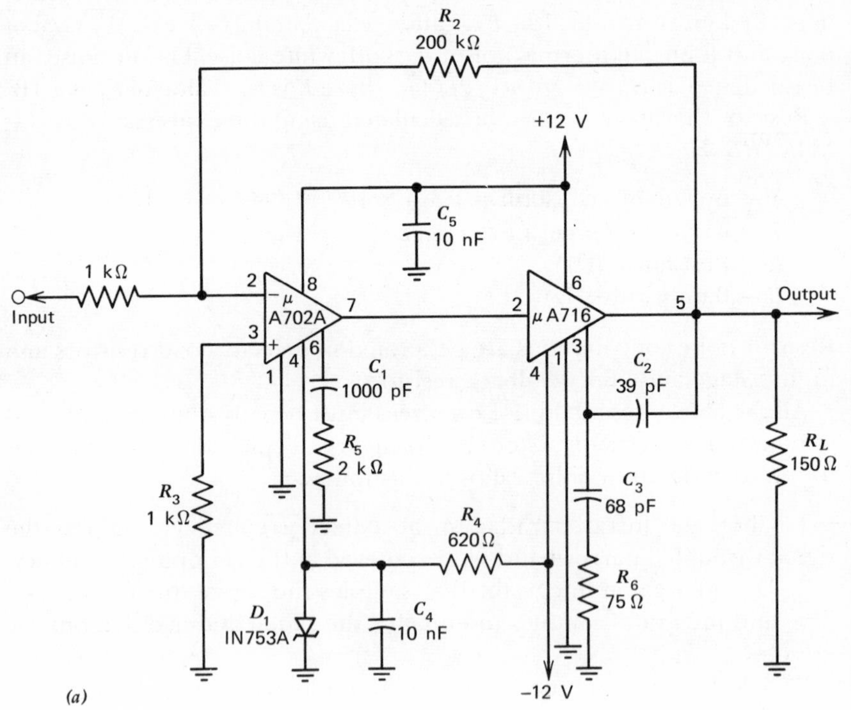

Figure 79. Fairchild linear preamplifier. (Courtesy Fairchild Semiconductor Company.)

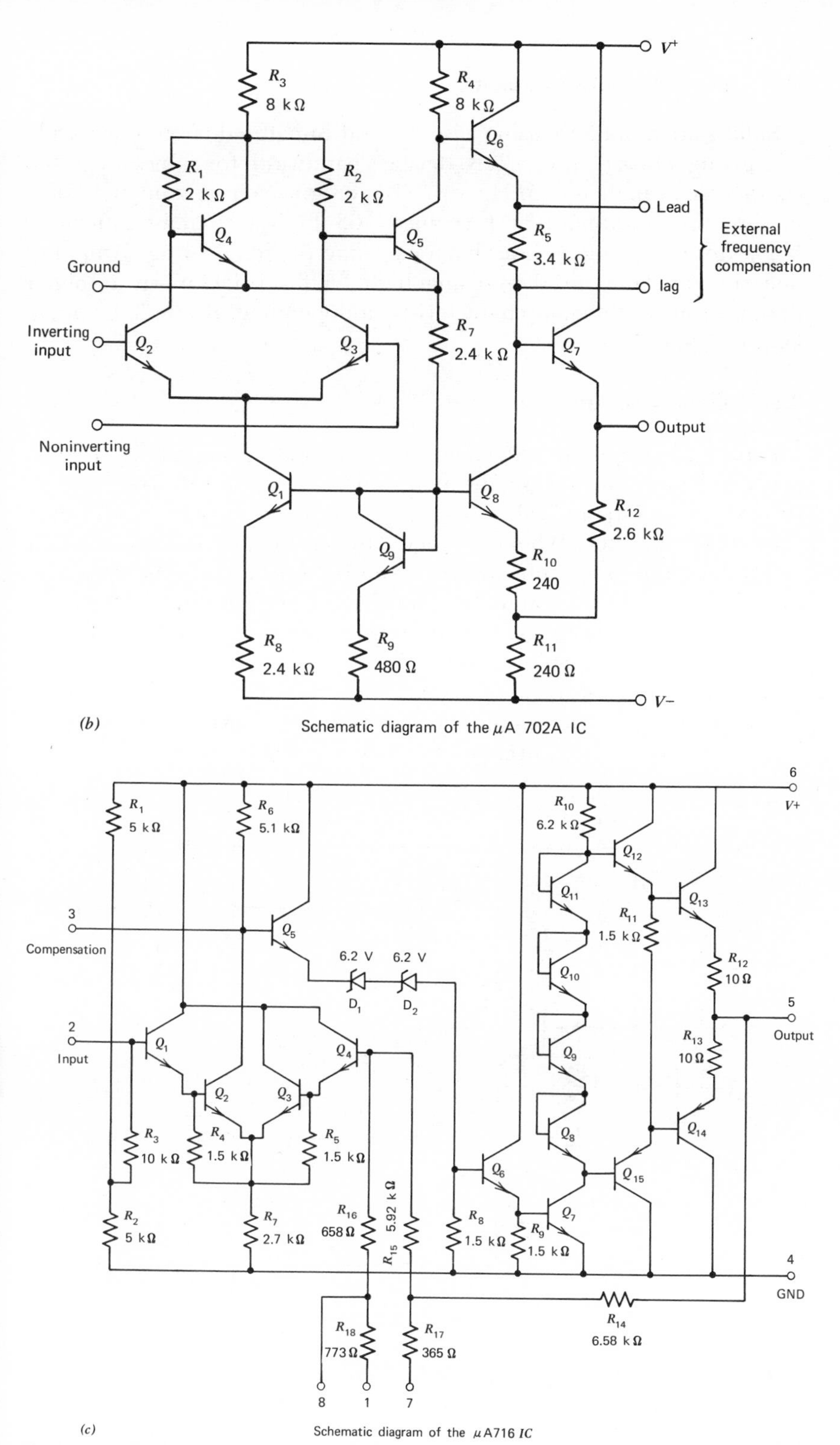

Figure 79. (Continued)

Solid-state amplifiers using discrete and integrated circuits are available giving a broad choice. Zero decibels bandwidth for some integrated circuits are as high as 65*M* Hz with a large signal voltage gain of 10,000, minimum. Typical noise performance is 63 db signal to noise ratio at 10 V *p-p* output giving a wide dynamic range of signal swing. Amplifier linearity can be assured by using high feedback factors in amplifier design to make the gain characteristic independent of amplifier open-loop nonlinearities.

Preamplifiers and Amplifiers

Figure 79 shows the schematic of a solid-state preamplifier used to process the output of a Ge:Hg photoconductive detector. To minimize stray signal pickup and cable loss, the preamplifier must be located very close to the detector. Whenever impedance matching between detector and preamplifier is a problem, an emitter follower unit can be used. Figure 80 shows the schematic of an emitter follower circuit.

When a very wide range of signal amplitudes is expected without the possibility of gain switching, a logarithmic gain preamplifier may be used to provide high gain for weak signals and lower gain for strong signals. In this manner, signal amplitude inputs ranging over 30 db may be accommodated without preamplifier saturation.

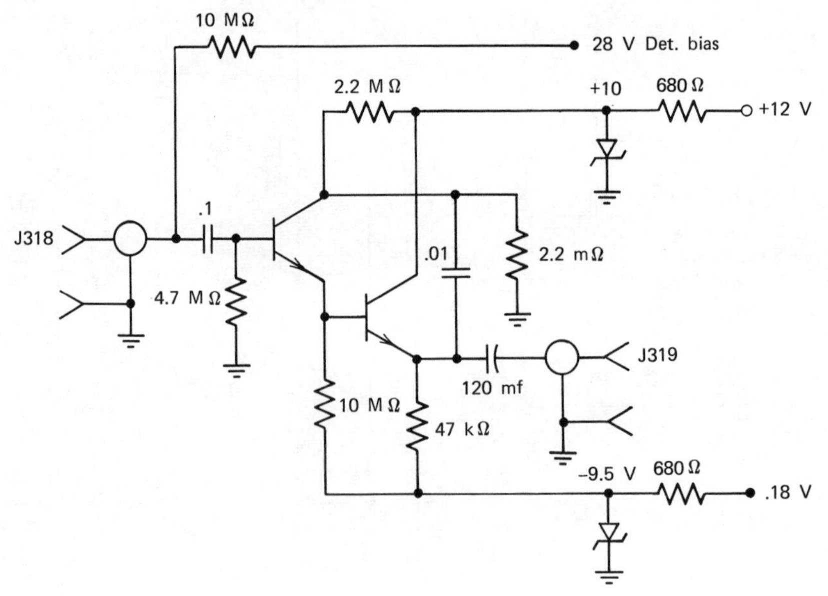

Figure 80. Emitter follower circuit. (Courtesy J. P. Ward.)

Its output is particularly suited for CRT displays of amplitude modulated scan traces. Figure 81 shows the schematic of a typical log amplifier. In this circuit the logarithmic characteristic is provided by transistor Q_1 which acts essentially as a diode.

The preamplifier drives a power or shaping amplifier that must provide the usual amplifier characteristics of large dynamic range, adequate bandwidth, and such. Its noise characteristics are, however, not as critical. It must also provide the interface match with the desired display medium and measures for attenuation in systems having very wide signal input variations.

The Display System

This is an area where a variety of techniques is applied. The major ones are as follows.

1. Oscilloscope trace display.
2. TV picture display (black and white, or color).
3. Photographic imaging (black and white, or color).
4. Digital conversion and printout display.
5. Digital conversion and generation of isothermal map.
6. Three-dimensional representation.

A detailed description of these systems can be found in Chapter 4. It will suffice here to say that every one of these systems tries to solve the problem of displaying the electrical output of the detector with the approach best suited for the evaluation of the thermal behavior of the chosen target. In general, this involves the problem of representing on a two-dimensional display a three-dimensional function that might even vary with time, hence the great variety of solutions, none of which is totally satisfactory.

Recording Systems

The most common recording devices used with infrared instrumentation are as follows.

1. Chart recorders (pen recorders, visicorders, electrocorders).
2. Magnetic tape recorders.
3. Photographic recorders.
4. Facsimile recorders.
5. Digital printout equipment.
6. Punched tape recorders.

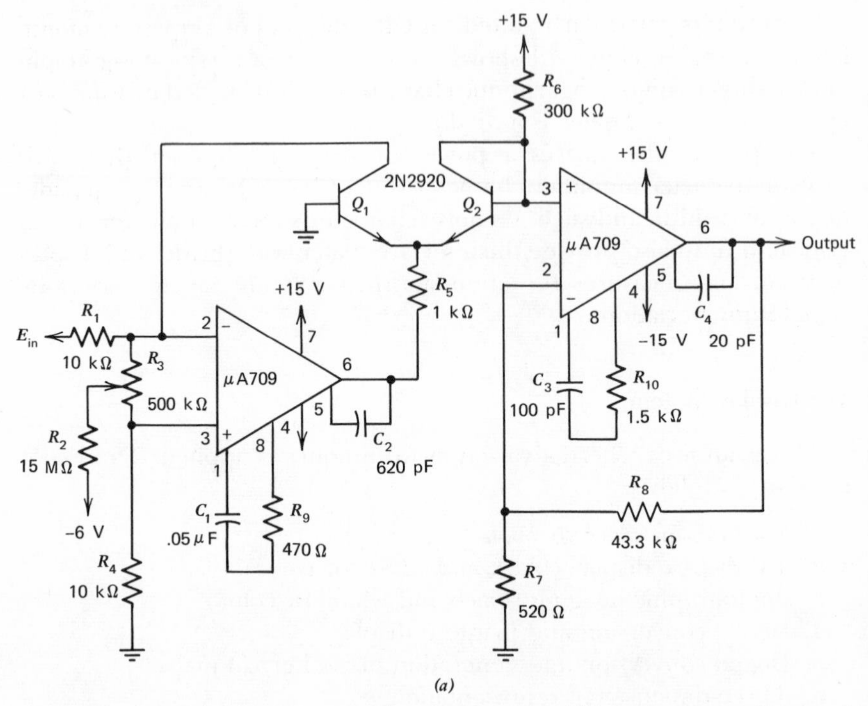

(a)

Figure 81. Fairchild logarithmic amplifier. (Courtesy Fairchild Semiconductor Company.)

Again here, the choice is dictated by the system's performance characteristics. Chart recorders are used for point detectors or for slow scanning systems, while high scanning speed requires the use of magnetic tape recorders or special photographic systems. Punched tape, facsimile, and printout equipment all require preliminary conversion of the analog output into digital information.

INFRARED RADIOMETERS

Definition

Generally speaking, all infrared measuring equipment can be divided in two classes: *spectrometers* and *radiometers.* The spectrometers are capable of measuring the discrete energy content of the radiation emitted by the target at every frequency of the spectral band; the radiometers instead measure the total energy content of the radiation emitted by the target within a predetermined spectral band.

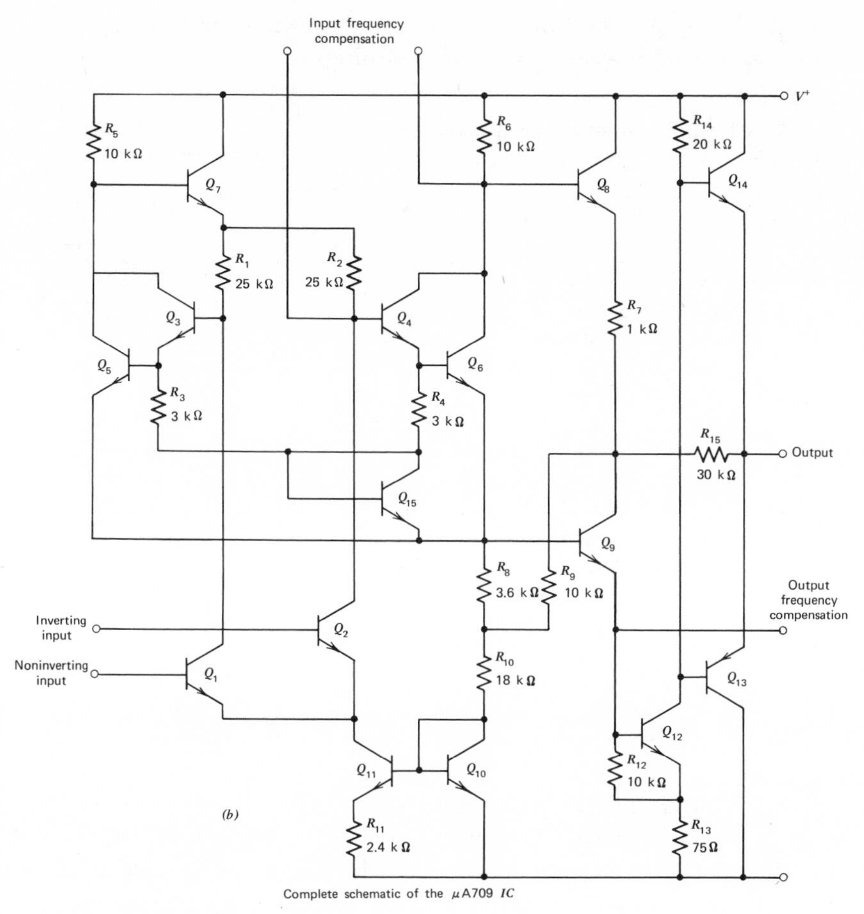

Figure 81. (Continued)

Spectrometers are widely used for analysis of the molecular structure of chemical compounds, since most of the vibrational and rotational frequencies of the molecules happen to lie in the infrared spectrum. As already mentioned elsewhere in this book, spectroscopy is outside the scope of the present work.

Radiometers are those systems whose output can yield either absolute or relative measurement of the magnitude of the infrared radiation emitted by the area viewed by the detector. The output signal is usually an analog function, whose value can be precisely measured either directly or through translation into digital information.

Radiometers can be divided in various groups or classes, according to their principle of operation, the "mode" of viewing the target, and the optics type. Table 5 lists the major groupings.

Table 5 Classification of Infrared Radiometers

Radiation Detected	Field of View	Optics Type	Description
Incoherent	Stationary	Telescope	Point detector
	Stationary	Microscope	Infrared microscope
	Stationary	Single optical fiber	Point detector
	Line-scan	Telescope	Line-scanner
	Surface-scan	Telescope	Raster-scanning radiometer
	Surface-scan	Microscope	Scanning microscope
	Stationary wide field	Telescope	Imaging device
Two-frequencies	Stationary area	Telescope	Two-wavelength radiometer
	Stationary area	Fiber optics bundle	Small area detector
Recombination	Stationary area	Single optical fiber	Point detector

These instruments can be equipped with cooled or uncooled detectors of the various types that have been described earlier in this book, and they all have an optical system more or less sophisticated, either using conventional optical elements or fiber optics. The spatial resolution, that is the ability to resolve two adjacent elements emitting different levels of infrared radiation, varies widely between very large areas (for radiometers designed to survey clouds or earth regions) to a few microns (for infrared microscopes). Logically, the spatial resolution requirements are dictated by the size of the smallest element whose infrared radiation must be measured.

Likewise, the thermal resolution requirements are dictated by the smallest temperature gradient to be detected. Also, in view of the fact that radiometers measure infrared radiation and not temperature, emissivity factors must be taken into account. Temperature resolution varies from a small fraction of 1°C (for targets around 300°K) to several degrees centigrade for infrared monitors of welding operations.

Finally, the time response requirements are dictated by the speed of the thermal transients to be detected, which for semiconductors can be very fast. Further requirements are set, for scanning systems, by the scan speed, which in the fastest systems might reach 1600 lines/sec. The limiting factor, in the time response of systems using quantum detectors, is to be found in the processing electronics.

Radiometric systems with performance characteristics useful for infrared evaluation work are made by several manufacturers, among which are the following.

A.G.A.
Lidingo, Sweden

Automation Industries
Boulder, Colorado 80302

Baird-Atomic Corporation
Cambridge, Mass. 02138

Barnes Engineering Company
Stamford, Conn. 06902

Dynarad, Inc.
Norwood, Mass. 02062

Infrared Industries, Inc.
Santa Barbara, California 93102

Mac-Iris, Inc.
Mountain View, California 94040

Ray-Tek, Inc.
Mountain View, California 94040

Servo Corp. of America
Hicksville, L.I., New York 11801

Vanzetti Infrared & Computer Systems, Inc.
Canton, Massachusetts 02021

In the following sections, examples of infrared radiometers are described for each major group listed in Table 5.

Point-Detector

A point-detector, sometimes called a "staring" detector, is the simplest radiometric system. Figure 82 shows a unit of this type made by Raytek, Inc., trademarked Circuit Ryder. It is a small, portable instrument, battery-operated, and provided with a built-in "light spot" finder that visually points out the area viewed by the detector. A single reflecting optical element, 2 in. in diameter, is the collector, and a thermistor bolometer is used as the sensor, with ambient temperature compensation provided by a twin shielded element. It is battery powered, and its smallest spot size is 0.1 in. at a 1 in. distance. This 1 : 10 distance-to-size ratio is, of course, constant, so that at 4 in. the spot size is 0.4 in. and so on. Emissivity correction can be introduced by adjusting a gain control knob located on the side of the instrument. Scanning a number of elements can be carried out by sequentially pointing on them the light beam of the unit, and by recording the meter's indication for each of them.

The temperature range in which the instrument operates varies with the model. The near-ambient region from 20 to 120°C is covered by the

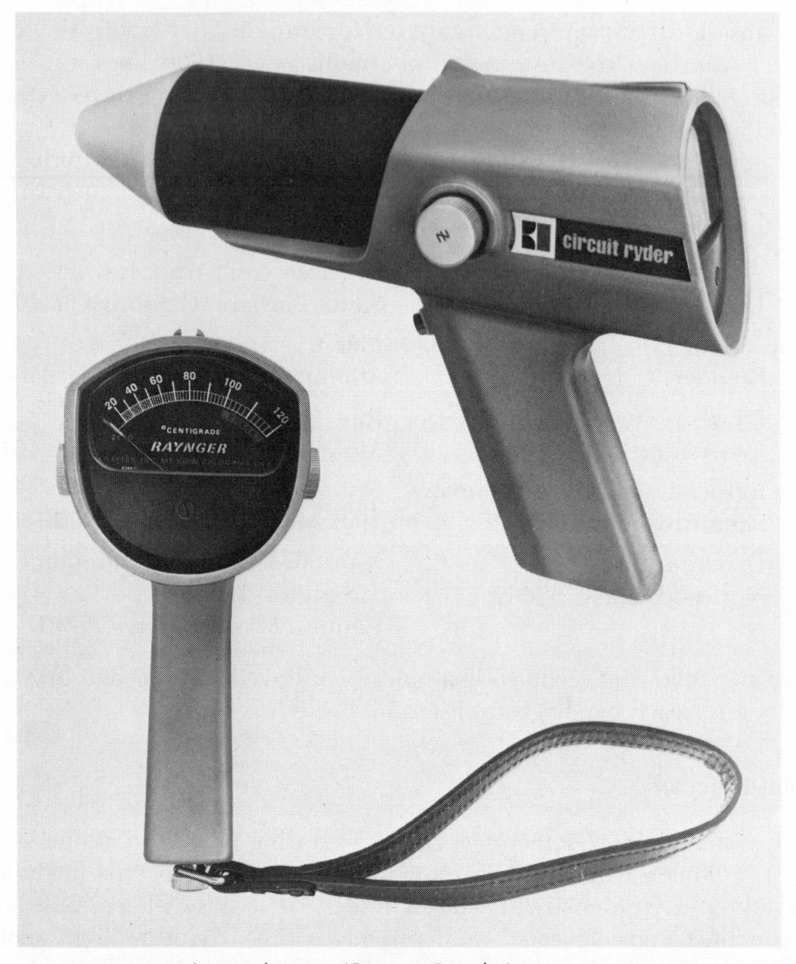

Figure 82. Infrared point detector. (Courtesy Raytek, Inc.)

model R38E-A, which was designed specifically for the evaluation of electronics. However, the instrument accuracy, given by the manufacturer as 2% of full scale reading, does not allow the fine resolution needed for today's solid-state electronics, and the instrument's usefulness is probably limited to the detection of serious conditions of component overheating.

Point-Detector with Movable Target

An instrument that utilizes a point-detecting radiometer to scan a two-dimensional surface has been designed and built by Servo Corporation of America.

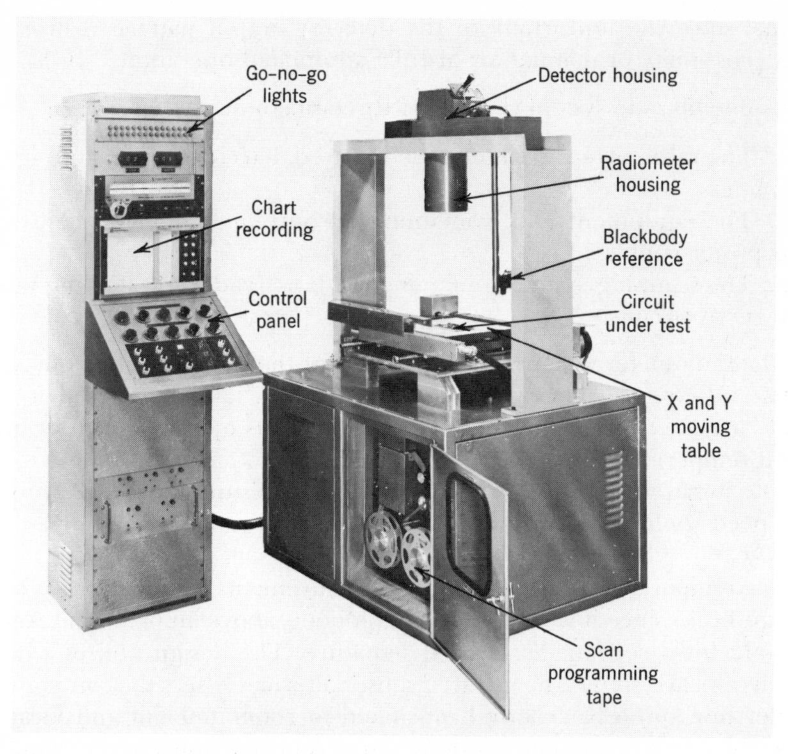

Figure 83. Infrared automatic test station for printed circuit boards and electronic components. (Courtesy Servo Corporation of America.)

In this instrument, (Figure 83) the detector is stationary, while the target, mounted on an *x-y* table, is moved in the horizontal plane according to a program punched on a control tape, and designed to bring into the area viewed by the detector every element of interest, in a predetermined sequence.

The system has the following advantages:

1. Fixed focus; for a planar target the focal distance is always the same, since every point is moved into the same viewing area.
2. No detector time wasted looking at background or at areas between components.
3. Visually identifiable target area: a spot of visible light is projected onto it during the operation.
4. Viewing of the components can be programmed in any desired sequence: for instance, a most desirable scanning sequence is one where the radiation level steadily decreases (or increases) from the first to the last element. This distribution makes it easy to detect defective compo-

nents, since the uniformity of the detector output pattern is broken.

5. Possibility of manual, or of fully automated operation.

Essentially, the system consists of three major subsystems:

1. The radiometer containing the infrared detector, optics, and signal amplifiers.
2. The equipment rack containing all output equipment, controls, and power supplies.
3. The scanning system containing the tape reader, positioning table and servo circuits.

The radiometer was specially designed for the testing of electronic circuits and components. The spot size of 1 mm^2 and the sensitivity of 0.1°C are adequate for testing small components operating at near ambient temperature.

The infrared detector is an immersed thermistor bolometer. It does not need cooling and is stable over the thermal range from ambient to 1000°C.

The temperature reference for all measurements is *room ambient.* This is significant since it is the rise of components above ambient that is important for a repeatable infrared signature. The design utilizes a dual beam optical system. The infrared sensor alternately sees the component under test and a blackbody heat-sinked to room ambient and located near the component. The resulting signal is proportional to the temperature difference between the component and its immediate environment.

The infrared optical system uses germanium lenses. A concentric, doughnut-shaped additional lens projects a small visible light spot on the detector's field of view to aid the operator in aligning the target.

In the *manual mode, X* and *Y* position data are fed in via knobs located on the control panel. Upon pressing a button the table will advance to the selected point. Another button then starts the infrared test, whose data is recorded on the chart located on the control rack.

The manual mode can also be used for the generation of a punched tape, since *X* and *Y* coordinates of components can be established by the use of the projected light spot. When testing printed board assemblies, minimum and maximum control limits for any electronic component can be established by recording the radiation value of the measurements taken for said component on a number of boards of the same type.

In the *automatic mode* the station is tape controlled and as many as 3000 measurements can be made without operator intervention.

Upon pressing a button, the automatic run begins and each test programmed on tape is executed. The run can be stopped either by manual control or by a "hold" code programmed on the tape. After the run the operator can read the go, no-go indicators and thus identify the components which failed the test.

The devices available at the output are the following:

1. A logarithmic scale voltmeter calibrated in degrees Celsius above ambient.
2. An *X-Y* recorder which records temperature in a bar-chart format thus generating an infrared "signature" for each circuit.
3. A set of go, no-go lights indicating whether the temperature reading falls between the preprogrammed minimum and maximum for the particular point or components.

As an alternate, the output equipment can be modified by the addition of a digital printer which replaces the *X-Y* recorder as the display medium.

Line-Scanner

Instruments of this type move the detector field of view back and forth along a stationary line. Most of the time they are used to monitor moving targets, such as a ribbon of material (paper, for instance) moving along rollers. In other applications, the radiometer itself is moved either lineally or radially to allow coverage of a two-dimensional target. Often they are designed and developed for special purposes, such as the high speed line scanner built by Lockheed Missile and Space Corporation for the U.S. Naval Weapon Station of Concord, California, and designated as the Mod. VI Scanner. Figure 84 shows its optical schematic: the radiation from the target is collected by the off-axis elliptical mirror after having been folded by the scanning plane mirror. This radiation is chopped by an element rotating at 4000 Hz at the primary focal point, from where it is transferred by a 1 : 1 relay lens system to the infrared detector. When the chopper is in such a position as to block the radiation from the sample, it reflects the radiation from the blackbody reference source. In this manner the detector continuously compares the target's radiation with the reference.

The optics have three interchangeable off-axis ellipsoidal mirrors which allow focusing targets located at a distance of 6, 10, and 17 in. The spot-size is correspondingly 0.6, 1.0, and 1.7 mm in diameter, while the scan line length is 0.35, 0.59, and 1.0 in., respectively.

The scanning mirror rate is 100 lines/sec obtained by oscillatory excursions of 3° of the mirror at its resonant frequency: this makes it possible to use relatively low power for the motion (about 200 mW), in spite

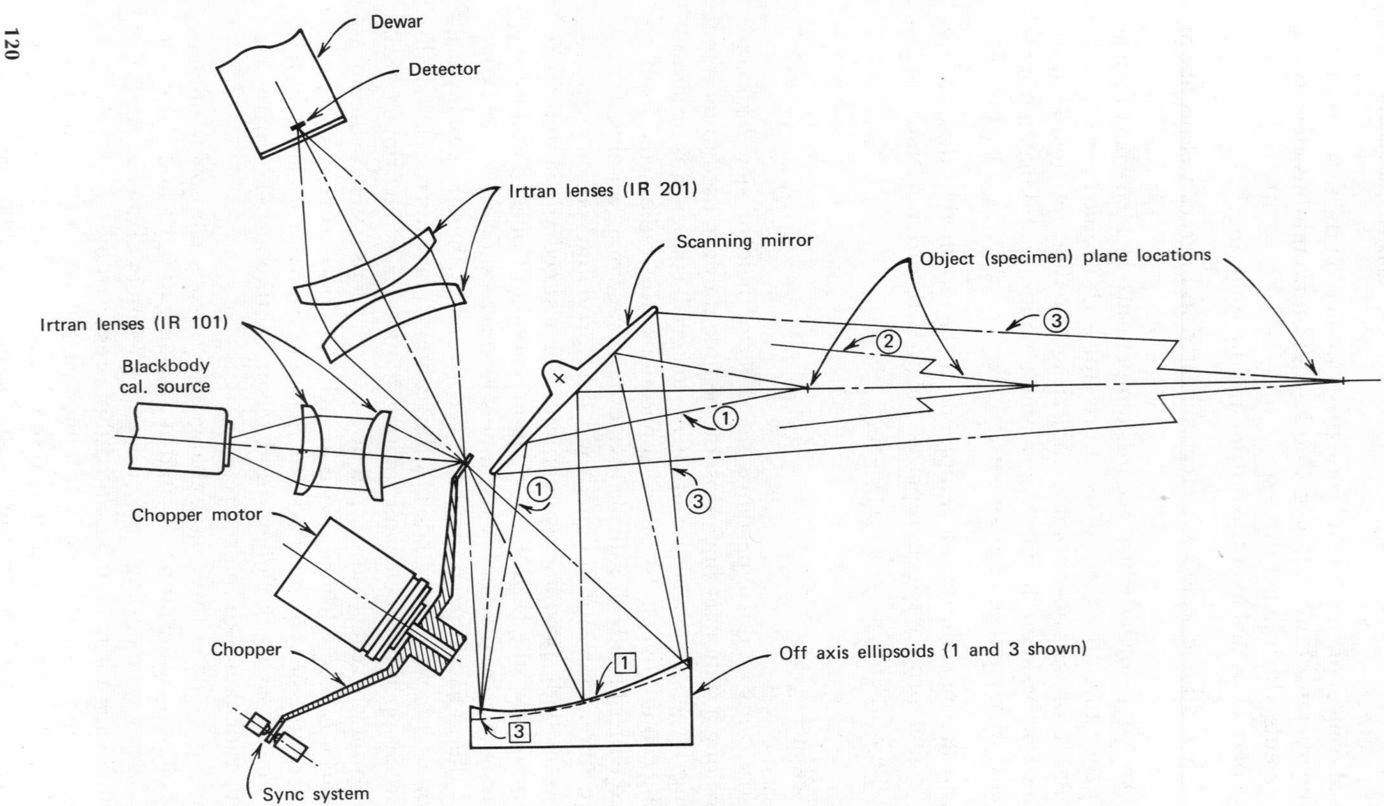

Figure 84. Optics schematic for model VI scanner. (Courtesy J. W. Patterson and R. A. Wallner.)

Figure 85. AGA thermovision Model 665 infrared camera. (Courtesy AGA Corporation.)

of the fact that the physical size of the mirror is rather large (2.5 in. diameter). The detector is a gold-doped germanium photoconductive detector cooled to liquid nitrogen temperature inside the Dewar flask. The noise equivalent temperature of the system is 0.5°C with a target temperature of 50°C.

Raster-Scanning Radiometer

While line-scanning equipment finds use in specialized applications, raster scanning radiometers lend themselves to a wide range of applications in medicine, earth sciences, night reconnaissance materials, and electronics evaluation. Among the most interesting and recent equipment in this group is the AGA Thermovision Model 665 area-scanning radiometer built in Sweden by AGA Corporation. The system shown in Figure 85 is remarkable for its fast scanning speed, for the use of a silicon polygon whose rotation performs the line scan, and for the use of refractive optics in addition to reflective elements.

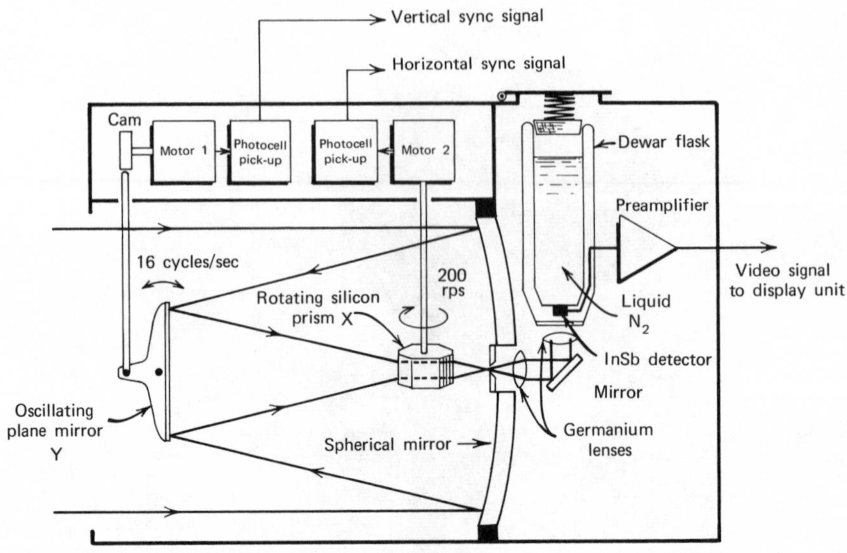

Figure 86. AGA thermovision schematic diagram of Model 665. (Courtesy AGA Corporation.)

As shown in Figure 86, the detector is an InSb cell housed in a 4-hour Dewar for liquid nitrogen. The focus can be adjusted anywhere between 20 in. and infinity. Scanning in the vertical direction is performed by the front mirror *Y*, oscillating 16 times/sec, while the rotating silicon prism *X* scans 1600 lines/sec. The combined action thus yields 16 full frames having 100 lines each. The optical resolution is about 100 points per line.

The video signal generated by the system is forwarded to a crt viewing system where an intensity-modulated display is generated, either as a black and white, or a color "picture" of the target, or a quantized thermal map.

Other characteristics of the system are as follows.

Field of view: 11° × 11°
Temperature resolution: 0.2°C
Thermal range: maximum sensitivity 1°C
minimum sensitivity 200°C
Target thermal range: −30 to +200°C
Capability of superimposing isotherms onto the Crt intensity-modulated display

The Infrared Microscope

When the physical size of the target is very small, such as in the case of the semiconductor "chips" on which transistors and integrated circuits

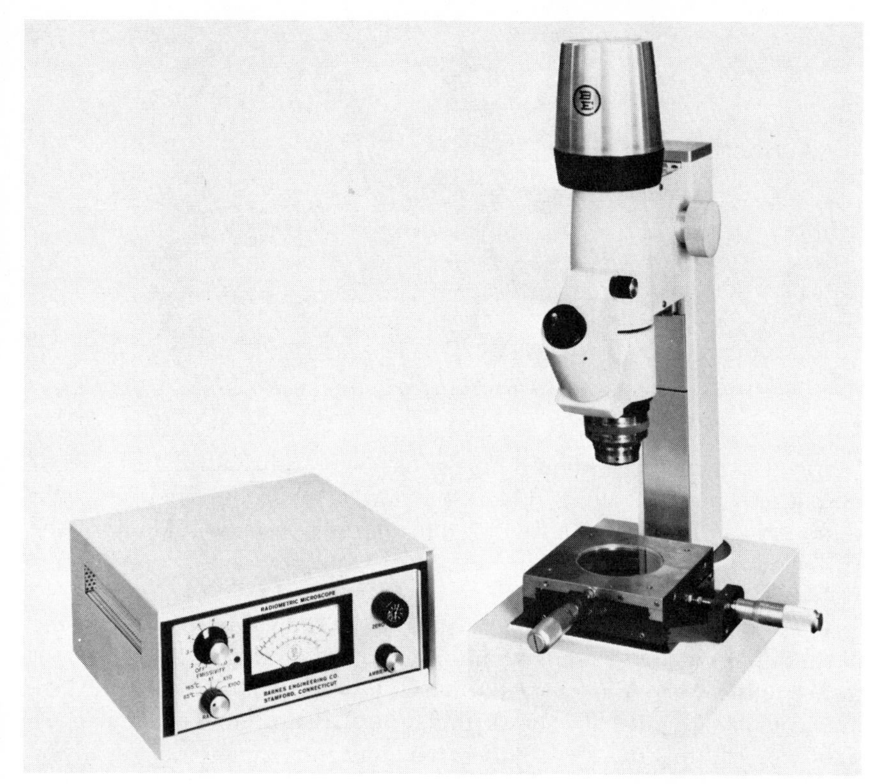

Figure 87. Barnes infrared microscope Model RM-2A. (Courtesy Barnes Engineering Company.)

are deposited, the spatial resolution of the radiometers so far described becomes inadequate, and a change in the optics must take place, from telescope to microscope.

Furthermore, the reduction in surface of the elementary area upon which the detector is being focused, causes a corresponding reduction of the infrared signal reaching the detector.

This can be in part compensated by decreasing the /*f* number of the optical system, so that a larger cone of energy is captured by the optics and forwarded to the detector.

Figure 87 shows the infrared microscope made by Barnes Engineering Company. It is designated as their Mod. RM-2A and it consists of two major elements: the detector unit, which includes the microscope, substage, detector cell, and coolant reservoir, and the control unit, which contains all of the solid-state circuitry for amplifying, filtering, and demodulating the signal from the detector and displaying it on a meter.

The infrared detector is an indium antimonide cell mounted in a

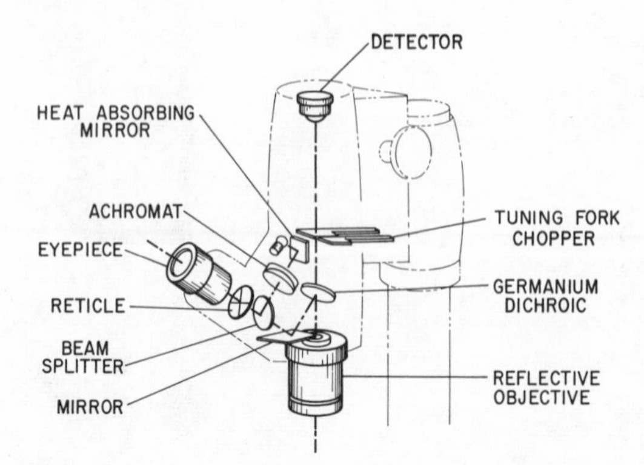

Figure 88. Optical diagram of RM-2B microscope. (Courtesy Barnes Engineering Company.)

Dewar flask which has a liquid nitrogen capcity sufficent for 12 hours of operation without refilling. The instrument is capable of operating continuously for an indefinite time, requiring only periodic refilling of the liquid nitrogen reservoir. The reservoir may be refilled without interrupting operation.

The reflective optical system permits visual observation and focusing of the subject with the aid of a built-in vertical illuminator. Reflective objective lenses with nominal resolution of 0.0014, 0.0007 and 0.0003 in. are available. Figure 88 shows the optical diagram of a similar instrument, equipped with a noncooled detector, namely, a thermistor bolometer.

The solid-state control unit provides external connections for an *X-Y* plotter, strip chart recorder, or oscilloscope. The target is mounted on a micrometric substage whose operation can be manual, semi-, or fully-automatic. Since the microscope is essentially a point-detector, the scanning takes place by moving the target, turning the micrometers controlling the *X* and *Y* position of the substage. Figure 89 shows one of the substages available for use with the instrument: it offers the options of manual or motor-driven operation, with a wide choice of scan speeds. The finest spatial resolution attainable by the instrument is 0.0003 in. Thermal resolution at 30°C is 0.5°C. The chopping frequency used is 400 Hz, but the instrument can also be used without chopping, in the so-called transient measurement mode. The spectral range covered by the unit equipped with an In-Sb detector is from 1.8 to 5.5 μ in wavelength.

Fast-Scan Infrared Microscope

Beyond a certain speed, physical movement of the target cannot be achieved without the danger of damage to it. But the extremely close

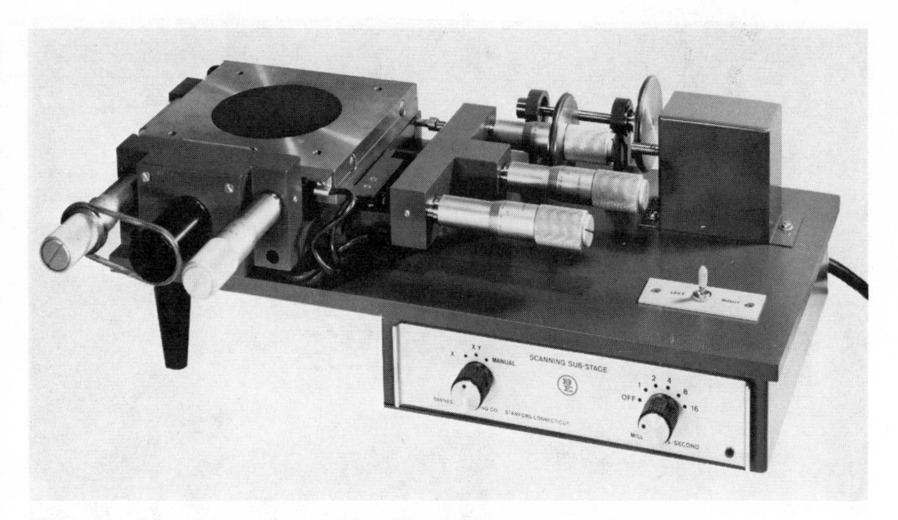

Figure 89. Micrometric substage. (Courtesy Barnes Engineering Company.)

spacing of the active elements deposited on a semiconductor chip makes it desirable to scan the target during warmup time, before thermal interaction between these elements takes place. It is at this point in time that electrical, physical, and mechanical properties of each and every component of an integrated circuit can best be evaluated by infrared, while thermal contrast is still good. After the heat emitted by the power dissipating elements has thermally flooded the unit, much detail disappears and only the major gradients are still visible.

Upon these considerations, the fast-scan infrared microscope was developed by Raytheon Company for the National Aeronautics and Space Administration. The instrument, mainly intended for evaluation of microelectronic circuitry, is shown in Figure 90. Clearly visible at upper left is the coaxial optical focusing system that allows the operator to view the target and to identify the exact location of the line being scanned. Also visible are some elements of the display and recording equipment located in the console at right.

Figure 91 is the block diagram of the instrument and Figure 92 outlines the optics and the scanning elements. Magnification ratio is 1 : 7.6, obtained through the use of a single aspherical reflecting element of special design. The $/f$ number of 1.1 allows the capture of a large cone of energy radiated from the target, so as to still have a measurable signal, in spite of the extreme reduction in the area being viewed by the detector. The maximum performance specifications of the instrument are as follows:

Figure 90. Fast scan infrared microscope.

Area resolution: 100 μ
Temperature resolution: 2°C
Number of scan lines per frame: variable, up to 200
Number of frames per second: variable, up to 4
Area viewed in one frame: 1 mm^2
Detector response speed: <1 μsec

To meet these requirements, the design of the instrument had to reach the very limits of the state-of-the-art in the areas of optics, scanning efficiency and detector sensitivity. As a result, the optics are diffraction limited, the scanning employs the unique system described in p. 98 that reaches close to 100% efficiency, and the detector is an In-Sb cell. Cooling of the detector to 77°K takes place in a Dewar containing an 8-hour supply of liquid nitrogen.

To eliminate the inconveniences connected with the use of liquid coolants, an optional feature of the instrument is a cryogenic system equipped with a Ge-Hg detector whose sensitivity approximates very closely the performance requirements in the thermal range of interest.

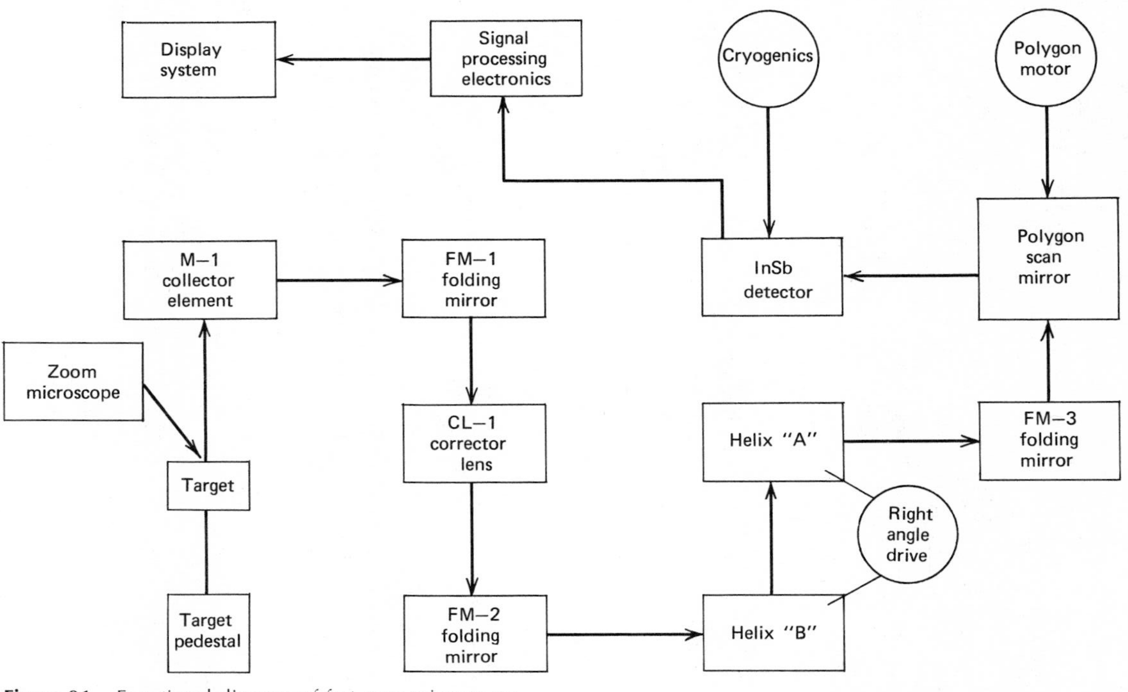

Figure 91. Functional diagram of fast scan microscope.

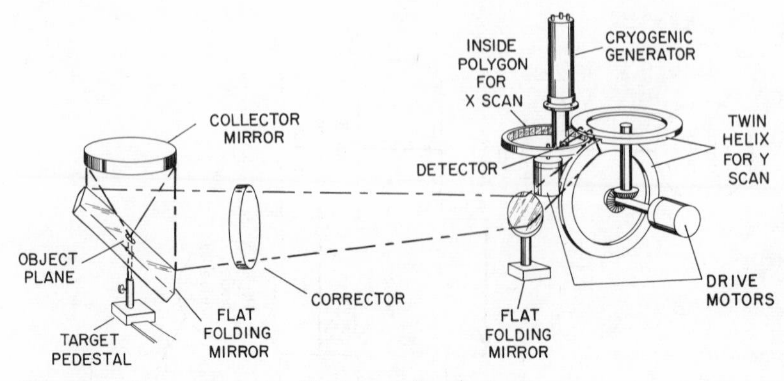

Figure 92. Optics schematic of fast scan infrared microscope.

The use of the instrument is thus simplified, since the detector cooling takes place automatically by simply turning the AC power switch on. The system's output is an analog signal composed of frequencies as high as 100 kHz and having a dynamic range of 60 dB, complete with line and frame sync pulses.

Two-Wavelength Radiometer

This instrument was invented in order to eliminate errors due to emissivity variations. Its operation is based on the assumption that the incoherent radiation emitted by a nonblackbody surface follows a "graybody" distribution, which is identical to the blackbody curve, although it is located at a lower level in the power scale.

For many surfaces this assumption can be taken as true in first approximation, and consequently every temperature of the target is precisely identified by a unique value of the ratio of the power radiated at two different wavelengths, A and B, conveniently chosen, independently from the level at which the corresponding graybody emission curve is located.

Figure 93 illustrates this concept better than any words. In (*a*), three radiation curves: 1, 2, and 3, are shown respectively for a blackbody at 1000°K, 2000°K, and 3000°K. In (*b*), three "graybody" curves: I, II, and III, are showing the radiation from a surface of 0.1 emissivity respectively at 1000°K, 2000°K, and 3000°K. If we choose the two wavelengths A and B at a safe distance from the radiation peaks, we can see that the ratios of the values $A_1:B_1$; $A_2:B_2$; and $A_3:B_3$ are identical to the ratios of the values $A_{\mathrm{I}}:B_{\mathrm{I}}$; $A_{\mathrm{II}}:B_{\mathrm{II}}$; and $A_{\mathrm{III}}:B_{\mathrm{III}}$.

In the former example, for simplicity, we have chosen the value of 0.1 for the emissivity of the graybody surface, but the law is equally valid for any value of the emissivity. For instance, in Figure 93 (*c*) the graybody emissivity is made equal to 0.5 simply by sliding the radiation curves down one half order of magnitude to indicate that the graybody emits exactly half the power radiated by the blackbody. Still the ratio of the power emitted at the chosen wavelengths *A* and *B* remains unchanged for each of the emission curves.

In conclusion, the two-wavelength radiometer is emissivity-free, since it measures the temperature of a target independently from the emissivity value of its surface. This is quite a convenient characteristic, in those instances where the value of surface emissivity is not known, or when it varies during the measurement process due to changes in surface characteristics such as oxidation and crystallization.

The two-wavelength principle is currently used in processes conducted at high temperature where the radiation emitted by the target contains a significant portion of the visible spectrum. The instruments used for this purpose are called "two-color pyrometers" and are often used to control metal melting and forging operations.

Separation of the radiation emitted at the chosen wavelengths is made by using "notch-filters" that reject all the incoherent radiation contained in the rest of the spectrum.

However, for targets whose radiation is located in the infrared spectrum only, the two-wavelength principle is difficult to apply because of the low level of the power radiated by the target's surface. This difficulty is further increased by the fact that only the radiation emitted at the chosen wavelengths *A* and *B* is measured, and this is just a very small fraction of the total graybody emission which is already small. Consequently, the lower the target's temperature, the more difficult becomes the task of measuring it with the use of the two-wavelength principle.

One instrument whose operation is based on the two-wavelength principle is the Weld Monitor described in the following pages.

Weld Quality Monitor

When the target is located in such a way as to preclude direct viewing by the detector, the use of fiber optics instead of conventional optical systems allows the detector to "see" around corners or inside opaque enclosures. The optical principle upon which the use of optical fibers is predicated was previously described in this same chapter.

The principle of two-wavelength radiometry is used in the Weld Quality Monitor, an instrument developed by Raytheon Company for process control of welds of small size, such as those used to make interconnections in wiring electronic components. An optical fiber bundle is used to

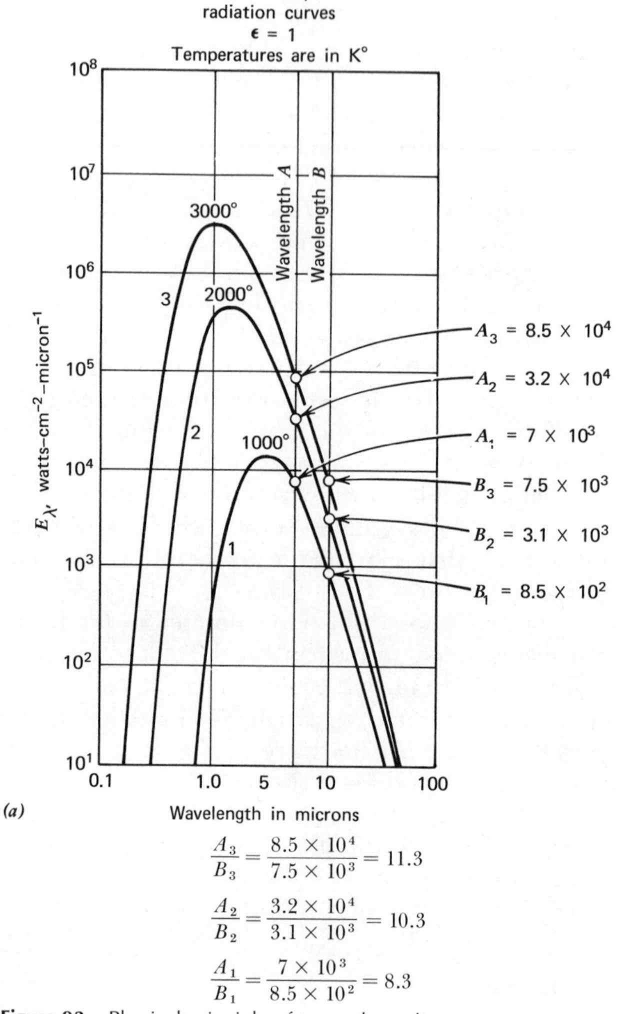

$$\frac{A_3}{B_3} = \frac{8.5 \times 10^4}{7.5 \times 10^3} = 11.3$$

$$\frac{A_2}{B_2} = \frac{3.2 \times 10^4}{3.1 \times 10^3} = 10.3$$

$$\frac{A_1}{B_1} = \frac{7 \times 10^3}{8.5 \times 10^2} = 8.3$$

Figure 93. Physical principle of two-color radiometer.

pick up the infrared radiation from the weld area, while the weld is made. The fiber bundle is split in two equal branches that transmit to two separate detectors the radiation, after this has passed through two narrowband filters. This eliminates possible errors due to emissivity variations of the surface observed. The instrument is capable of giving temperature measurement of the weld nugget while it is being made, regardless of weld size, and with a high degree of precision, namely ±15°C at the melting point of steel.

This information is displayed in real time through a set of colored

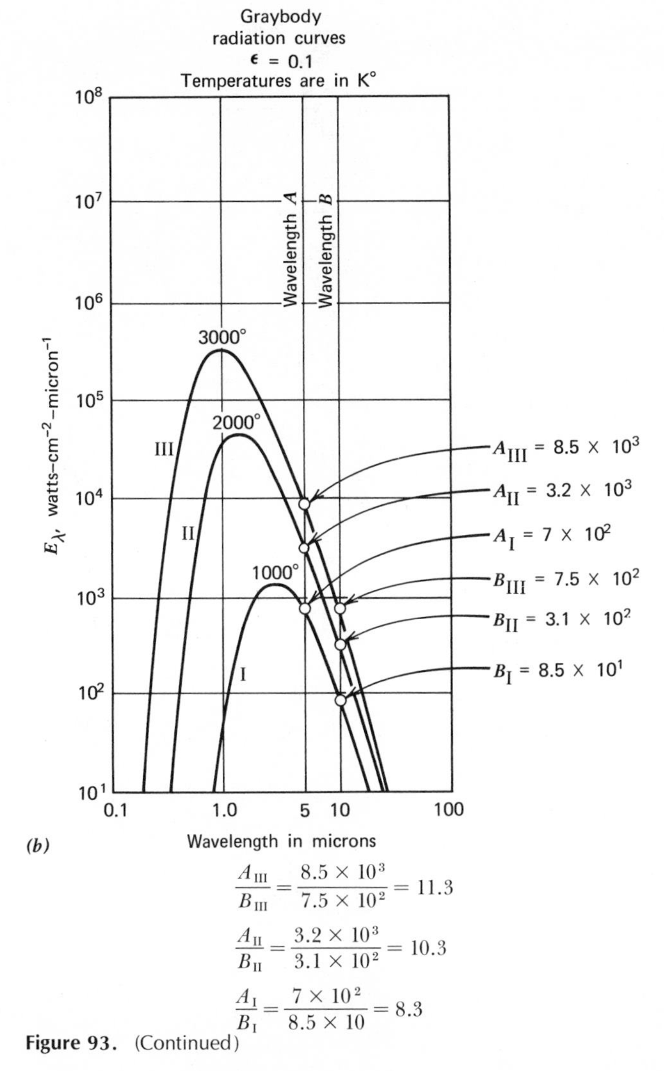

$$\frac{A_{\text{III}}}{B_{\text{III}}} = \frac{8.5 \times 10^3}{7.5 \times 10^2} = 11.3$$

$$\frac{A_{\text{II}}}{B_{\text{II}}} = \frac{3.2 \times 10^3}{3.1 \times 10^2} = 10.3$$

$$\frac{A_{\text{I}}}{B_{\text{I}}} = \frac{7 \times 10^2}{8.5 \times 10} = 8.3$$

Figure 93. (Continued)

lights that turn on according to the temperature to which they have been preset by the operator. Figure 94 shows the instrument: the low setting level is indicated by a yellow light, the optimum weld temperature by a green light, and the high setting by a white light.

Since weldments take on a variety of geometries, it is necessary to taylor the sensing system to each application individually. Normally, welding engineers routinely accomplish this as a part of electrode and joint design, and the flexibility of the fiber optic system provides a simplification to the normal problems of focusing and accessibility.

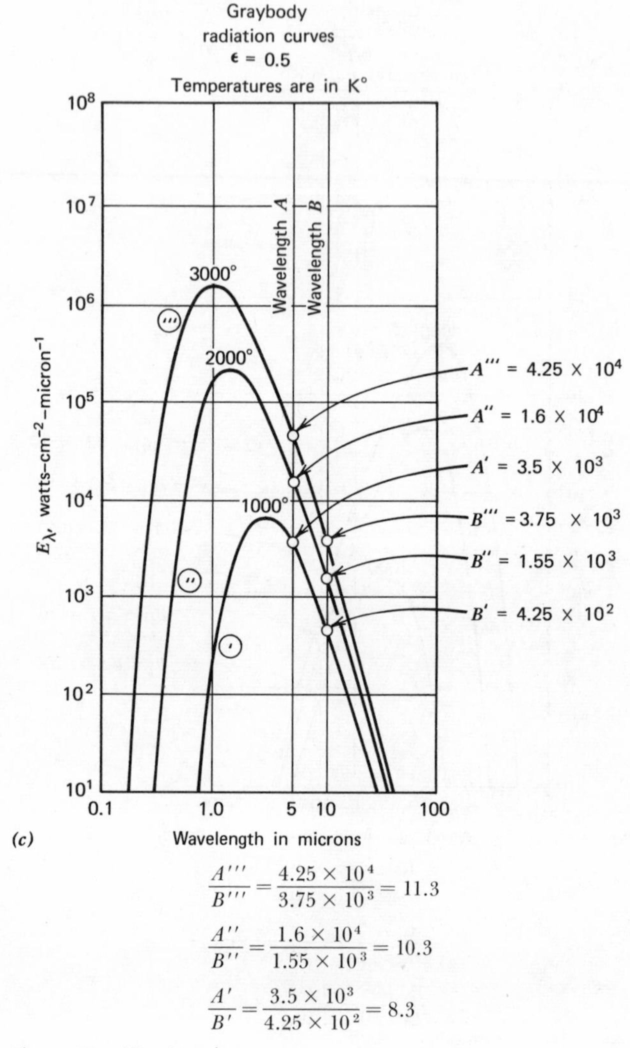

$$\frac{A'''}{B'''} = \frac{4.25 \times 10^4}{3.75 \times 10^3} = 11.3$$

$$\frac{A''}{B''} = \frac{1.6 \times 10^4}{1.55 \times 10^3} = 10.3$$

$$\frac{A'}{B'} = \frac{3.5 \times 10^3}{4.25 \times 10^2} = 8.3$$

Figure 93. (Continued)

According to Raytheon[6] it is possible to evaluate almost any type of weld using the fiber optic weld sensor, providing the sensing tip geometry is worked out.

Thermal Monitor

This instrument also uses fiber optics instead of conventional optical systems. However, in this case, a single optical fiber is used to transmit the infrared radiation from the target to the detector.

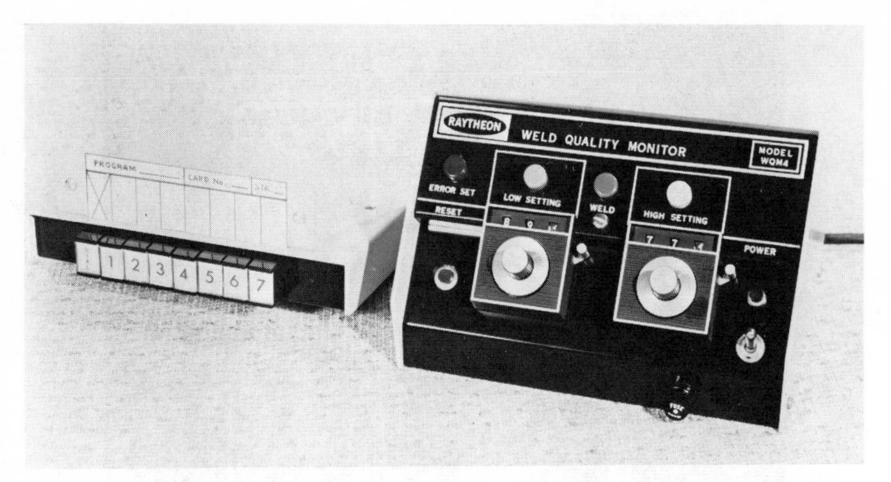

Figure 94. Weld quality monitor. (Courtesy Raytheon Co.).

The instrument is essentially a simple system, composed of an optical assembly (fiber in a stainless steel jacket) a detector unit (chopper, detector, and bias supply), and an electronic console for control and display of the signal. Figure 95 shows these three basic components designated as *A*, *B*, and *C*.

A typical application of this instrument is the process control of the die-attach operation in semiconductor manufacturing. This is achieved by inserting the front end of the optical fiber into the vacuum collet that holds the semiconductor chip. This insertion is done through a tiny hole especially drilled through the upper part of the collet, so that the front end of the fiber faces the semiconductor chip from a short distance.

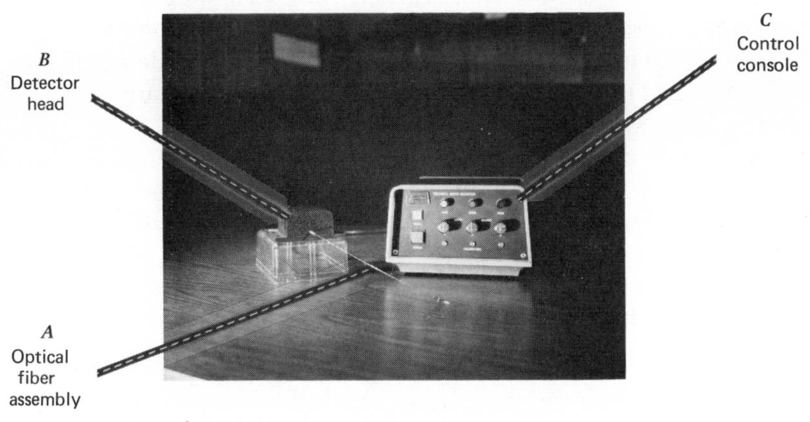

Figure 95. Thermal monitor.

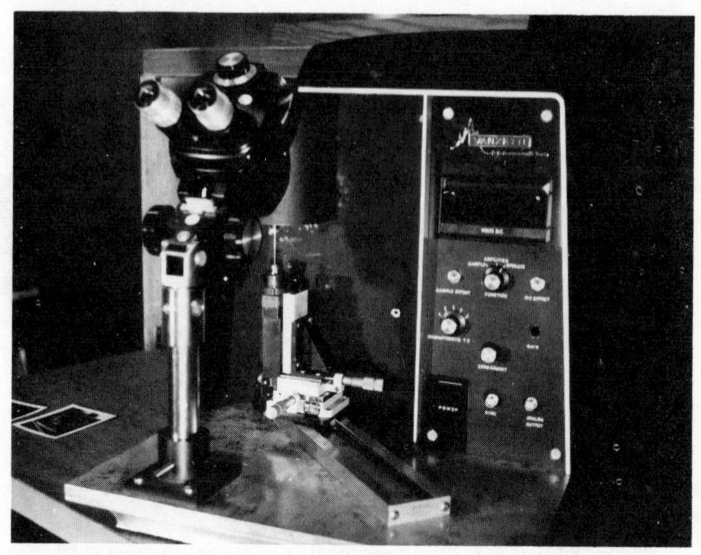

Figure 96. Semiconductor junction analyzer.

Typical dimensions are as follows:

Optical fiber: 14 in. long; 0.006 in. diameter

Fiber-chip distance: min. 0.020 in., max. 0.050 in. (according to chip size)

During the bonding operation, the detector output is a voltage signal whose magnitude is directly correlated to the chip's temperature, and to its variations in real time.

The control console of the system uses this voltage to sequentially turn on three warning signals: the first, a yellow light to tell the operator that bonding temperature has been reached and "scrubbing" should begin; the second, a green light to signal that a good bond has just been made; the third, a red light that indicates overheating, thereby identifying thermally damaged units that might otherwise go undetected and cause early failures.

The system, developed by Vanzetti Infrared & Computer Systems, Inc., can also be used to run through a feedback loop a thermocompression bonder, thus turning the die-attach process into a fully automated operation.

Semiconductor Junction Analyzer

This instrument, shown in Figure 96 was developed to allow noncontact measurement of the current flowing through a semiconductor junction in discrete devices and also in integrated circuits.

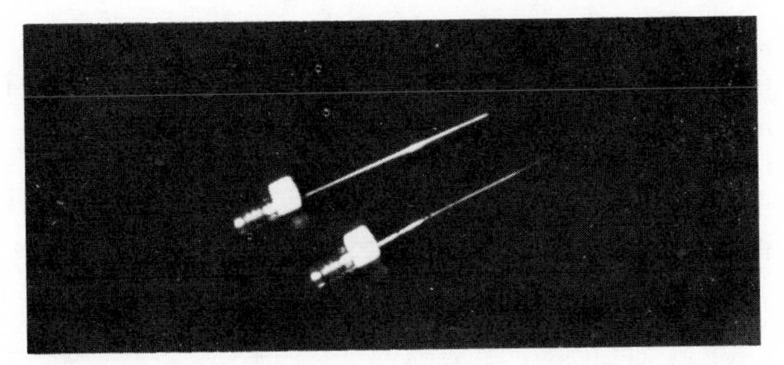

Figure 97. Plug-in fiber optics.

This is done not by measuring the incoherent infrared radiation of thermal origin emitted by the junction, but by measuring the intrinsic recombination radiation emitted by it, as already discussed in Chapter 1.

In view of the extremely low level of power emitted as recombination radiation, state-of-the-art solutions had to be found and employed in the following areas: optics, detectors, signal processing, and noise cancellation.

A single optical fiber 2 in. long having peak transmissivity at the wavelength of the radiation emitted (1.1 μ for silicon devices) constitutes the optics. Its diameter controls the spatial resolution of the system. It is quickly interchangeable to accommodate different spatial resolution requirements, from a maximum of 0.040 to a minimum of 0.001 in. Two of these fibers, respectively, of 0.006 and 0.002 in. in diameter, are shown in Figure 97.

Figure 98. Operator's view.

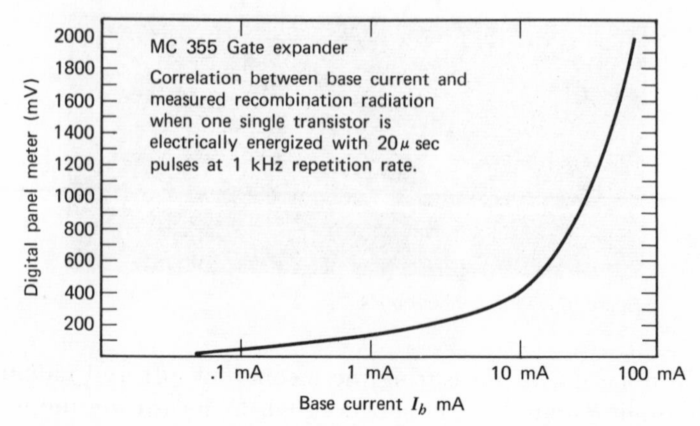

Figure 99. I_b-DPM correlation.

The infrared detector has the highest D^* so far attained in the 1.1 μ spectral region. Its response is on the order of approximately 40 nsec, and it has no cooling requirements.

The signal processing electronics have a total gain of 150,000 and a noise canceling capability that can extract and make measurable a signal "buried" under electrical noise that is one hundred times greater.

Target aiming is done visually, by means of a stereo microscope through which the operator can view the target, while adjusting a micromanipulator holding the pedestal where the target is mounted. Figure 98 shows the operator's view through the microscope: it is clear that the junction under test is located in the area visible between the tip of the fiber and its reflection onto the chip's surface.

A response characteristic of the system is shown in Figure 99. The curve depicting the correlation between base current flow and radiation magnitude shows a "knee" in correspondence of the saturation threshold of the device under test (in this case, a transistor incorporated into an IC type MC 355).

IMAGING DEVICE (THE EVAPOROGRAPH)

The radiant power accumulated during a certain length of time can be used when a relative, instead of absolute, radiation measurement is desired. Just as in photography the duration of the exposure controls the average tone of the picture, so in the evaporograph the length of the time exposure controls the colors of the display picture, which is the conversion, in the visible range, of the infrared image of the target.

The colors, however, do not bear a univocal correlation to the radiance of each point of the target, since the duration of the exposure has also an effect on their formation. In other words, the same object, always having the same radiance, will appear in different colors in the visible display, according to the length of the exposure. However, radiance gradients will appear as color differences, and whenever absolute values are not required, the evaporograph will display these gradients in bright colors.

Actually, of all imaging systems, the evaporograph was the first one capable of giving color displays. Made by Baird-Atomic Corporation, the instrument employs the oldest method of infrared imaging: a method first used in 1840, by John Herschel, the son of Sir William Herschel, to demonstrate the existence of absorption bands in the infrared portion of the solar spectrum—evaporation produced on a thin film of oil by the infrared radiation impinging on it.

In the same way that an "oil slick" on water reflects different colors of the spectrum according to the thickness of the oil layer, the evaporograph's oily membrane, when illuminated by white light, reflects different colors related to the thickness of the oil film left on it after the partial evaporation produced by the infrared image focused upon it.

The operation of the instrument is shown schematically in Figure 100. The optical system gathers infrared radiation from the field of view and focuses an image of the field on the membrane. A special coating on the

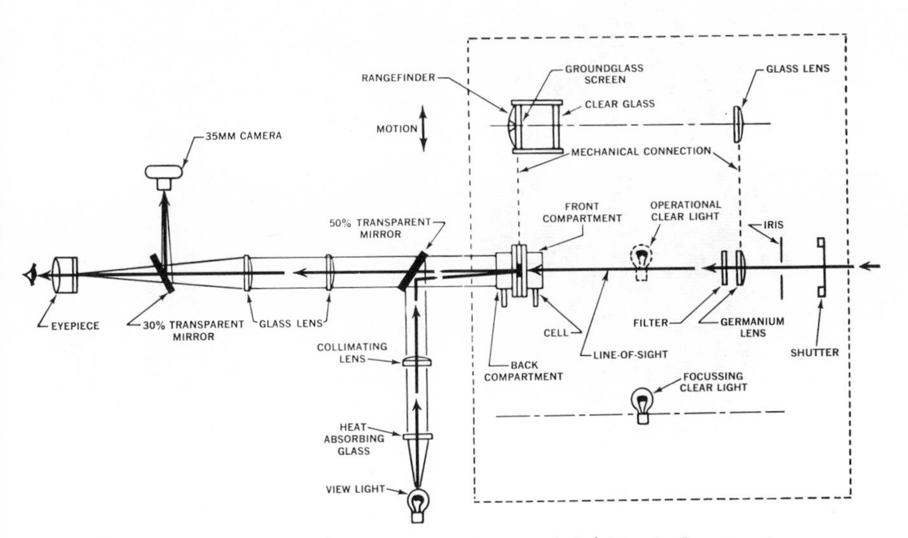

Figure 100. Diagram of evaporograph system. (Courtesy Baird-Atomic Company.)

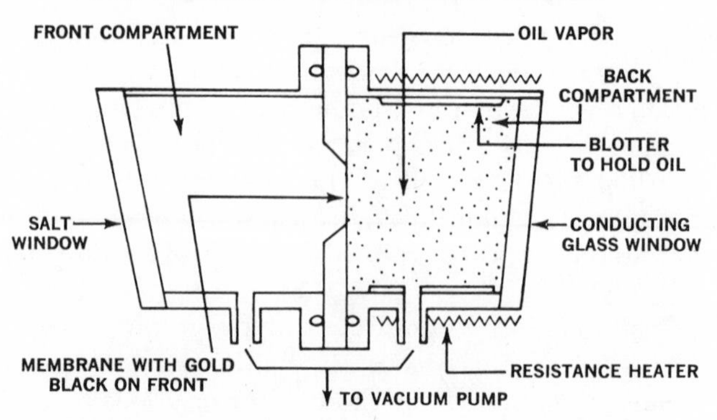

Figure 101. The evaporagraph cell. (Courtesy Baird-Atomic Company.)

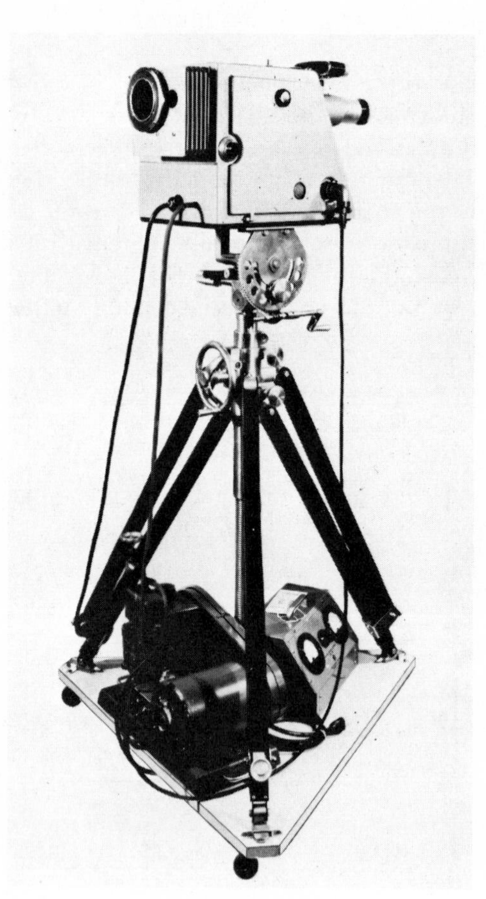

Figure 102. Evaporograph Model KR-1. (Courtesy Baird-Atomic Company.)

front surface absorbs this radiation and changes temperature from point to point in accordance with the amount of radiation received by each portion of the membrane. These point-to-point temperature variations alter the thickness of an oil film condensed on the rear side of the membrane. Thus temperature differences are resolved into differences in oil thickness, which cause white light to be converted into interference patterns of different colors, giving a visible, colored thermal image of the entire field of view.

Figure 101 shows the heart of the instrument: the evaporograph cell, containing a nitrocellulose membrane blackened on one side. Oil vapor becomes deposited on the other side, and infrared radiation from the target causes differential evaporation of this oil film, forming a visible image.

Finally, Figure 102 shows the evaporograph model KR-1 mounted on a tripod and the accessory hardware, which includes a vacuum pump needed to evacuate both compartments of the evaporograph cell.

The performance characteristics of the instrument vary with the lens used and with other fabrication details: the fastest model meets the following specs:

Optics	$2\frac{1}{4}$-in. diameter, $f/1.5$ germanium
Detector	Evacuated "membrane" cell
Readout	Optical viewing or 35 mm black and white or color transparencies
Minimum detectable temperature difference	0.5°C for blackbody at 25°C
Angular resolution	1.0 mrad
Maximum time to obtain thermal image	20 sec*
Maximum spatial resolution	10 lines/mm with a 10°C temperature difference at 6 in. distance from lens
Size, optical head complete	18 × 14 = 11 in.
Weight, optical head complete	48 lb.
Power	115 Vac 800W

* Approximate image forming time can be given by $T = 20/\Delta t$ where T is time in seconds and Δt is temperature difference in degrees Fahrenheit.

Chapter 4 Information Processing and Display

The information developed by the infrared-detecting equipment can be of value only if it is available in a form that our senses can perceive and our minds can understand. This is achieved by displaying it in the following ways:

Transient image, intensity modulated
Permanent image, in black and white or in colors
Oscilloscope traces, amplitude or intensity modulated
Film or chart recordings, amplitude modulated
Isothermal maps, digital notation
Digital printout

INFRARED-TO-VISIBLE CONVERSION

This type of information display is the most "natural" one. The equipment turns the infrared radiation emitted by the target into a visible image or picture presenting it to the human eye in a two-dimensional display where the radiation intensity at each point is measured by the intensity or by the color of the light.

The infrared-sensing equipment capable of producing such a conversion was reviewed in Chapters 2 and 3—infrared photography, phosphorluminescent coatings, liquid crystals, ferroelectric converters, photoconductive-electroluminescent devices, infrared cameras, the evaporograph, and the infrared vidicon tube. In this chapter we discuss these conversion systems in more detail.

INFRARED PHOTOGRAPHY

Due to the fact that the available infrared photographic plates and film are only sensitive to radiation in the near-infrared band, the pho-

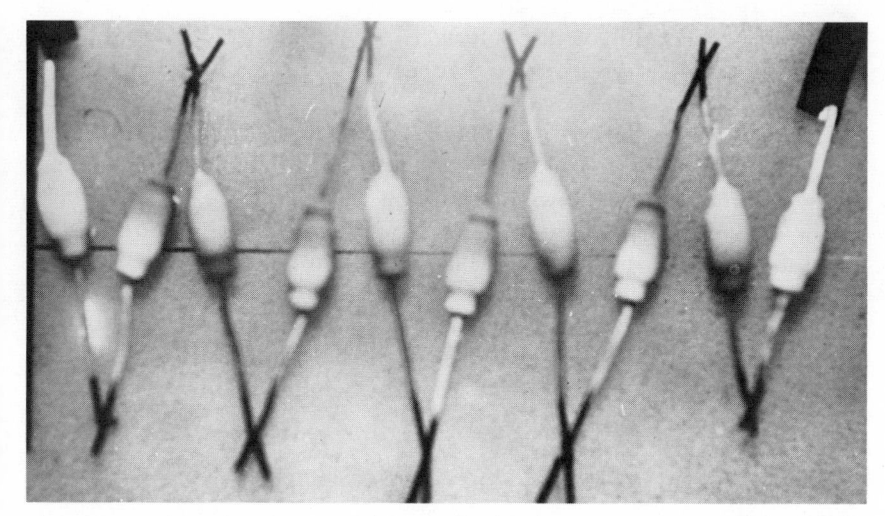

Figure 103. Thermal evaluation of diodes with phosphor paint. (Courtesy H. O. Frazier.)

tographic process can be used in the conventional way only when the target is at high temperature. However, we know that the blackbody emission curve covers the whole infrared spectrum at amplitude levels increasingly lower further away from the radiation peak. It should therefore be possible to take pictures of the intermediate infrared radiation by using very long time exposures. Although some examples of this are known to exist, the length of time needed to obtain them makes this technique impractical.

PHOSPHORLUMINESCENT COATINGS

When surface contamination of the target can be disregarded, the use of phosphor paints can give a quick, live display of the thermal condition of the object, and even of its transient variations, provided that the response time of the phosphor compound is not exceeded. Most of the paints presently available are of the emission-quenching type, which means that temperature and light emission are inversely correlated.

Figure 103 is an example of this technique: the diodes in the picture are all connected in series, but their forward resistance is different, so that their operating temperature is different: the ones having lower voltage drops appear brighter because their operating temperature is lower. Different amount of quenching is apparent on the surface of different diodes, since most of the heat is dissipated at the *N* end of the unit, where the semiconductor disc is located. The P end of the diodes holds the whisker, and therefore is lower in temperature and brighter in appearance.

The picture was taken by H. D. Frazier of Pacific Semiconductors, Inc., using a Radelin phosphor, and is a good example of the possibilities of the thermographic phosphor technique.

The information generated by this type of conversion can be photographically recorded either in black and white or in color. A movie camera loaded with color film was successfully used to record transient thermal conditions. Quantitative measurement of temperature is difficult to obtain by this system, because of the need to exactly measure the intensity of the light emitted by every elementary point of the target.

FERROELECTRIC CONVERTERS

The display produced by a ferroelectric converter plate of the type described in Chapter 2 is a visible, monochromatic image where the brightness of every point is directly proportional to the intensity of the infrared radiation emitted by the corresponding point of the target.

There, too, the image can be recorded by photographic means, using black and white film, and brightness can be measured with a light meter. Again, it is difficult to turn this information into absolute quantitative measurements.

The advantage of using a converter plate is mainly due to its very limited space requirements: the device can be made very thin (less than $\frac{1}{4}$

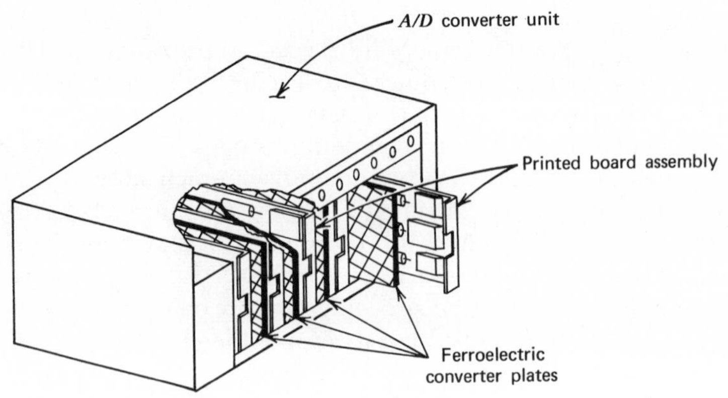

Figure 104. Component thermal measurement with ferroelectric converter plates.

in.), and formed in any shape. Its front surface can be coated in such a way as to accept almost only radiation incident on its surface at 90° angle while reflecting back all oblique rays. This feature eliminates the need for an optical focusing system whenever the target is located at close distance in front of the converter plate. Figure 104 shows such an arrangement, designed to check the operating temperature of every component-part mounted on the panels of an *A/D* converter unit.

With this setup, temperature measurements can be taken when the unit is all enclosed and operating in any desired environment, including the vacuum simulating high altitude or outer space condition.

LIQUID CRYSTALS DISPLAY

The thermal information yielded by liquid crystals solutions is very colorful and can best be assessed by visual observation of its display, which can only be recorded in color if it is to keep its significance. We have already seen that for every type of compound the correlation between color and temperature is unequivocal.

Figure 105 shows in a schematic representation how a poorly bonded area between two aluminum stiffeners appears when a liquid crystals solution is used on the surface to be inspected: the hot air flowing from below the assembly produces a temperature rise in the lower aluminum element. The heat is transmitted to the top element, but a void in the center area creates an obstacle to the heat transfer, so that the temperature of the top layer is lower in the corresponding area, than in its immediate surroundings. This condition shows up as a change in color: namely from blue to red if the temperature difference is at least 1 °C when the liquid crystal compound has the thermal range indicated in the illustration (blue = ~30°C; red = ~29°C).

Since the colors indicating the temperature are created by refracted light, strong illumination of the target is essential. Figure 106 shows a typical setup for detection of hidden flaws in honeycomb sandwich structure: floodlights provide light and radiant heat at the same time, while cooling air blown from another direction introduces the necessary temperature differential. Any difference in the heat transmission from the surface toward the core of the sandwich structure produces a color variation clearly detectable.

Basically, the system used makes it possible to observe visually the heat flow at the surface of the target. This is a dynamic condition best suited to visual analysis and evaluation. Color movies are thus far the best recording and display method.

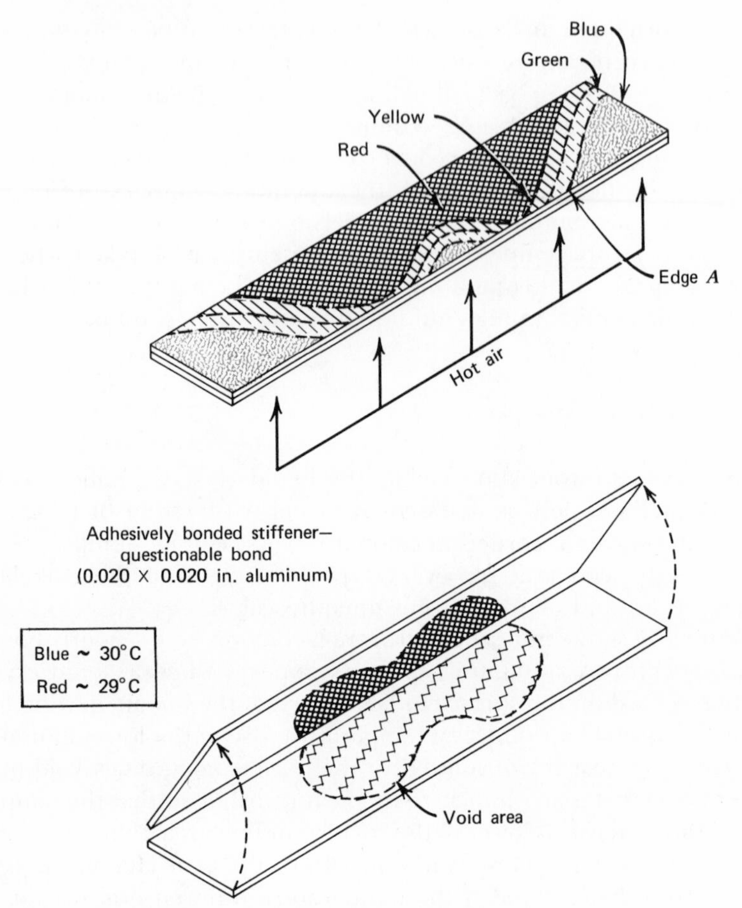

Figure 105. Bond evaluation with liquid crystals. (Courtesy W. E. Woodmansee.)

CATHODE RAY TUBE DISPLAY

The analog signal representing the amplitude of the infrared radiation emitted by every point of the target can be displayed as intensity modulated lines. This evidently is the natural choice for raster-like imaging. Precise quantitative measurements are impossible, but the visual presentation allows overall, instant evaluation at a glance. Figure 107 is an example of such display.

Imaging devices such as the infrared Vidicon supply an electrical output that is already conditioned for use as the video signal in a TV monitor system. Other scanning systems require the use of synchro-

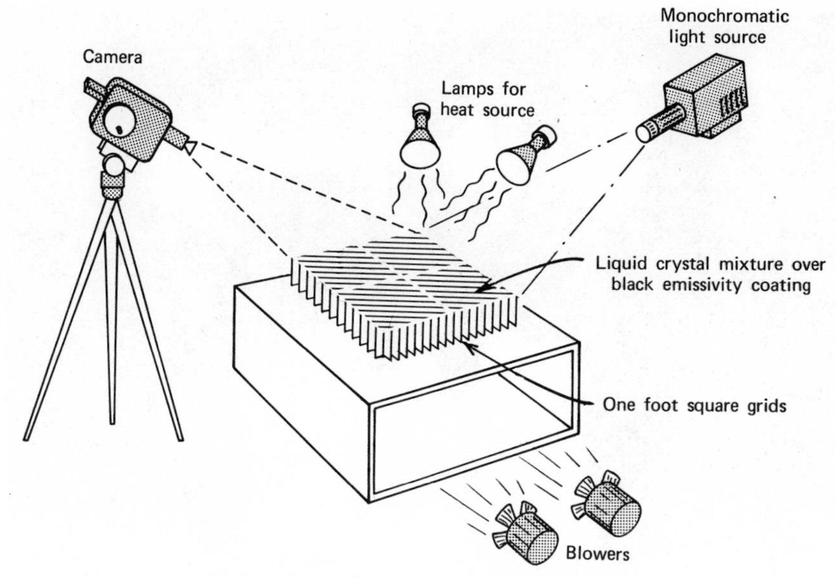

Figure 106. Typical liquid crystal inspection station.

nizing pulses to have all the horizontal lines start from the same edge and orderly stacked in the vertical direction. Whenever the scan speed is slower than the rate used in conventional TV systems, cathode ray tubes with long-persistence phosphors can be used to hold the image as long as needed.

The main problem in the CRT display system is the limited dynamic range of these devices: 10 to 15 db as opposed to the 20 or 30 db covered by the radiometer's electrical output. This difficulty can be reduced by a log amplifier and by a zero level offset capability. The log amplifier "compresses" the signal into a narrower dynamic range, thus a given percentage change in sensor output appears as the same voltage change at the CRT's input regardless of whether the radiation level being sensed is high or low. The zero level offset feature allows one to limit the display of the signal to the desired range only, eliminating the levels that are below any chosen threshold. In this way, for instance, background radiation can be dialed out of view.

Nevertheless, if we compress 30 db of signal onto the CRT's screen, the intensity resolution that will be available at the peak signal will be 10 db at best.

Consequently, TV-type displays can best be used when a visual, direct presentation is more important than exact, quantitative data, and when

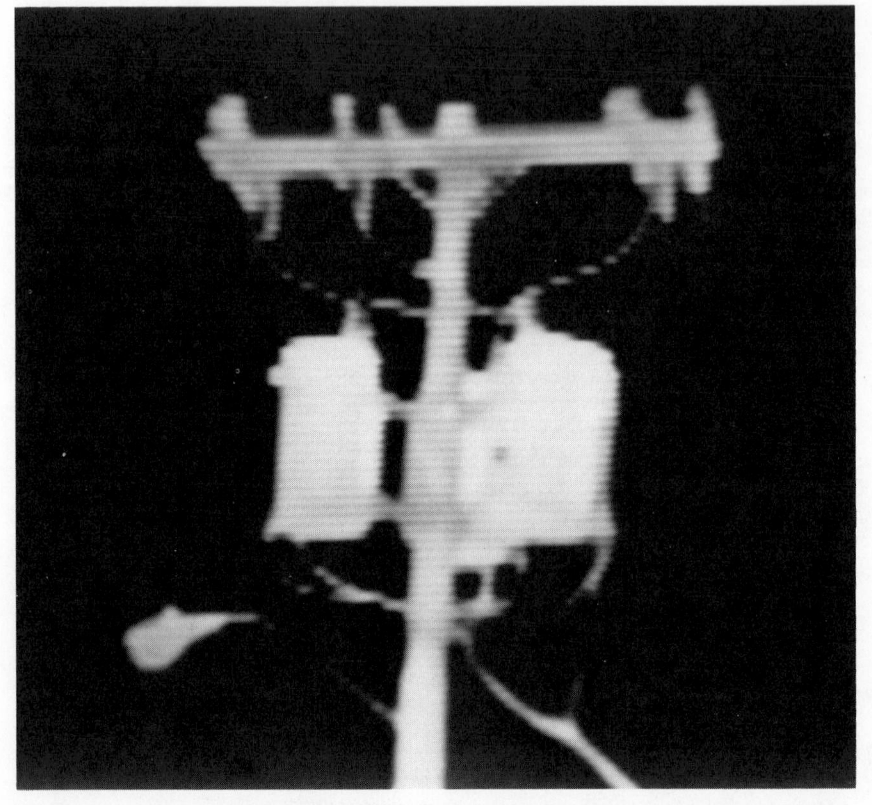

Figure 107. CRT intensity modulated raster display. (Courtesy Barnes Engineering Company.)

the dynamic range of the signal variations between different points of the target is moderate or low.

The infrared scanner made by A.G.A., and described in p. 121 has a display of this type. Barnes Engineering also uses this system for its Mod. T. 101 infrared camera, in addition to quantitative reading devices that can be used in conjunction or as alternates.

INFRARED CAMERAS

The two instruments just mentioned are sometimes called "infrared cameras," since they can yield Polaroid pictures of the infrared image of the target almost immediately. Actually, they are image converters, which are capable of turning the infrared radiation emitted by each

Figure 108. Visible picture of electronic module. (Courtesy Barnes Engineering Company.)

point in the field of view into an electrical signal, whose amplitude is proportional to the intensity of the infrared radiation emitted. This electrical signal can be displayed on the screen of a CRT, thus forming a visible image of the infrared target.

A Polaroid picture of such CRT display is often called an "infrared picture," and for convenience we can use this term. This book contains many different types of "infrared pictures". Figures 108 and 109 are examples of visible and infrared pictures of an electronic module.

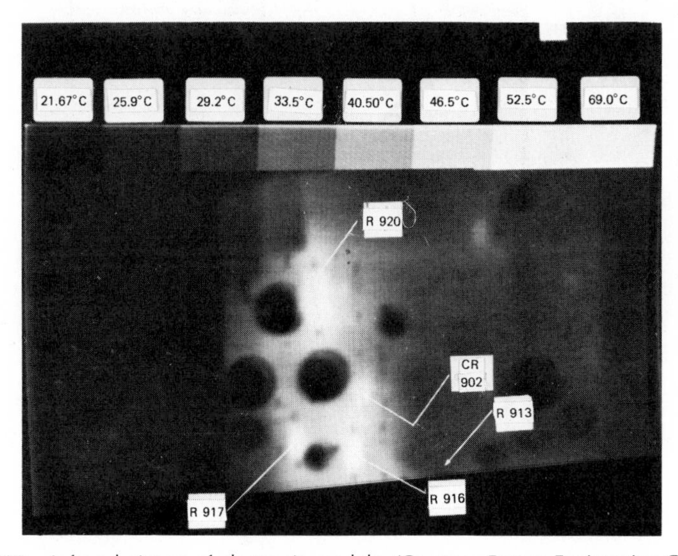

Figure 109. Infrared picture of electronic module. (Courtesy Barnes Engineering Company.)

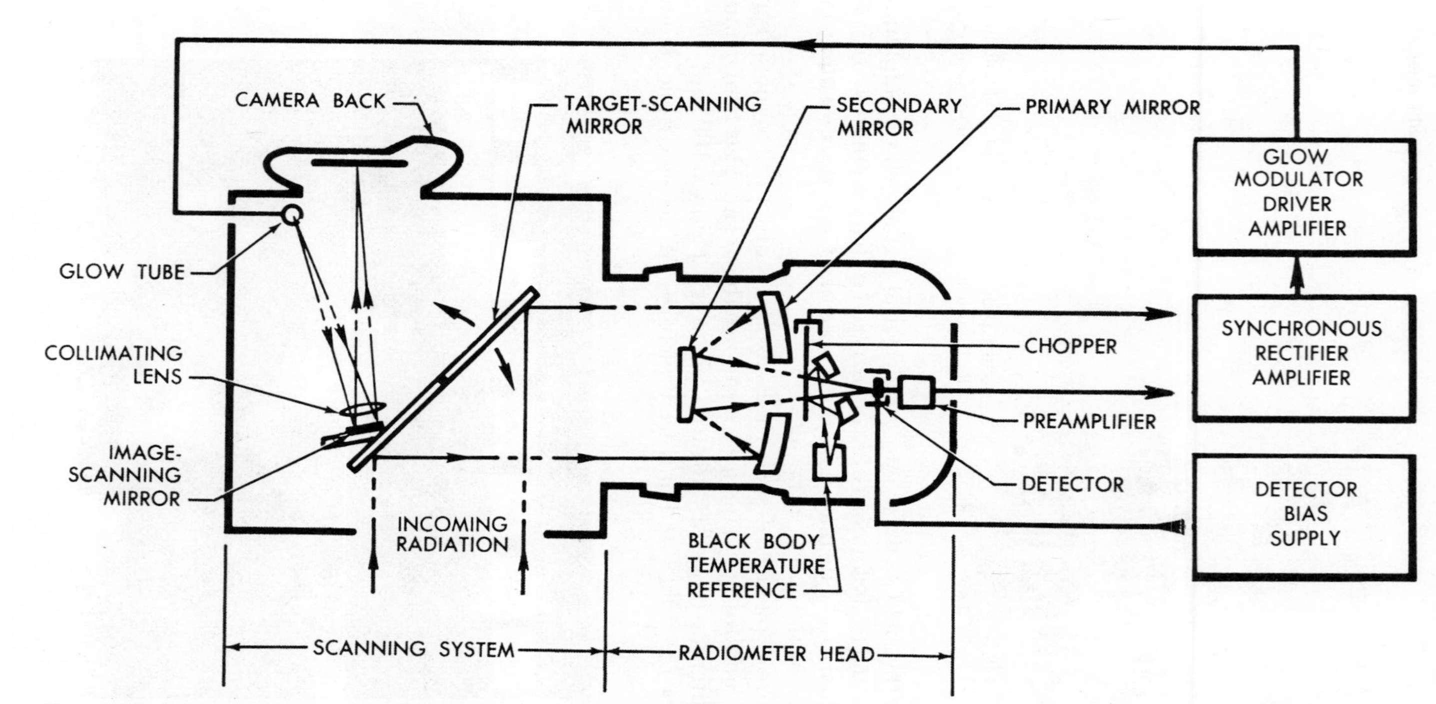

Figure 110. Schematic diagram of Barnes infrared camera. (Courtesy Barnes Engineering Company.)

Figure 109 was taken with the module electrically energized and at thermal equilibrium. The camera used was a Barnes T-4 model, which is composed of an 8-in. radiometer, with a scanning optical system. The detector output, after adequate amplification, energizes a glow tube whose light is driven across a Polaroid film in the same raster configuration used to scan the target. Figure 110 shows the solution devised to achieve simultaneous scanning for the two systems: the imaging-forming system is rigidly mounted on the back of the target-scanning mirror so that they move together. As a result, when the scanning of the target is completed, the picture is ready to be developed, which for a Polaroid film takes 10 sec.

The camera has a built-in blackbody for radiation reference purpose, and every picture is provided with an eight-step "gray" scale whose extremes (shown respectively as "white" and "black" steps) can be set at any predetermined thermal level. Radiant intensities therefore can be read as gray shades, either by visual comparison with the steps of the gray scale or with the help of a densitometer.

In conclusion, the infrared picture of Figure 109 is not only the intensity modulated display of the infrared radiation emitted by every point of the electronic assembly shown in Figure 108, but also is its permanent recording, which carries its own calibration scale, whose range can be preset to cover any desired temperature span expected at the target.

EVAPOROGRAPH DISPLAY

A similar display, although in color, is produced by the evaporograph instrument that was described in p. 136. Here the radiation produces different colors in relationship to its intensity; Figure 111 is an example of such type of display. A drawback of this system is the absence of a built-in calibration system; therefore reference levels must be located in the viewing area to appear in the picture as elements of the target. Despite the fact that this display appears very similar to the color displays yielded by some infrared cameras, we must not forget that they are produced using completely different methods. The evaporograph works very much as a photographic camera, by projecting on the membrane, simultaneously all the points of the image and holding this image fixed onto the membrane for the required length of time. The infrared cameras instead use a scanning method to view the target not in its entirety, but point by point. Its image is then reconstructed by aligning all the corresponding points of the detector output in the correct sequence, just as in a TV system.

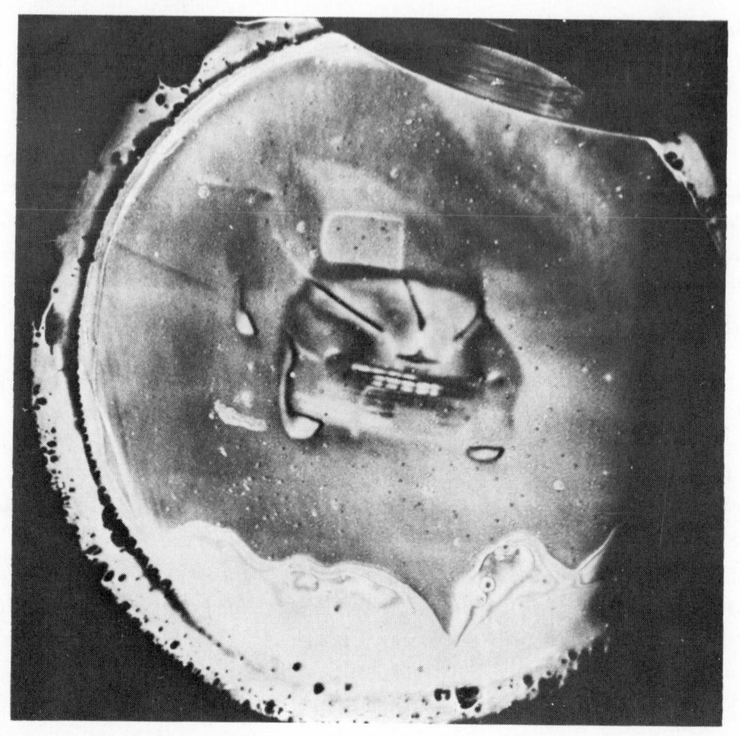

Figure 111. Evaporagraph picture of parked automobile at night. Note hot tires and radiator. (Courtesy Baird-Atomic Company.)

DISPLAY OF ANALOG INFORMATION

When the radiometer's output is displayed on a CRT in the amplitude modulation mode, a scan line of a nonuniform target will produce a variable-amplitude line similar to the one shown in Figure 112. This type of display allows precise measurement of the signal amplitude at every point of the scan, and a logarithmic amplification of the detector output will allow it to cover a wide range of signal magnitudes without losing sensitivity at the lower levels.

Permanent recording of these lines can be achieved by taking Polaroid pictures of them, and quick identification of anomalies can be performed with a simple visual system based on automatic cancellation of all those elements that are identical in the "standard" and in the "under test" signatures.

Figure 113 shows the principle upon which this visual comparison system is based: the traces of the "standard" unit are printed as a nega-

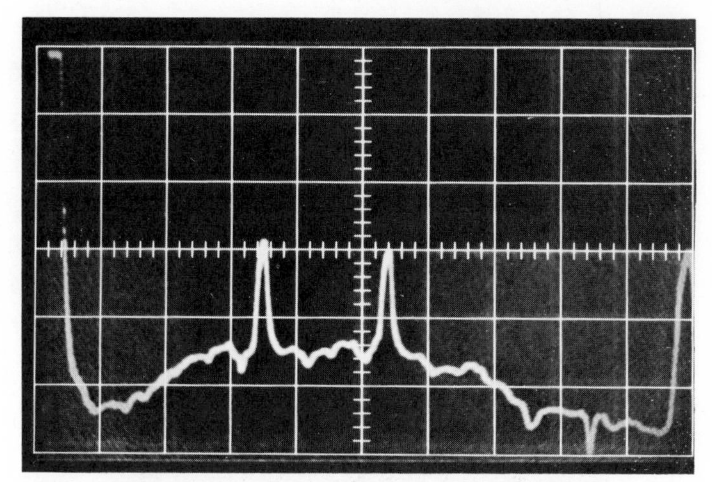

Figure 112. Oscilloscope analog display of single scan line.

tive, that is, black on a transparent background, as shown in *a*, where the thickness of the line covers the tolerance variations of the different elements; the traces of the unit to be compared against the standard are printed as positive, that is, white on a black background, as shown in *b*. The comparison process takes place by superimposing *a* over *b*: the black traces of *a* block out the white traces of *b* as long as these are contained within the tolerance limits. Any deviation beyond these limits, above or below the black line of the standard, stands out conspicuously as a white trace on a black background, as illustrated in *c*.

Discrete Points Display

We have discussed infrared sensing equipment where the detector thoroughly investigates an area by sequentially covering each and every point of it through a scanning motion along a certain predetermined pattern. The detector output reflects this continuous motion with its continuous variations in amplitude, proportional to the intensity of the radiation emitted by every point of the target.

However, other systems are so designed that it is possible to program the detector to "look" only at preselected points of the target; or in other systems, the detector output signal is accepted by the processing electronics in coincidence with preestablished "windows" that have been programmed in coincidence with the desired points of the target.

In both these cases, the output reaching the display section is not a continuously variable analog signal, but a sequence of pulses whose

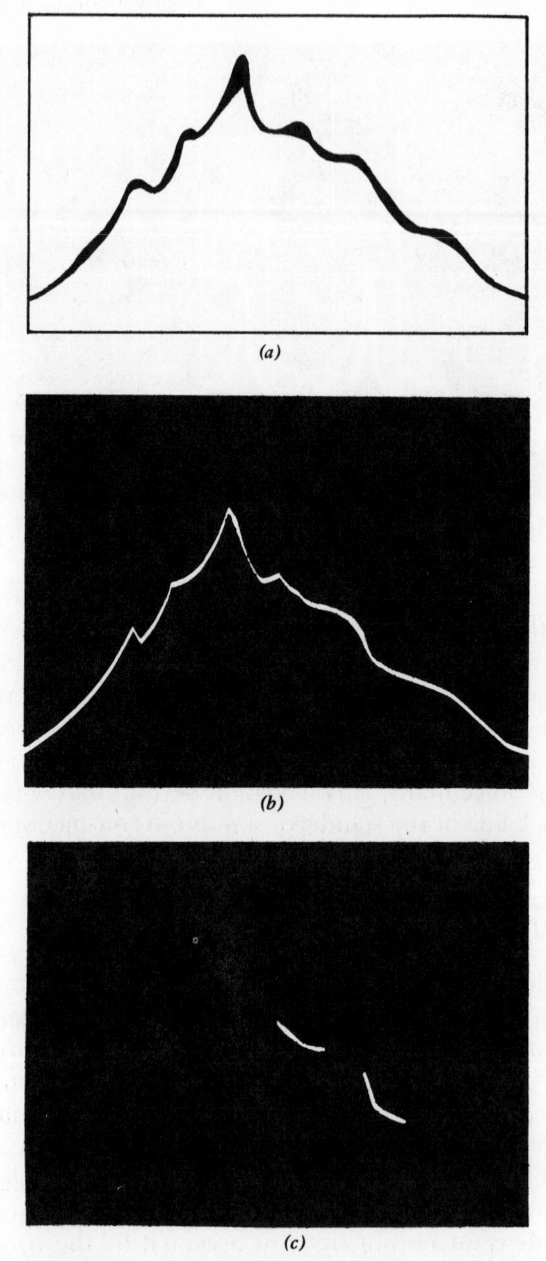

Figure 113. Trace comparison principle.

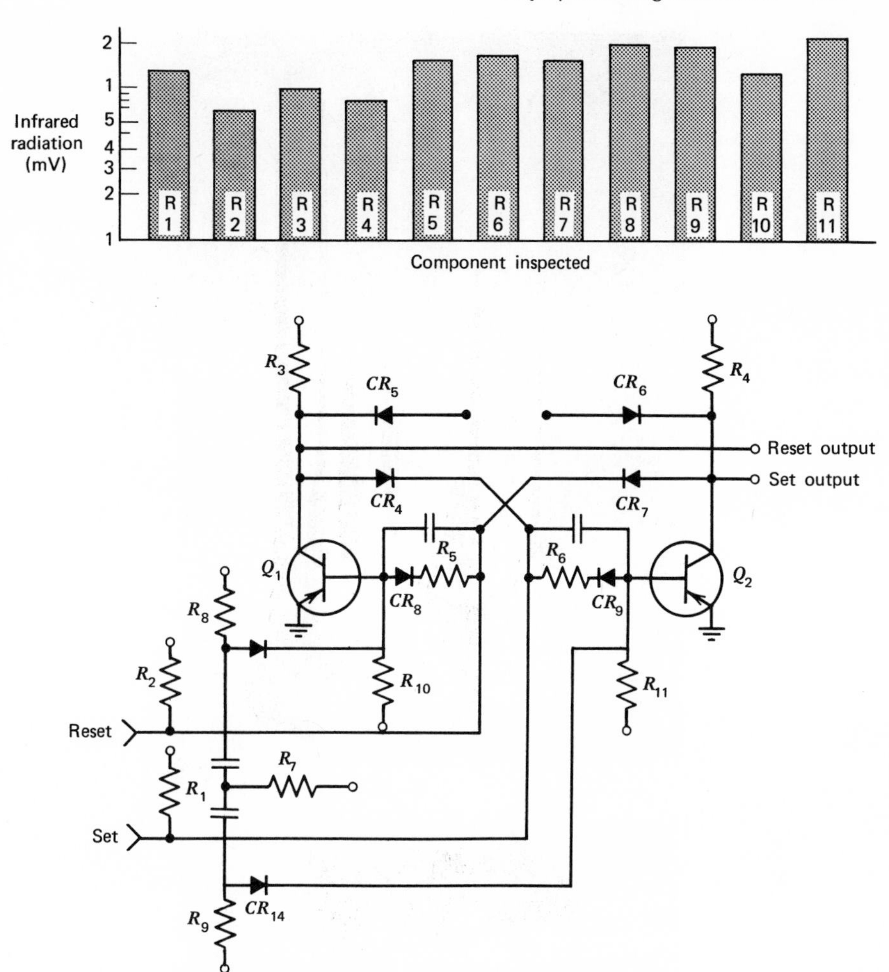

Figure 114. Bar chart recording system. (Courtesy Jose Pina, Boeing Company.)

amplitude corresponds to the power of the infrared radiation emitted by every observed point of the target.

A typical example of such a display, called a bar chart, is shown in Figure 114. This chart was generated by the Automatic Infrared Test Station built by Servo Corporation of America and described on page 117. Each bar corresponds to an active element of the circuit under investigation, and the system has the advantage of presenting only the information related to the elements of interest: background and nonessential areas are eliminated. As a consequence, evaluation and comparison are

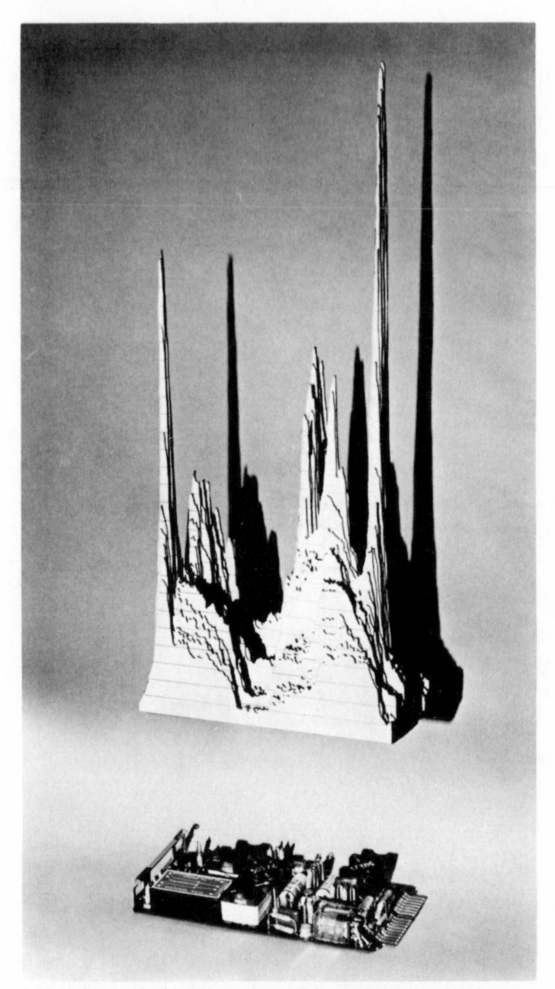

Figure 115. Three-dimensional model of radiation from electronic module.

remarkably simplified. Direct recording of the bar chart on paper was possible because of the relatively slow operation of the equipment.

Three-Dimensional Model

As mentioned early, most of the display systems are designed to present in a two-dimensional version a three-dimensional correlation—infrared power emitted by every point of a surface.

Figure 115 shows a three-dimensional model of this correlation: the vector representing the magnitude of the infrared radiation is pointing vertically upward from each point of the emitting surface. The locus of all the terminations of these vectors forms a three-dimensional surface that gives an immediate measure of the power emitted by each point. Whenever all these points have the same emissivity, the model can be calibrated in equivalent temperature by the traces of the intersection with horizontal planes representing given predetermined thermal levels.

In the example shown, perhaps the most remarkable feature is the wide dynamic range of the radiation emitted. This explains why the logarithmic amplification of the detector output is often adopted, so that compression of the large signals will not cancel all detail at the lower levels.

Thermal Map

The concept here is the same as in geographical maps or weather maps: only the "quantity" to be plotted is temperature instead of altitude or barometric pressure. The contour lines in this case are called "isothermes" and they connect all the points having the same temperature.

An isothermal map can easily be displayed on the face of a CRT, such as the one shown in Figure 116*a*, while the CRT trace of the scan line *A-A′* is shown in Figure 116*b*. The analogy with the geographical map is complete, if we only replace "altitude" with voltage."

Figure 117*a* shows the circuit that can be used for this purpose: the radiometer's output feeds a number of Schmitt triggers, S, S_1, S_2, and so on. These are high-gain operational amplifiers, each of them biased in such a way as to generate an output when the radiometer signal reaches a certain voltage. This output, which is a voltage step, is differentiated so

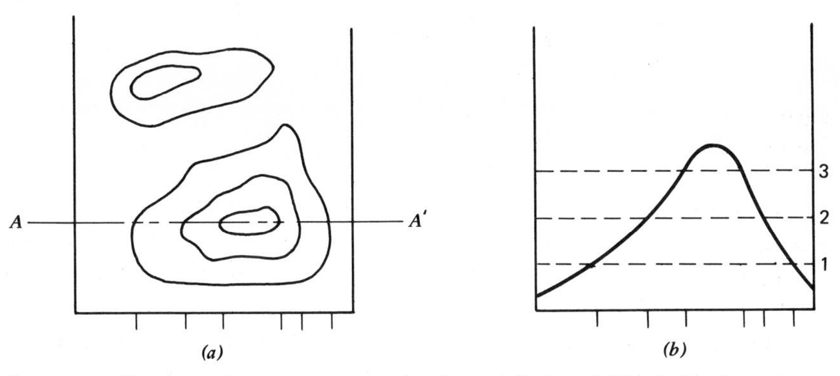

Figure 116. Correlation between analog and isothermal displays. (*a*) Typical isothermal map. (*b*) Analog scan trace along line *AA′*.

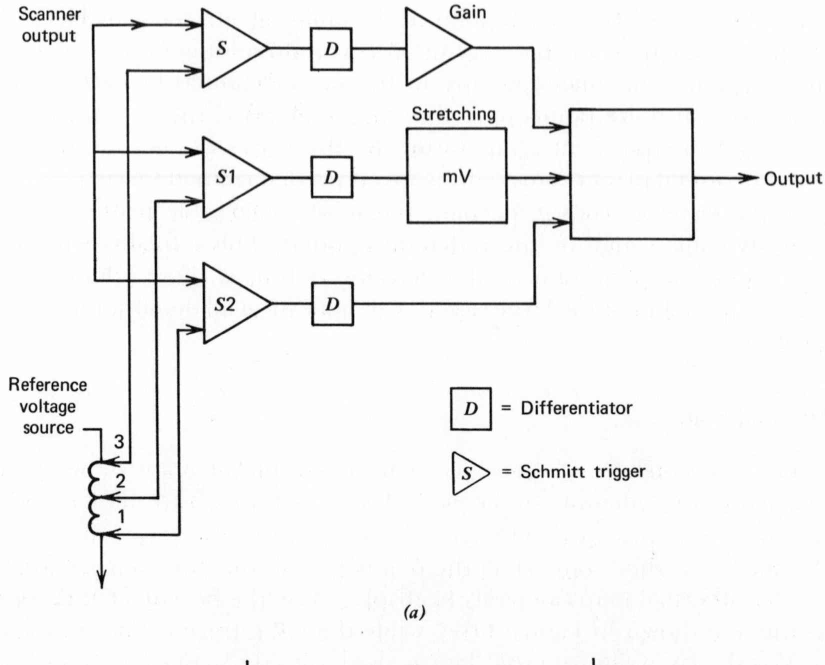

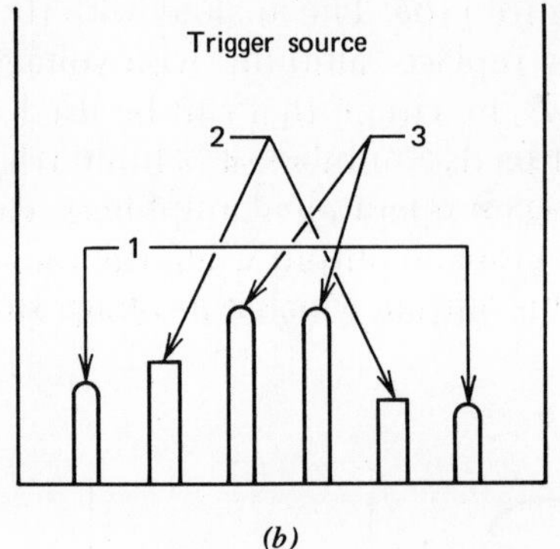

Figure 117. Circuitry of mapping system. (*a*) Functional diagram. (*b*) Schmitt trigger output (input to display).

as to form a sharp pulse, which is used to mark a point at the contour line on the scope. Figure 117*b* shows how these pulses can be shaped to originate contour lines of different width, so that their identification might be possible.

A display instrument capable of presenting on a CRT isothermal maps, thermal plateaus and intensity-modulated images was made by Philco-Sierra. Called the Model 711A Quantizer, it was designed to convert the output of the infrared detector into six discrete levels, each of which can be independently adjusted by the operator.

Figure 118*a* shows the CRT display depicting the distribution of the isothermes on a sheet of plastic, one end of which is heated. This display in itself does not tell us which end is up and which end is down as far as temperature is concerned, but the "thermal plateau" display of Figure 118*b* solves this problem.

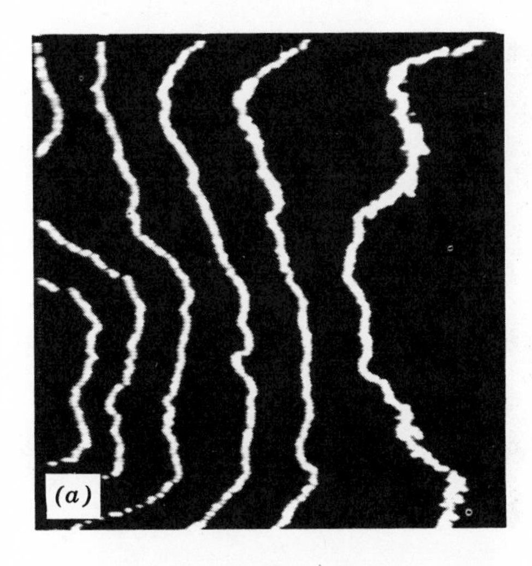

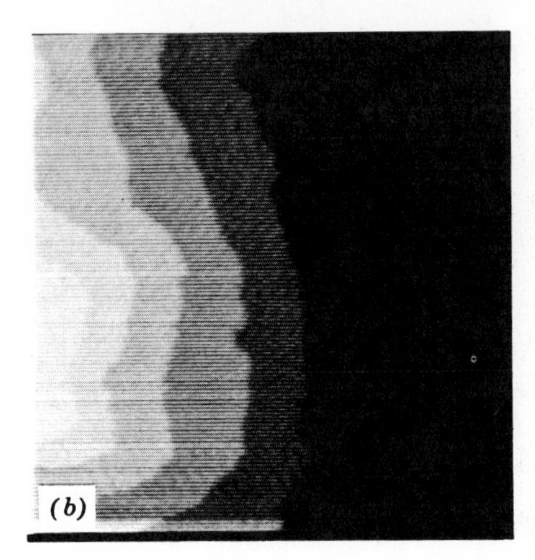

Figure 118. Oscilloscope map displays. (*a*) Lines only. (*b*) Bands only. (Courtesy Philco-Sierra Division.)

For more complete, compounded information, the thermal map picture can be combined with the intensity-modulated display of the plateaus, as in Figure 119*a* or of the video output of the detector, as in Figure 119*b*.

Multicolor displays based on the same principle but offering a greater range of thermal levels are available when using the T-101 camera, or the AGA Thermovision system. The use of polacolor film makes it possible to record and develop these color maps in a rather short time. How-

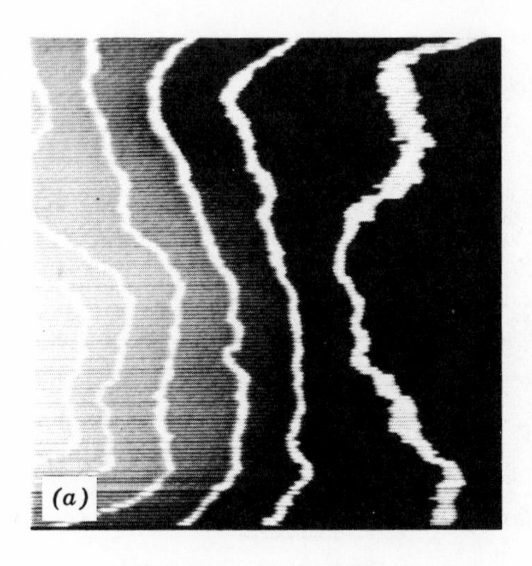

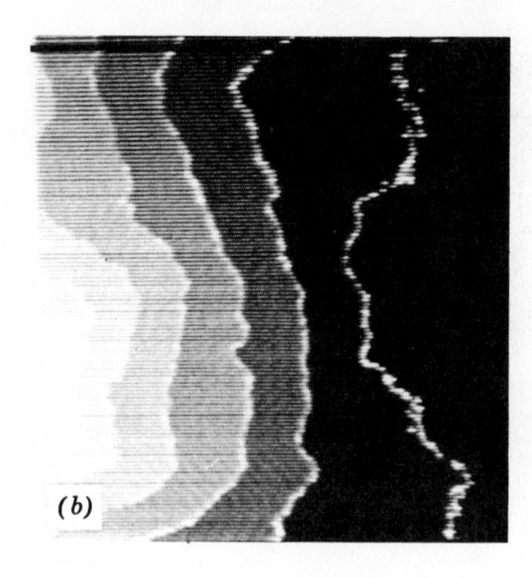

Figure 119. Oscilloscope combined map displays. (*a*) Lines plus video. (*b*) Bands plus lines. (Courtesy Philco-Sierra Division.)

Figure 120. Picture of lamp shade. (Courtesy AGA Corporation.)

ever, the map is not generated simultaneously, but a different color is assigned to each different thermal plateau, and each of them is recorded onto the same polacolor film in a sequence synchronized with the insertion of color filters in the path of the white light used to expose the film. This technique limits the use of color maps to targets that are stationary and in thermal equilibrium.

Figure 120 shows a good example of such a display. The object is a lamp shade, and the color-temperature scale appears at the bottom of the picture taken with AGA thermovision system.

Figure 121 is another example of color display; two hot irons show different temperatures and different heat distribution. This picture was taken with the Barnes T-101 camera, and the color-temperature correlation is different from the preceding picture. There, the lower tempera-

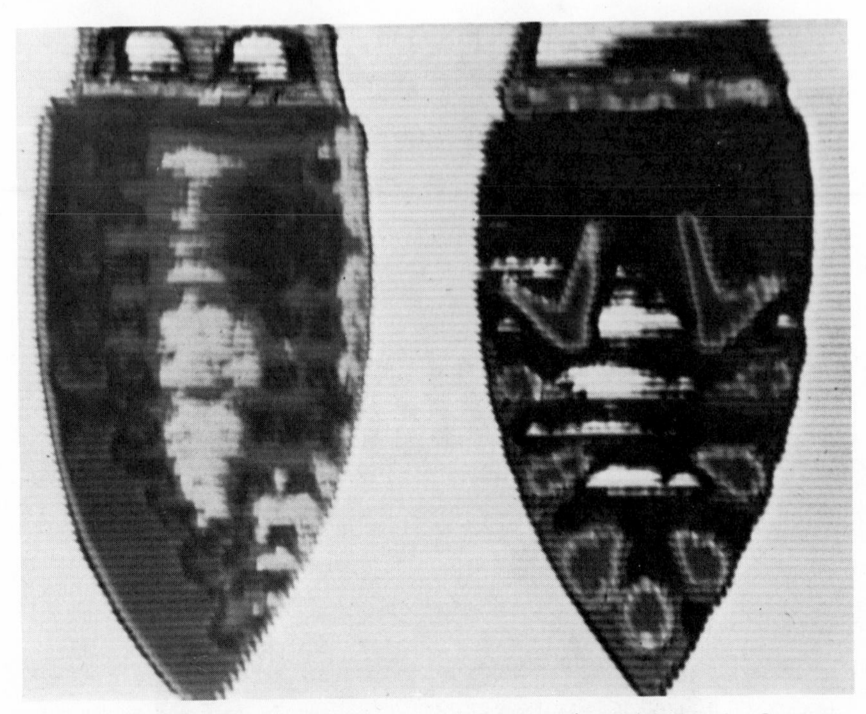

Figure 121. Picture of hot irons. (Courtesy R. W. Astheimer and Barnes Engineering Company, *PS&E Journal*, Vol. 13, 1969, P. 129.)

ture appears blue, the higher red, and the intermediate yellow. Perhaps some day in the future infrared color reudition will be standardized?

Other color displays now in experimental phase use color TV tubes that can either be viewed or photographed, usually with polacolor film.

Deviations from a Standard

Besides being useful in displaying information directly related to the intensity of the infrared radiation emitted by every point of the target, intensity-modulated systems, contour systems, or combination of both can be used to display *only the deviations* from the standard.

Figure 122 shows the schematic configuration of such a system: the "standard" infrared profile is recorded on magnetic tape and can be read out at the same speed and in synchronism with the "live" infrared profile as generated by the radiometer while it scans the target. Due to the presence of the subtractor unit, the two profiles will cancel each other if they are identical. If they differ, only the differences will emerge

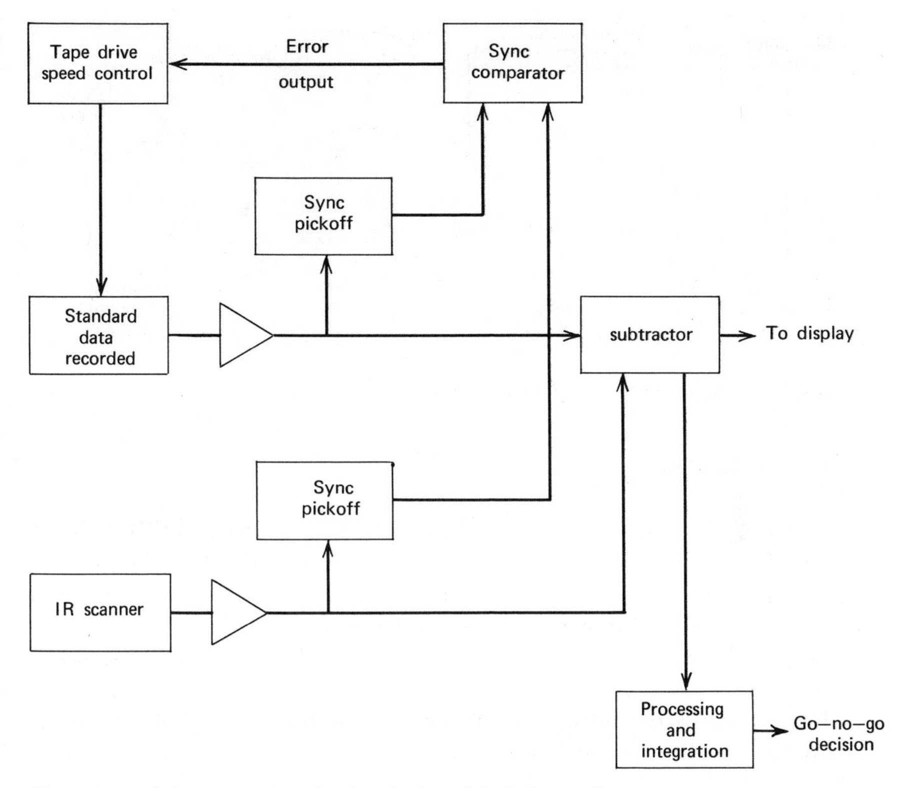

Figure 122. Subtractor system for the display of deviations only.

from the subtractor and these will appear in the visible display systems mentioned above.

This approach greatly simplifies comparison, since all conforming elements are canceled out, while only the deviations are shown. The major difficulty is the synchronization of the magnetic tape with the recorder's output. The use of a buffer unit offers a solution to this problem.

Another system that can yield the same result of displaying only the deviations is shown in Figure 123.

Here a two-gun storage tube is used with information written in one side of a storage surface in real time, read out on the other side repetitively at a high TV-type frame rate, and displayed on a conventional TV monitor. The dynamic range problem is still not solved here, but it does become possible to automate the comparison process, since reference data can be stored on another tube and read out synchronously with the sample's data; a point-by-point comparison is made by subtracting the electrical outputs of the two read scans and displaying their difference.

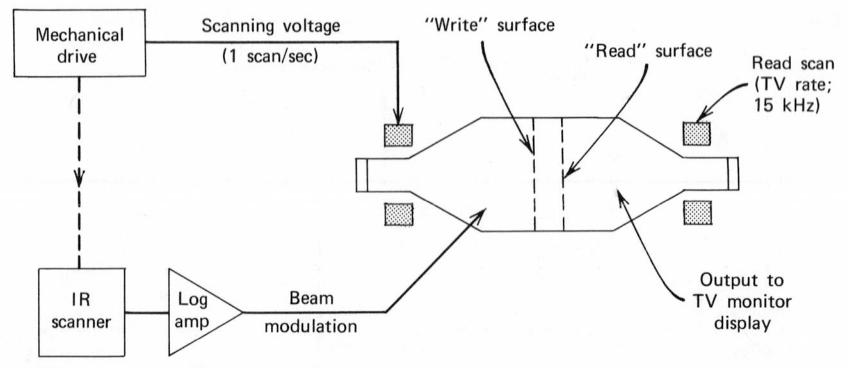

Figure 123. Indirect view storage tube display.

A/D CONVERSION

Probably the most sophisticated processing systems are those where the analog signal produced by the radiometer is turned into digital information, which in turn can be processed, stored, read, compared, and displayed in many different ways. Figure 124 shows such a system, where the output of the infrared scanner is compared against a stored reference pattern.

The sync pulses generated in the clock generator by the motion of the scanning system cause the following:

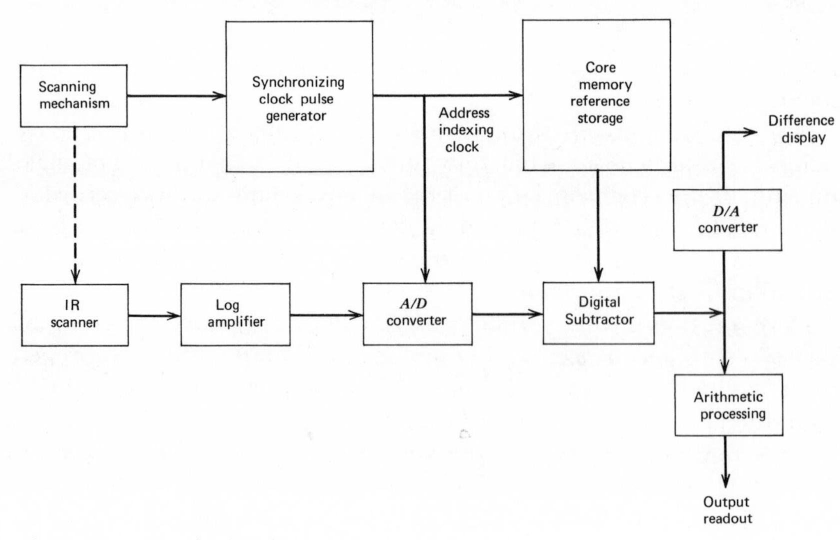

Figure 124. Digital infrared scanner processing.

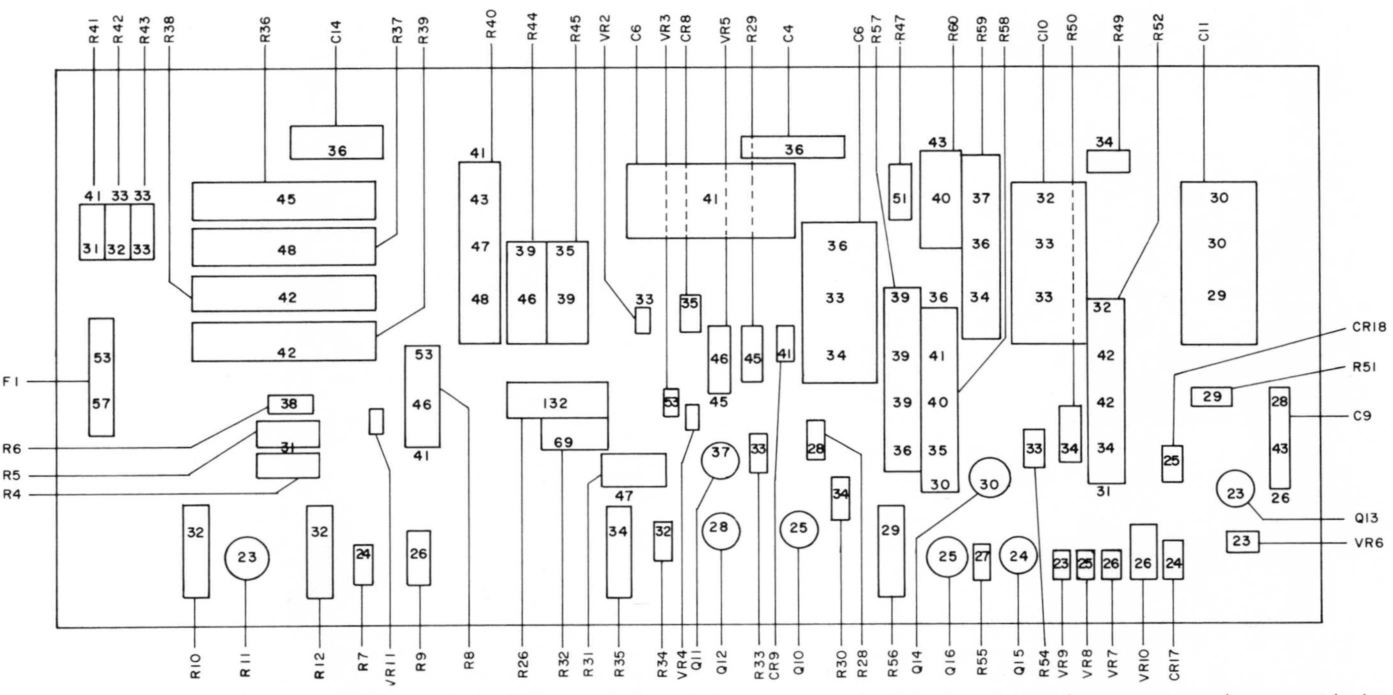

Figure 125. Printout display: temperatures in degrees Celsius are marked in the components' areas. Components' designations are shown outside the panel's outline.

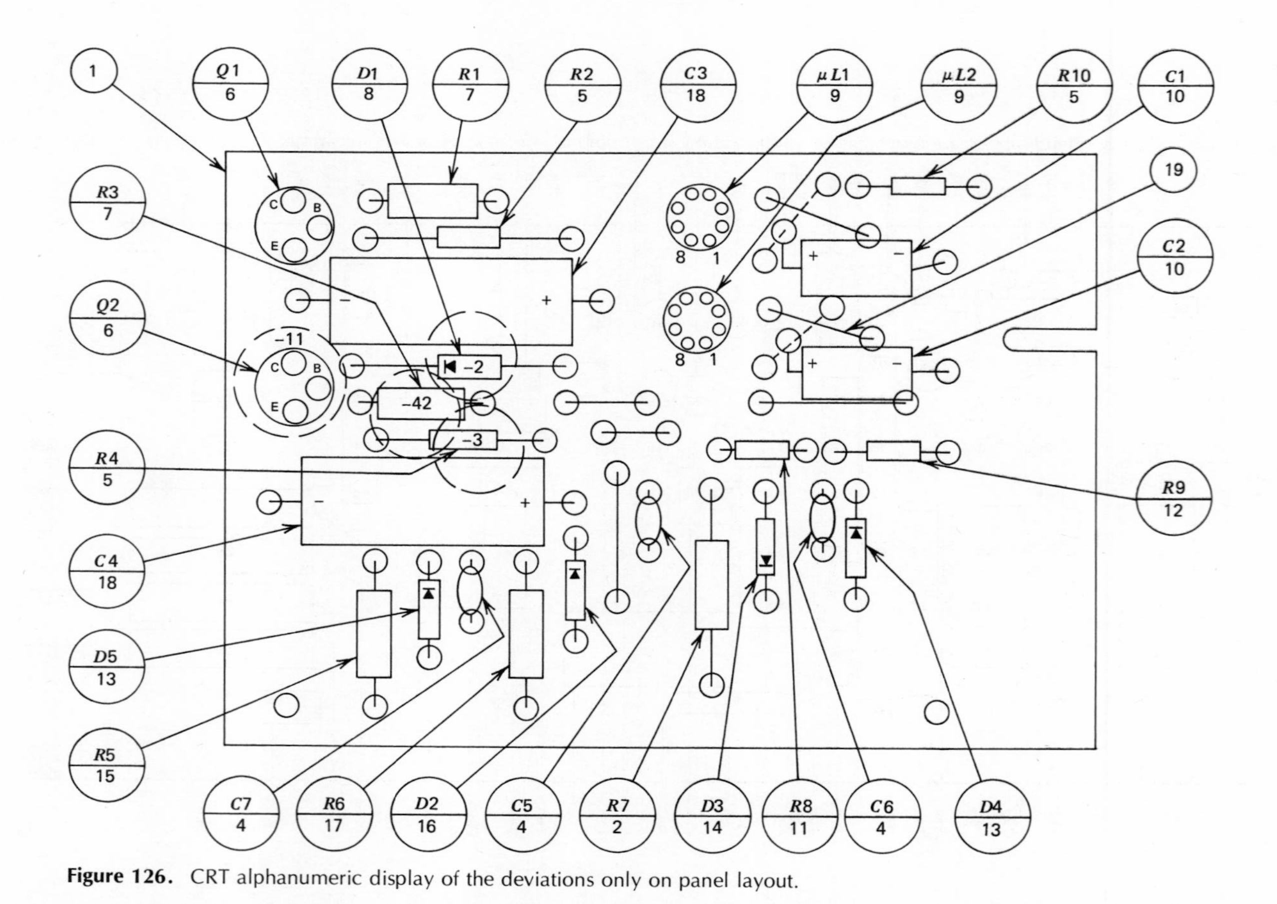

Figure 126. CRT alphanumeric display of the deviations only on panel layout.

1. The *A/D* converter to sample the radiometer's output and to read it out.
2. The memory to index to its next position and to read out (nondestructively) the contents of this position.

These two readings are then substracted from each other in the "digital subtractor" unit, and this result is made available at the output readout end for printed display, such as numerical information typed out on a target layout drawing. The same information can also be reconverted from digital to analog and presented in visual display on CRTs for a qualitative look at the deviations.

Figure 125 shows how the temperatures of operating component parts of an electrically energized electronic assembly are printed out within the components' outlines on the assembly layout. This can be done by a flexowriter and it constitutes one of the preferred display techniques. Of course, in place of the actual temperature values shown in the illustration, only the deviations from the "standard" can be typed out on the same layout drawing. In this way, only those elements that are affected by a failure mechanism will stand out.

Instead of the flexowriter, a CRT capable of alphanumeric display can

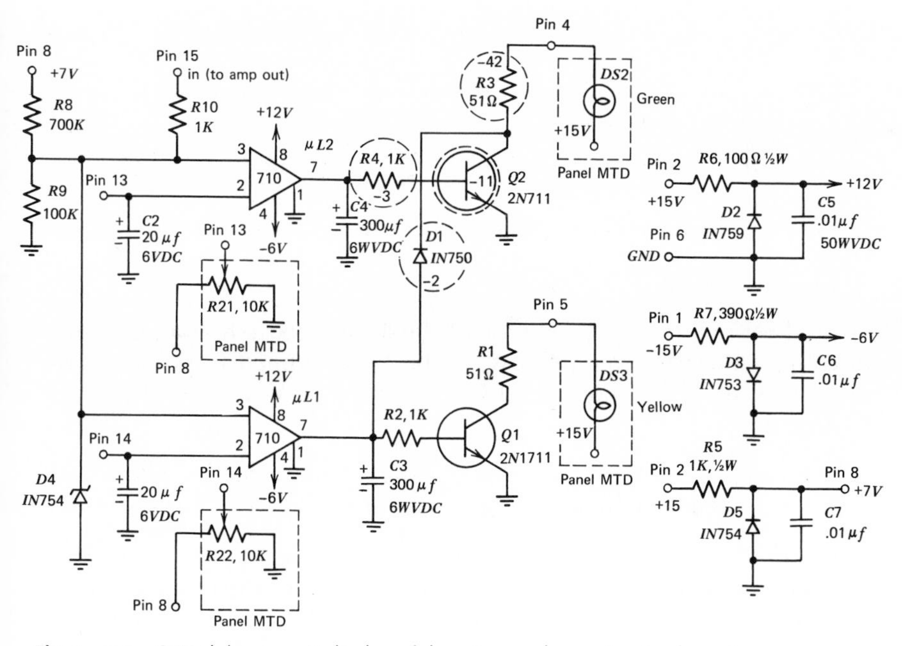

Figure 127. CRT alphanumeric display of deviations only on circuit schematic.

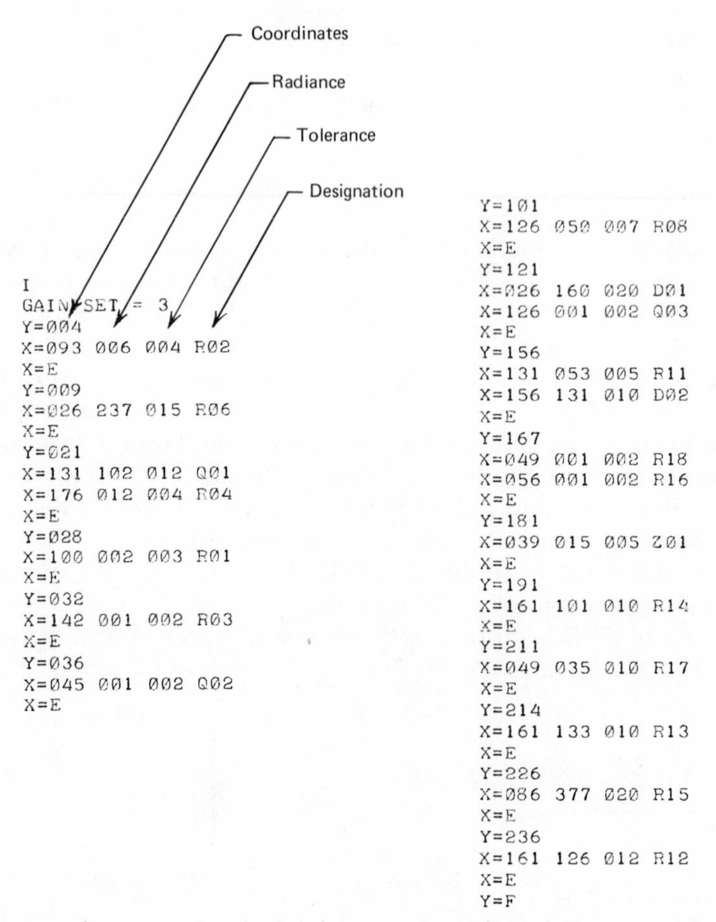

Figure 128. Teletype printout of infrared signature (partial) of electronic printed circuit.

be used for real-time presentation of the deviations. The computer used in conjunction with this process can be programmed to point out the deviations in sign and magnitude, located in such a way as to do the following:

1. Fit a panel layout outline, such as the one illustrated in Figure 126.
2. Fit a circuit schematic, such as the one illustrated in Figure 127.

The display in Figure 125 is preferred by design engineers, product designers, and reliability engineers, because it shows the actual temperature of the components including the effects of all the thermal and radiative interactions.

```
S

COMPONENT   DIFFERENCE

   Q05          +005
   D02          +005
   R03          +004
   R24          +001
   R26          +000
   R30          +001
   D01          +003
   Q06          +000
   D03          +013
   D04          +014
   R22          +016
   R04          +001
   R31          +002
   R17          +000
   R49          +003
   R38          +001
   Q09          +000
   R46          +002
   R16          +000
   R53          +004
   R56          +006
   D17          +006
   R55          +005
   R05          +003
   D18          +001
   R52          -005
   Z01          +001
   Q13          +031
   R58          -041
   R50          -005
   D07          +001
   Z02          +003
   D20          +002
   R29          +042
   Q12          +035
   RE3          -003
   Q14          +024
   Q11          +045
TEST COMPLETE
```

```
S

COMPONENT   DIFFERENCE

   D01          -042
   Q04          -000
   Q10          -000
   Q12          -000
TEST COMPLETE
```

```
S

COMPONENT   DIFFERENCE

   D01          -045
   Q04          -000
   Q10          -000
   Q12          -001
TEST COMPLETE
S

COMPONENT   DIFFERENCE

   D01          -044
   Q04          -000
   R16          -017
   R53          -054
   R56          -044
   D17          -043
   R47          +001
   Q14          +000
   Q11          +012
TEST COMPLETE
```

```
S

COMPONENT   DIFFERENCE

   Q05          +002
   D02          +001
   R03          +003
   D01          -037
   R53          +000
   R56          +004
   D17          +001
   R55          +001
   D16          -017
   R59          -025
   R58          -105
   Z02          +002
   Q11          +015
TEST COMPLETE
```

Figure 129. Printout of deviations from standard.

The display in Figure 127 is preferred by test engineers, trouble-shooters, and maintenance engineers, since it points out the effects of a failure or of an anomaly that will eventually cause a failure.

Both these displays are available as options for the Inspect System that was mentioned in p. 101 under "computer-controlled scanner." However, the basic display of this system is a teletype printout, which upon operator's request will do the following:

1. List the infrared radiance value of every programmed component on the board under test, along with its designation, its coordinates, and the tolerance assigned to it (see Figure 128 as a partial example).
2. List the designation of every component whose radiance differs from the preestablished standard, along with the magnitude of such deviation, if it exceeds the assigned tolerance. Figure 129 shows examples of such printouts.

Chapter 5 Infrared Emission by Electronic Equipment

BASIC CORRELATION

Whenever an electric current flows through a resistive element, a certain amount of heat is generated. The magnitude of the heating power so produced is

$$P = I^2R$$

where R is the ohmic resistance of the element.

In this process, the electrical energy of the current has changed its physical state by becoming an increment to the energy level of the atomic and subatomic particles of which physical matter is made. In turn, this added energy manifests itself as an increase of the temperature of the conductive element where the current flow takes place. The rate of temperature rise is simply

$$\frac{dT}{dt} = \frac{P}{mc}$$

where c = specific heat
m = physical mass.

If we assume that the element was initially at ambient temperature, as soon as its temperature starts rising above said level, three heat dissipation processes will take place: conduction, convection, and radiation. Ordinarily, conduction losses primarily depend on the physical-mechanical configuration of the element and its supports, and are linearly dependent on the temperature gradient from element to support. Convection losses primarily depend on the flow of a heat transfer medium in proximity to the element, and the magnitude of convection loss is ordinarily linearly dependent on the temperature gradient from element to medium. We have seen that radiant losses depend on the fourth power of element temperature, and this power loss manifests itself as infrared radiation. In other words, the dissipated electrical power is proportional to radiated power, or to the fourth power of absolute temperature. Thus a measurement of the temperature of a radiating element can provide a useful correlation with dissipated electrical power.

All the passive infrared measurement techniques described in this chapter are based on this concept. In other words, every conductive element increases or decreases the infrared power emitted in direct proportion to the variations of the electrical current flowing through it. The emitted infrared signal is amplitude modulated, and when duly complemented, and correctly interpreted, contains all the information necessary for determining the characteristics of the thermoelectrical stress level imposed on the component.

HEAT TRANSFER MECHANISMS

As already mentioned, convection, conduction, and radiation are the three physical mechanisms through which the heat can be transferred from an object whose temperature is above the ambient. Generally, for objects mechanically mounted or secured to a holding fixture, the conduction process is predominant and can carry away 70 or 80% of all heat being lost by the object.

The quantity of heat removed by convection can vary greatly between the extreme of zero in the vacuum of outer space to over 90% in a forced stream of a cold gas or fluid. However, for "normal" conditions such as uncooled still air at sea level, convective removal of heat is around 15% of the total. This condition is defined as "free convection."

Radiation losses make up the balance. Again, in "normal" conditions, the heat removal by radiation is around 10% of the total, but the emissivity factor can make this figure vary within a wide range. Only seldom, such as is the case of hardware floating in outer space, can radiation become the only heat transfer mechanism in existence.

It might be worth mentioning that negative heat removal can sometimes take place, such as when the power dissipating element is connected to a heat-sink at a higher temperature, or when the "cooling" fluid is warmer or the impinging radiation introduces more energy than is emitted.

The chart in Figure 130 shows how greatly the temperature of a conventional electronic component operating in a "normal" environment is affected by the ratio between conduction and the other two heat transfer

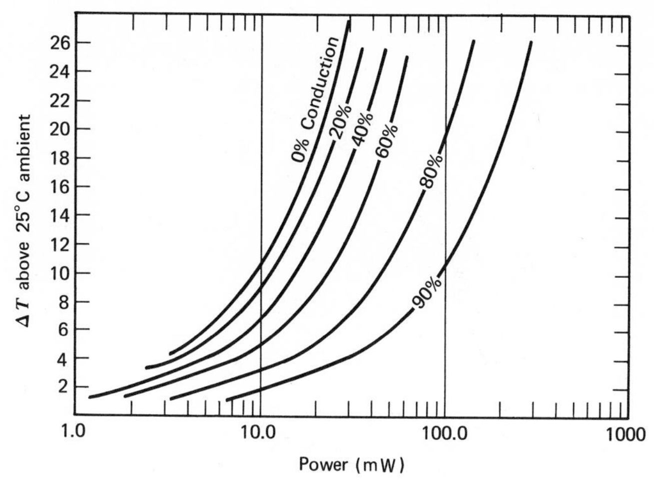

Figure 130. Correlation of component temperature and conductive heat loss.

mechanisms. The same chart, read in a different way, shows what increases in temperature are taking place to step up convection and radiation losses to compensate for reduced levels of heat carried away by conduction.

The relationship between convection and radiation losses, again for "normal" environment, is shown in Figure 131. It is apparent that their magnitude is similar, although energy losses by radiation are limited by the maximum value of 1 for emissivity, while energy losses by convection do not have an upper limit, since they are a function of the temperature and of the speed of the circulating coolant.

In the great majority of the practical applications of thermal measurements by means of infrared, it is very important to keep in mind that thermal energy can only be emitted by physical matter when the ambient, or environment, is at a lower thermal level. As a consequence, a power-dissipating component must raise its own temperature above its immediate environment to dissipate said power under the form of heat.

This is why the charts in Figures 130 and 131 are calibrated in "degrees above ambient": because in effect the ambient temperature is the "zero level" of our thermal system. Furthermore, the ambient temperature of 300°K is indicated, because of the nonlinear characteristic of the expression for radiated power, which includes the T^4 factor. For all practical purposes the curves are usable in a limited thermal range around 300°K, but cannot be considered valid for large deviations from this level.

This is why all techniques attempting to measure electrical power dis-

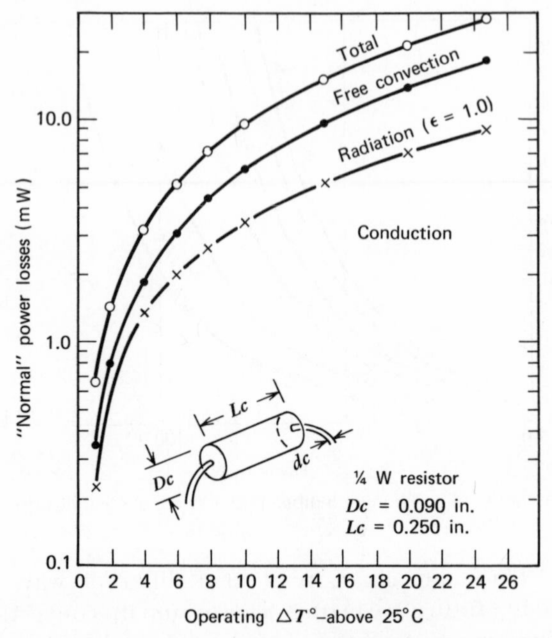

Figure 131. Correlation of component temperature with conduction, convection and radiation losses.

sipation and current through the measurement of temperature must take into account the local zero level, or, in other words, the ambient temperature in the immediate surroundings of the element being measured.

It is of interest to examine the effects of emissivity and background variations on an ideal model of electronics circuit assembly where interaction between components does not exist.

Such a unit is schematically illustrated in Figure 132 which shows six resistors wired to a printed board in whose ends two blackbody reference elements are located. These reference elements are assumed independent from ambient temperature variations, and can be designated as H_{RL} and H_{RH}, respectively, for the low radiation reference and the high radiation reference level.

The illustration also shows the background surface around the assembly, and we assume that its temperature is coincident with the ambient temperature at all times.

We also assume that the six resistors are all at the same temperature, and that this temperature is higher than that of the board on which they are mounted. Finally, we will assume that the emissivity of the resistors and the emissivity of the board is a fraction of unity.

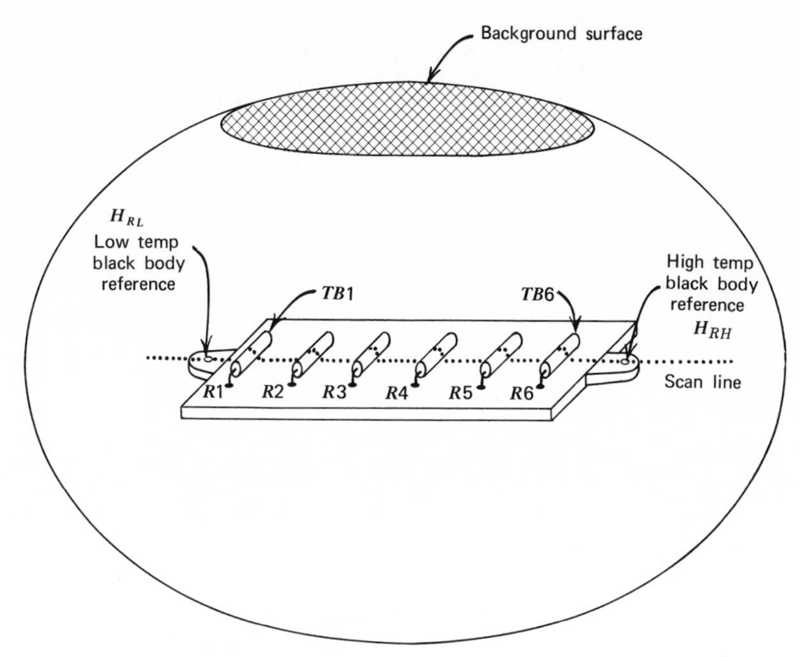

Figure 132. Theoretical circuit board.

We are going to examine what the infrared readings of a radiometric system will be in the following instances:

1. Uniform emissivity, uniform background temperature.
2. Variable emissivity, uniform background temperature.
3. Uniform emissivity, variable background temperature.

Figure 133 shows the radiance variations as the assembly of Figure 132 is being scanned from H_{RL} to H_{RH} along the dotted line, in the instance 1, where the six resistors are all at the same temperature and the mounting board is at ambient temperature.

In the instance 2, the emissivity of the surface of the resistors varies as follows: $\epsilon_1 < \epsilon_2 < \epsilon_3 < \epsilon_4 < \epsilon_5 < \epsilon_6 < 1$, while their temperature and the ambient (mounting board included) remain uniform. Figure 134 illustrates the radiance variations as the assembly is scanned from H_{RL} to H_{RH}. The radiation peaks are aligned along a line that depicts the emissivity increments of the resistors' surface, while the radiation lows are aligned along a horizontal line that indicates no change in the temperature of the board.

In the instance 3, Figure 135 illustrates what a scan of the six resistors may look like when all are operating at the same temperature with uni-

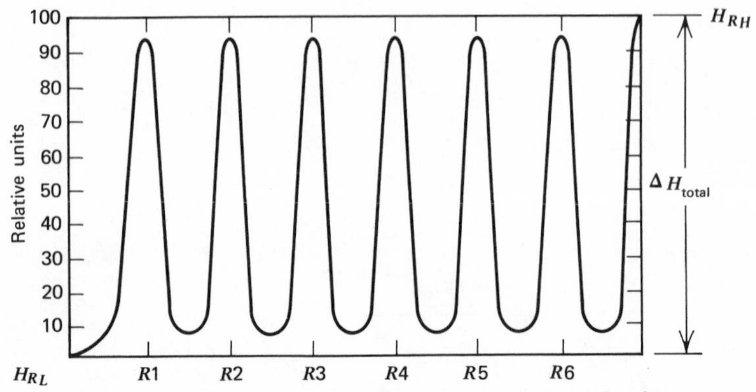

Figure 133. Infrared scan across resistor assembly, when emissivity and ambient are constant.

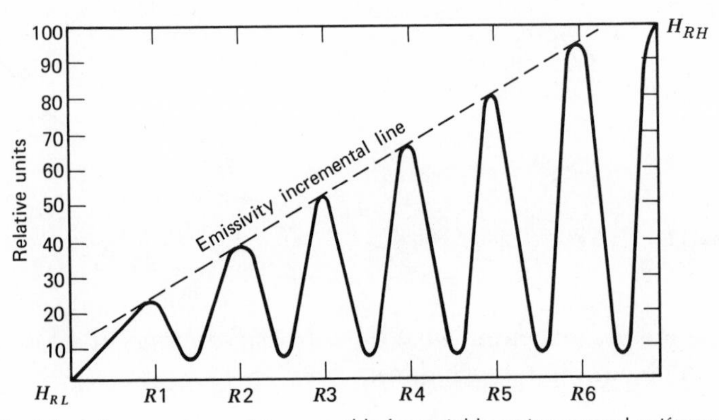

Figure 134. Infrared scan across resistor assembly for variable emittance and uniform ambient.

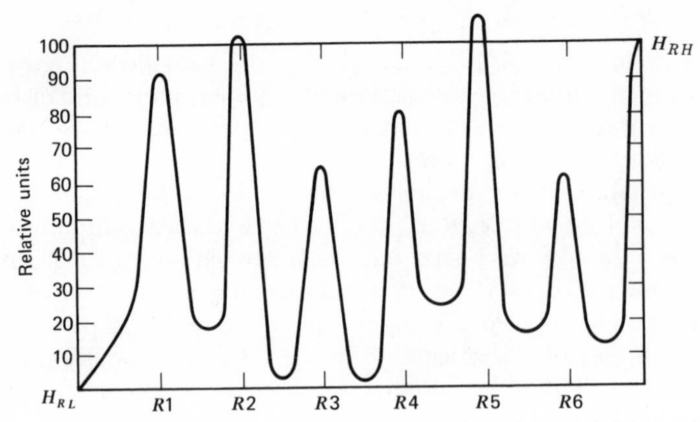

Figure 135. Infrared scan across resistor assembly, with uniform emissivity and variable background.

form emissivity but being exposed to a nonuniform background whose variations affect the radiation measured.

The variations become even more pronounced in the case that emissivity and ambient temperature are nonuniform at the same time.

In conclusion, the theoretical examples shown above, underline the importance of conducting infrared measurements under known and controlled conditions related to ambient temperature and target emissivity. It is in view of this that the most sophisticated infrared test equipment incorporates an Automatic Gain Control (AGC) feature that automatically compensates for any variation of the ambient temperature.

THERMAL INTERACTION

In actual applications, the simple, clear-cut assumptions that were formulated in the past chapter are conspicuously absent. In every heat transfer mechanism, there is a great deal of mutual exchange or interaction that changes the thermal pattern quite substantially.

The *conduction* of heat from parts at higher thermal level to parts at lower temperature can be quite large and can severely affect the operating condition of the heat-receiving component. This is especially true for nonlinear devices, such as semiconductors, whose performance can vary rather sharply with temperature. For instance, a power-dissipating resistor needs adequate heat-sinks into which to transfer the self-developed heat. As a matter of fact, both the design and the rating of resistors is based on the assumption that 20% of the developed heat is dissipated by the body, while 40% of the heat is transferred out of each terminal lead into adequate heat-sinks. If the thermal capacity of the designed heat-sinks is inadequate, all other elements directly or indirectly connected with the resistor will become additional heat-sinks as long as their own temperature remains below that of the heat source. Thus the operating temperature of a small diode or of a transistor located near the hot resistor could reach a level that will bring operation into an overstress condition.

Even more insidiously, this condition could develop as an unforeseen consequence of a seemingly innocent change, such as the replacement of a component part with a new version of equal electrical characteristics, but more convenient for style, price, and smaller size. What really happens is that the old-type component (e.g., a transformer) was acting as the heat sink for a power-dissipating element. The new transformer cannot fulfill this function because of its smaller size, and the result is unforeseen overheating of the power-dissipating element, or of some other temperature-sensitive element directly or indirectly connected to it, and now promoted to the rank of additional heat-sink.

The *convection* process, too, causes thermal interaction that often is ignored. Free-air convection develops along vertical lines, while forced convection follows the flow lines of the cooling fluid. The trouble is that sometimes the cooling fluid turns out to be a warming fluid, so that the end result might happen to be the opposite of what was originally intended.

The *radiation* mechanism is the only one that is always present, and is probably the most difficult to calculate because of its dependence on emissivity, reflectivity, and absorptivity of all the elements involved. Even the application of the basic rules can yield opposite results if some of the elements are disregarded. For instance, in opposition to the prevalent practice, the outside surface of semiconductor package envelopes should be black to enhance heat loss by radiation, so that the units might be operating at a lower temperature. Although in those particular configurations where the units are located near strong sources of radiation, a highly reflective coating might actually keep the unit cooler, since most of the impinging radiation will be reflected away.

In conclusion, the effects of thermal interaction are very complex and almost impossible to calculate exactly, except for the most elementary and simple configurations. Consequently, even in the best designed systems, there is always the possibility that some unsound thermal condition might be present to the detriment of the reliability and of the life expectancy of the system.

Until now, actual temperature levels were verified with the use of thermometers or the thermocouples. Besides being time consuming, of difficult implementation, and necessarily restricted to a limited number of points, these techniques cannot be used in high voltage areas or with elements that do not allow physical contact.

At the present time, the availability of infrared test equipment makes it possible to verify easily and quickly the true thermal condition of every element of an assembly, pinpointing the areas where the conditions are not as planned. And evaluation of complex assemblies is almost as easy as the evaluation of the simple ones, especially when emissivity of the target and ambient temperature can be kept constant.

THERMAL RESISTANCE

The heat flow along a thermally conductive path is directly proportional to the temperature gradient between the heat source and the heat sink at the other end, and inversely proportional to the thermal resistance of the path itself. This is expressed in the following equation:

$$\frac{dH}{dt} = -KA\frac{dT}{dx}$$

where K = thermal conductivity
A = surface area
dH/dt = time rate of heat flow
dT/dx = temperature gradient.

In other words, the heat transfer rate between two points that are kept at a given temperature gradient is directly proportional to the thermal conductance of the heat flow path, or inversely proportional to its thermal resistance. This thermal resistance can be evenly distributed along the transmission path, or it can be concentrated in one or more points, where discontinuities of the heat-carrying material are located. In the first case, the temperature variation along the heat flow path is uniform, while in the second instance a temperature drop is located in correspondence to every point where the thermal resistance is higher.

This concept is graphically illustrated in Figure 136: in the example, the corresponding ends of the two conductors A and B are kept at the same thermal levels: high at end 1, and low at end 2. Temperature probing along the two elements shows that conductor A has a constant

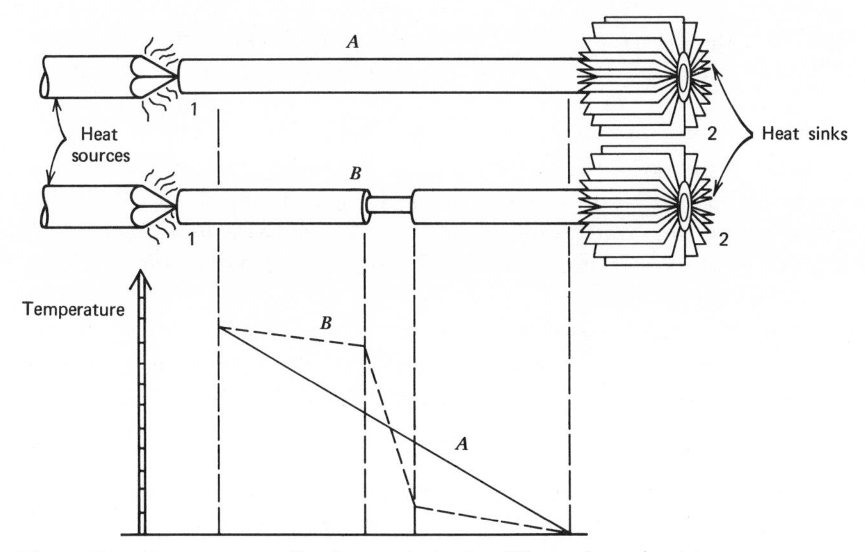

Figure 136. Temperature profile along paths having different thermal resistance.

thermal gradient along its whole length, while conductor *B* shows a sharp temperature drop in the area where its cross-section is greatly reduced.

This phenomenon, readily detectable by thermal surface mapping, is at the basis of most techniques designed to evaluate the integrity of materials and of physical bonds. Especially for large surfaces, thermal mapping by scanning them with an infrared detector is by far the most convenient, quick, and complete way.

EMISSIVITY CORRECTION

As already seen, surface emissivity is one of the elements controlling the magnitude of the infrared radiation. Variation of the emissivity coefficient between different points of the same surface or between different elements of the same assembly makes it impossible to obtain infrared readings that are readily comparable to each other. This difficulty has been largely responsible for the lack of confidence in the infrared approach, and for the long delay in the acceptance of the infrared measurement techniques.

Two approaches are devised to surmount the difficulty: emissivity compensation and emissivity equalization. The *emissivity compensation system* calls for measurement of the emissivity coefficient of every area where an infrared measurement is made. In this way, every reading can be corrected by dividing it by the emissivity coefficient so that the resulting values are expressed in terms of blackbody emissivity. Major difficulties of this system are the usually large number of points for which corrections must be made, and the problem of measuring the emissivity factor of every one of these points.

For this purpose, the target must be heated to known temperature levels and its radiation levels measured and compared with those of a blackbody at the same temperatures. Two or more of these readings are needed to plot correlation curves indicating the emissivity factor of each point being evaluated.

Another system makes use of the variations in reflectivity of the surface elements under evaluation. Since reflectivity and emissivity are complementary quantities, the emissivity coefficient is given by:

$$\epsilon = 1 - r$$

This system requires an exact way of measuring the reflected radiation, and this can be quite a problem, except in the case of geometrically simple, even surfaces.

Application of this principle is illustrated in Figure 137: radiation from the test surface is composed of emitted radiation due to its temperature and reflected radiation from a reference heater. When the total radiation from the test surface matches that from the reference heater, the heater temperature is proportional to the test surface temperature. Since the sum of emissivity and reflectivity is always unity, any change in test surface emissivity will cause a corresponding change in the amount of heater radiation reflected from the surface. Thus at a specific temperature the total test surface radiance (emitted plus reflected energy) remains constant regardless of its emissivity. This principle is used in a

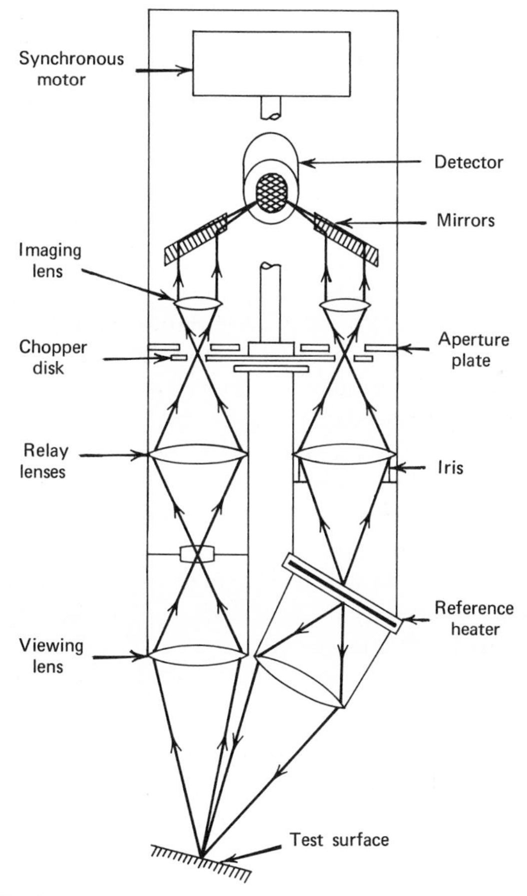

Figure 137. Reflection compensates emissivity variations. (Courtesy British Scientific Instrument Research Association, England.)

radiation pyrometer by British Scientific Instrument Research Association, Chislehurst, Kent, England.

The *emissivity equalization system* is based on the use of coatings that can be applied over the surface elements whose radiation must be measured. It is the equivalent, in the visible range, to painting with the same color all objects in sight. After this operation, it is clear that all readings are directly comparable, and if the coating compound has an emissivity coefficient close to unity, the measurements will also be comparable with those of blackbodies. A further advantage of the system is the increase in the energy level of the infrared radiation emitted by surfaces having low emissivity, so that thermal resolution is enhanced and measurements are more precise after the coating has raised the emissivity to a value that is comparable with that of a blackbody.

A large variety of compounds is available for emissivity equalization. For application purposes, they can be grouped as *removable* and *permanent.* Another grouping could be made with regard to their electrical characteristics—resistivity—thus we could speak of conductive or nonconductive coatings. Obviously only the latter ones can be used on electrical or electronic components.

The trend, at the present time, is to combine emissivity equalization with physical protection of the parts. In most instances, these parts must operate in environments that can damage them, through corrosion or contamination. Protective coatings are therefore a necessity, and many of them happen to have emissivity factors close to unity. Their use, consequently, offers the double advantage of physical protection and high, uniform emissivity factor.

Neville R. Burrowes of the U.S. Navy Applied Science Laboratory conducted a 2-year study of the problem, surveying the physical characteristics of about 50 coating compounds and completing the investigation on the 12 most promising materials. The great majority of these belong to the "thermosetting" family, liquid resins that solidify at moderate temperature. Table 6 lists the advantages and disadvantages typical of the four groups of this family.

Among the key properties taken into consideration for the selection of the best suited materials were the following.

Spectral transmittance. The coating must be transparent in the visible range to allow legibility of markings or codings of the circuit components. It must be opaque in the 3 to 10-μ range where most of the infrared emission takes place for components of electronic assemblies operating at near-ambient temperature.

Curing characteristics. Plastic materials are usually cured by the addition of a catalyst, a heat, or both. The addition of the catalyst causes an

Table 6 Comparative Evaluation of Thermosetting Compounds

Material	Advantages	Disadvantages
Polyesters	1. Good chemical resistance. 2. Low water absorption. 3. High dielectric strength.	1. Not suitable for high frequency applications. 2. Relatively high shrinkage. 3. Limited temperature range.
Epoxies	1. Outstanding adhesion to clean surfaces. 2. Excellent electrical properties 3. High mechanical strength 4. High thermal stability. 5. Good chemical resistance. 6. Low shrinkage.	1. High temperatures produced by exothermic reaction. 2. Some epoxies are highly toxic.
Silicones	1. Available in many states such as fluids, gels, elastomers, foams. 2. Good ozone resistance. 3. High operating temperatures. 4. Good electrical properties. 5. Relatively high thermal conductivity.	1. Poor adhesive properties. 2. Poor resistance to abrasion. 3. Low tensile strength.
Polyurethanes	1. Strong, tough. 2. High abrasion resistance. 3. High thermal resistance. 4. High surface adhesion.	1. Susceptible to moisture during casting process. 2. Low thermal conductivity.

exothermic reaction, which, in a potting operation, can generate enough heat to damage delicate components such as germanium diodes. However, for the thin coatings used to equalize emissivity, the exothermic heat generated is negligible.

Curing temperature. Normally the coating material must be baked at an elevated temperature to complete the cure. This cure temperature should be as low as possible to not damage components.

Shrinkage. It is desirable to minimize shrinkage of the coating material to maintain a good bond between the cured resin and the object coated.

Pot Life. The pot life or working life of a resin is the time interval between the initial mixing of all components and the polymerization of the resin to a viscosity degree beyond which it is impractical to apply.

Table 7 Properties of Thermosetting Coatings

Manufacturer	Coating Designation	Main Constituent	Pot Life	Cure[a]	Maximum Operating Temp. °C	Emissivity Index
Columbia Technical Corp. Woodside, N.Y.	Humiseal type x342(A-B)	Polyester	20 min	Overnight at room temp.	130	0.826
	Humiseal type 1H34	Silicone	6 mo	Room temp. or 1 to 2 hr at 200°C	180	0.876
3M Company Elec. Prod. Div. St. Paul, Minn.	"Scotchcast" brand resin no. 8	Epoxy	1–2 hr	Room Temp. or 2 hr at 60°C	130	0.896
Ciba Products Co. Fairlawn, N.J.	Araldite[c] 488E-32	Epoxy	1 year	Room temp.	130	0.871
Dow Corning Elect. Prod. Div. Midland, Mich.	Sylguard 182	Silicone	8 hr	Room temp. or 4 hr at 65°C	200	0.828
General Electric Silicone Prod. Midland, Mich.	LTV 602	Silicone	2–3 hr	Room temp. or 5 hr at 65°C	200	0.827
Hysol Corp. Olean, N.Y.	Hysol PC 12-007 (A-B)	Epoxy	1–2 hr	2 hr at 50°C	130	0.916
	Hysol PC 15(A-B)	Urethane	2–3 hr	10 min at 52°C	130	0.914
	Hysol PC 21(A-B)	Epoxy	1–2 hr	2 hr at 60°C	130	0.890
Sterling Varnish Company Swickley, Pa.	Sterling[b] E251-33	Epoxy	35 min	2 hr at room temp.	130	0.860
	Sterling[b] E 252-46	Epoxy	2–3 hr	Room temp.	130	0.866

Note. For those interested in temporary coating, the Sylguard 182 and the LTV 602 are easily removable at any time.

[a] Curing conditions flexible, check manufacturers literature.

[b] Found difficult to spray.

[c] Dielectric strength not determined.

Table 7 lists the main properties of the thermosetting coatings evaluated in the NASL's program. Besides the already-mentioned data, several other properties were investigated, such as spectral transmissivity in the 2 to 15-μ range, chemical and moisture resistance, dielectric strength, insulation resistance, thermal conductivity, and temperature tolerance. It was found that 130°C is the upper limit for more than 90% of available coatings.

Spectral transmittance was measured for different thicknesses of thin films of coating and was found to be essentially similar between the four groups of thermosetting compounds. Figure 138 shows the transmissivity spectra of "Scotchcast" brand resin no. 8, for thicknesses varying from 1 to 4.5 mils. It is apparent that in the 5 to 10 μ range, any coating thicker than 1 mil is close to blackbody characteristics.

Figures 139 and 140 illustrate the effects of the coating technique. In Figure 139 are recorded the measurements of infrared energy emitted respectively by:

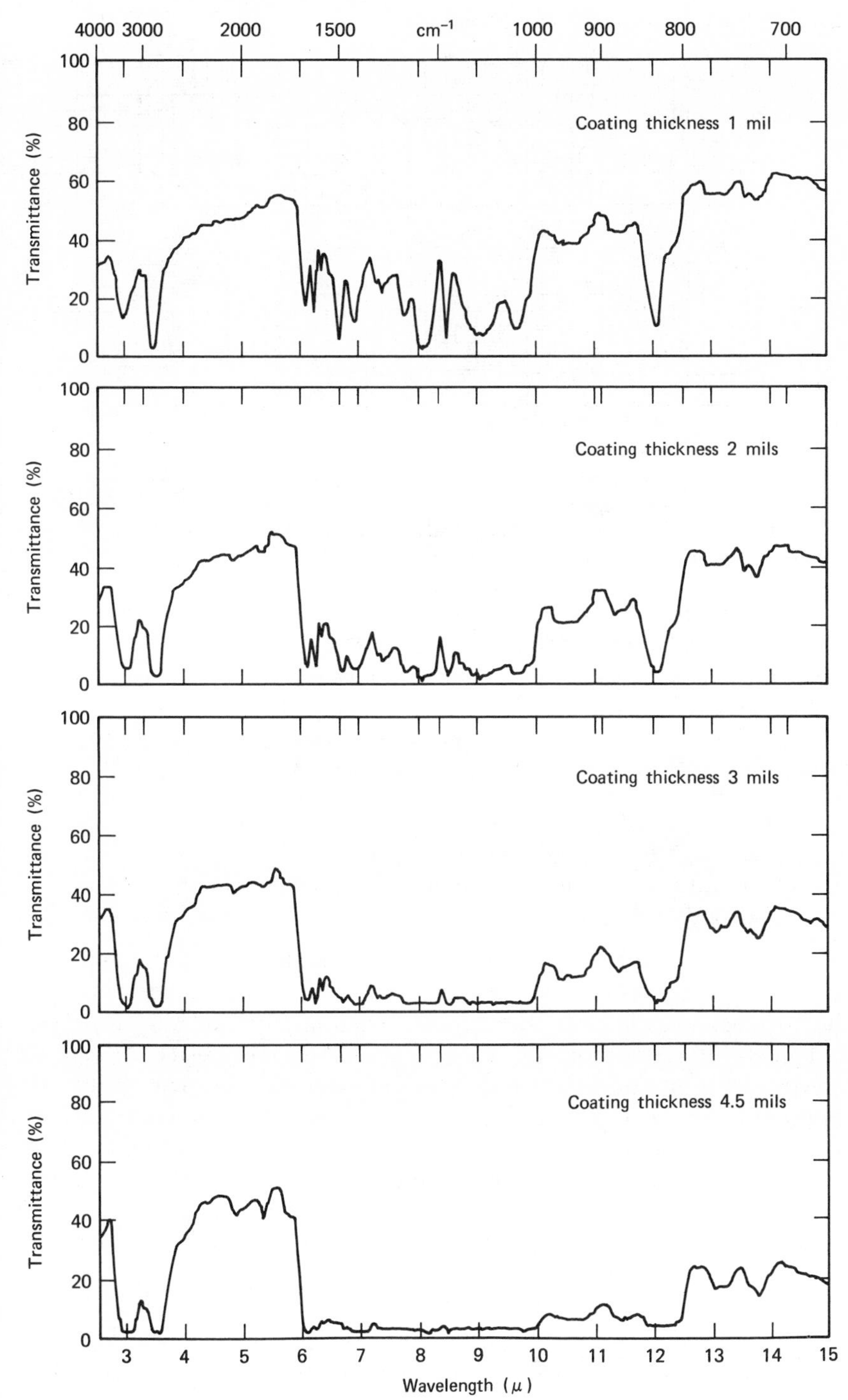

Figure 138. Variation of transmittance with coating thickness, for "Scotchcast" brand resin 8. (Courtesy N. Burrowes.)

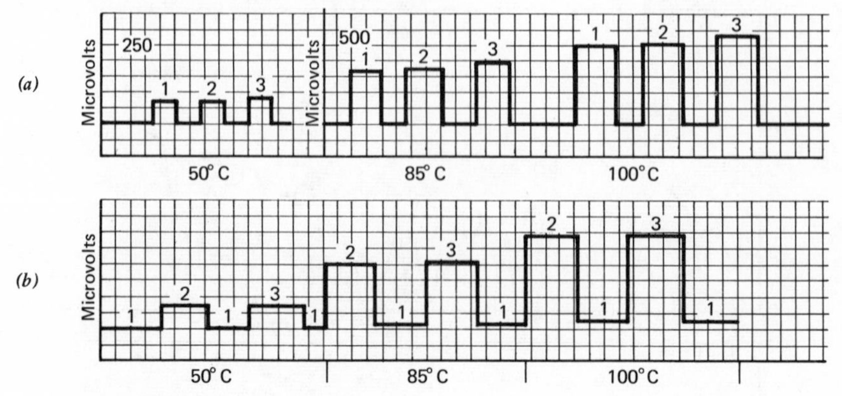

(a) Comparison of radiant emittance for:
(1) Aluminum foil tape, coated with Dow-Corning Sylguard-182 Silicone Resin.
(2) Aluminum foil tape coated with black lacquer.
(3) Barnes Engineering Co., Infrared Radiation Source-Model 11-101.
(b) Comparison of radiant emittance for:
(1) Aluminum foil tape (uncoated).
(2) Aluminum foil tape coated with Dow-Corning Sylguard-182 Silicone Resin.
(3) Aluminum foil tape coated with black lacquer.

Figure 139. Radiant emittance versus temperature for coated and uncoated aluminum foil surfaces. (Courtesy N. Burrowes.)

Aluminum foil uncoated
Aluminum foil coated with a silicone resin
Aluminum foil coated with a black lacquer
Blackbody reference source

The readings were taken respectively at 50, 85, and 100°C. It is apparent that the low emission of the uncoated aluminum foil has been brought up to blackbody level by the applied coating with the silicone resin.

Figure 140 shows how the same phenomenon takes place for various electronic components having different surface materials: after the application of a conformal coating of a silicone resin, the level of infrared radiation emitted has practically been equalized for all units.

Removable coatings offer the advantage of good reworkability of the assemblies: component measurement and replacement is easy, as opposed to the thermosetting compounds, that turn rapidly into a hard, strong material.

The easiest way to remove an emissivity-equalizing compound is by washing, possibly with water, and some soap-based compounds have been developed by Raytheon to this effect. By controlling the amount of

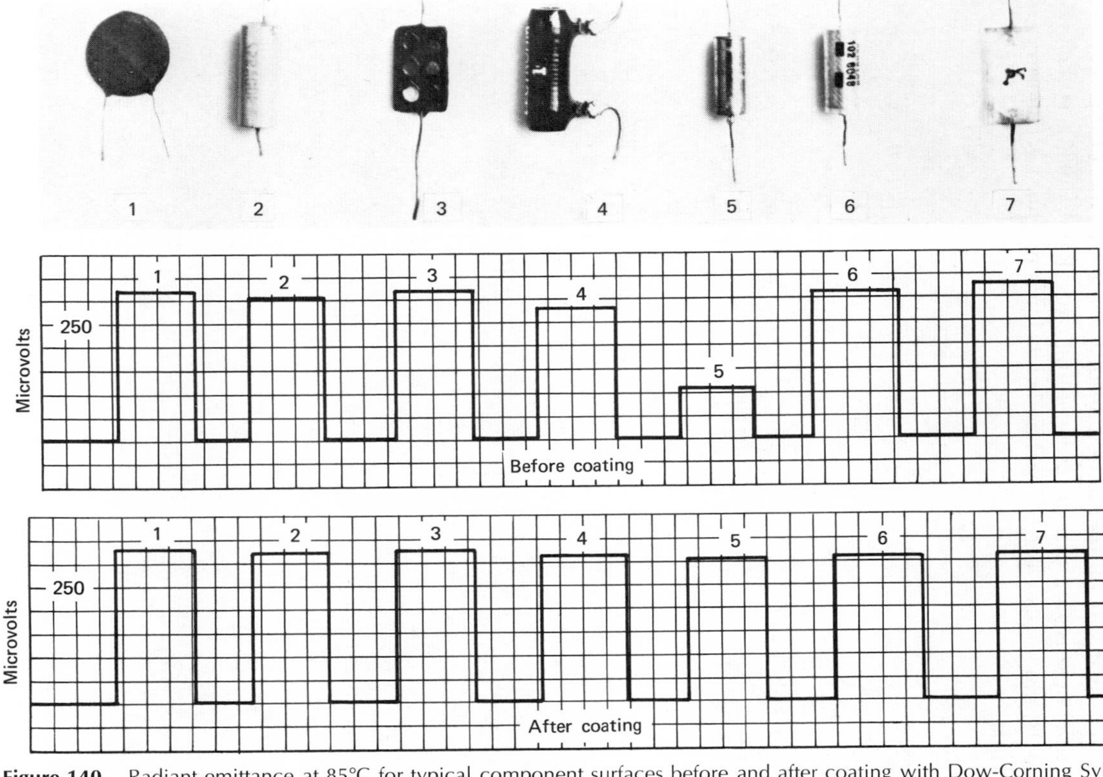

Figure 140. Radiant emittance at 85°C for typical component surfaces before and after coating with Dow-Corning Sylguard-182 Silicone Resin. (Courtesy N. Burrowes.)

pigment in colloidal suspension, the absorptivity factor can be raised close to the blackbody level, and the presence of soap helps the thorough cleaning of the exposed surfaces during the washing cycle. Also for these coatings, as it was already pointed out for the thermosetting materials, the thickness is not critical: any coating over 1 mil thick is sufficient to raise the emissivity close to blackbody level, within the spectral range of 3 to 10 μ.

Part II Applications

"Daddy, oh, Daddy," a jubilant Nelly announced. "Tiggy is going to have kittens! Can I keep them all?"

"Wow!" said I, with a faint smile. "Let's not rush things so fast. How do you know that?"

"She told me," Nelly replied with total assurance, "and she never lies to me."

"I believe you, Nelly. Unlike humans, most animals say mainly the truth. But *things* are even more sincere. They tell the truth all the time, to the whole world," said I, trying to steer away from the kittens.

"Things, Dad? Oh, come on. They don't even have a mouth. How can they talk? And what do they say, anyhow?"

"Oh, a lot of things about themselves, Nelly dear. They tell you what they are and what they are doing, and how they were made, and whether they have invisible faults, and how long they will keep working and the time when they will die. They keep repeating these things over and over, to everybody who cares to listen. But their language has no sound and cannot be seen or felt by our senses. Fortunately, today we have machines that can listen to this language and that can translate it into information that we can understand. You see, Nelly, their language is not English. It is made of invisible colors and it is called INFRARED."

"Infrared, infrared," sighed a disappointed Nelly. "It's not even a good name for a kitten. And I have to find at least five nice names for them."

WHO CARES TO LISTEN?

Of the 3 billion people on earth, approximately 2,999,900,000 are as interested as my daughter, Nelly, in listening to the infrared voice of physical objects. However, the remaining 100,000 (plus or minus X) should be interested in listening. They are the physicists and the engineers, the astronomers and the intelligence agents, the physicians and the chemists, and in general all the technical specialists working in such diverse fields as the automotive and the microelectronics, the steel manufacture and the crystal growing, the atomic reactors and the plastics, the communications and the agriculture, the glass manufacture and the aircraft.

To all these people the infrared voice of the objects of their work can supply significant, sometimes vital information that might enable them to achieve goals not otherwise attainable with conventional techniques.

In the following chapters I will endeavor to outline some of the current applications of passive infrared detection and measurement techniques, so that no one will have to reinvent the wheel. However, this should be only the starting point, intended to equip the reader with knowledge adequate to view his technical problem in the infrared light, and to devise his own solution with the help of this new capability.

Chapter 6 Infrared for Reliability

Reliability is defined as "the probability that a certain piece of equipment will operate in a given environment, within given performance limits, and for a certain length of time."

Only a perfect piece of equipment would have a 100% chance of performing perfectly, that is, it would have a reliability of 1. Since perfection does not exist in this world, this figure is never attained. All we can do is to try to come closer and closer to it.

THE FAILURE RATIO CONCEPT

It is my intent to describe in this book a new tool that can help the designer to assess the real extent of the stresses active in the component parts, so that the reliability figure might approach the unreachable limit of 1 even closer. As already mentioned, this tool is the infrared radiation emitted by all physical objects. This radiation carries with it significant information about the physical structure and the operating characteristics of the emitting elements. Whenever physical structure changes cause changes in the failure ratio of an element, a correlation between infrared emission and reliability can be established. Furthermore, whenever operating characteristics variations cause variations of the failure ratio of an element, then a correlation linking infrared radiation with reliability can be determined.

For instance, the probability of failure for a soldered joint is inversely proportional to the cross-section area at the narrowest point of the fused metal bonding the two elements. Since we can, by infrared measurements, determine the thermal drop located at this point and consequently the area of this cross-section, correlation between infrared data and failure probability can be established.

Similarly for changes in active characteristics, the probability of failure of a resistor operating at the conventional loading levels is directly proportional to the amount of power dissipated by it (assuming constant ambient temperature). Since power dissipation can be measured by the amount of infrared radiation emitted, correlation between infrared radiation and failure rate can be established.

For electronic components such as semiconductors, resistors, capacitors, coils, and transformers, the general rule is that the failure ratio increases in direct proportion with the increase in the temperature of the component. This correlation is clearly shown in the failure ratio charts of the current reliability manuals such as the *Mil-Handbook 217A*, entitled *Reliability Stress and Failure Rate Data for Electronic Equipment.*

Figure 141 shows one of these charts for precision wirewound resistors. We can see that an increase in the resistor's power stress ratio (indicated as the ratio operating wattage/rated wattage) causes an increase

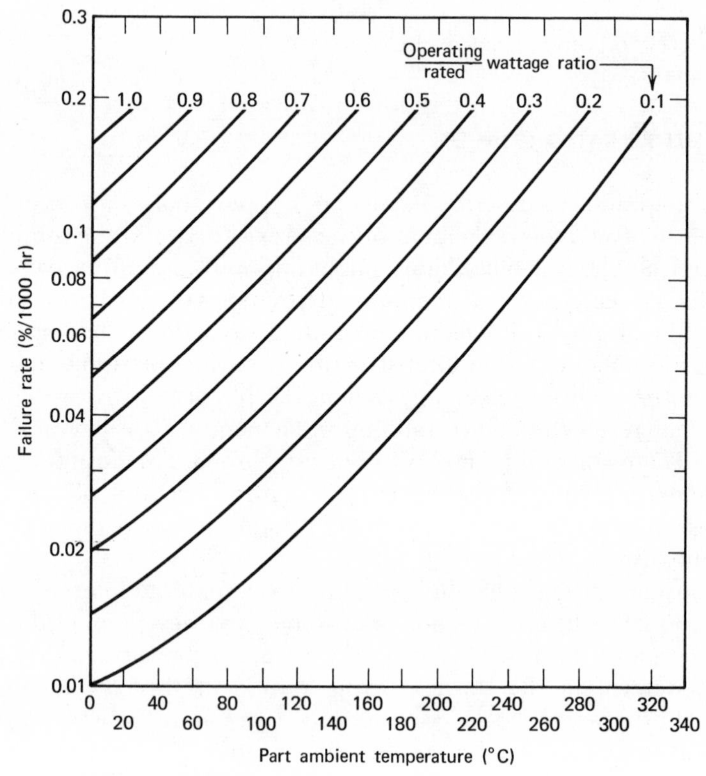

Figure 141. Failure rates for MIL-R-26C quality power wirewound resistors conventional chart. (From Mil. Handbook-217A.)

of the failure ratio index. We can also see that an increase in the ambient temperature causes an increase of the failure ratio index.

Consequently, the chart is made of a family of curves, one for every power stress ratio; as this ratio increases, the corresponding curve is displaced toward the higher failure ratio region of the chart. Since the effect of power dissipation is heat, the chart merely depicts the damaging effect of heat on the reliability of the resistors. Whether the heat is generated by electrical power dissipation or by the temperature of the surrounding ambient, the effect is the same—higher failure ratio or lower reliability.

At this point, a question comes to mind: since both mechanisms bring the resistor to its operating temperature, would it be possible to plot a chart where the failure ratio would be solely a function of the operating temperature of the component?

The idea seems to offer some advantages: in the case of nonlinear devices, such as semiconductors, it is difficult sometimes to assess the power dissipation level; furthermore, the term "ambient temperature" is rapidly losing its meaning: the present trend toward tighter packaging and miniaturization increases the effects of thermal interaction between components, so that quite often these elements are located in regions where the temperature is much higher than the so-called "ambient" surrounding the equipment. This makes it very difficult to determine the real ambient temperature for each and every component part, and the use of the failure-rate charts can become misleading and conducive to error.

Worse yet is the case of space hardware: outside the earth's atmosphere the very concept of ambient has no meaning, and the charts cannot be used, thus every reliability engineer must use his own ingenuity.

Still, the correlation between component's operating temperature and reliability is valid, ambient or no ambient, and this is the rationale for replacing the failure-rate charts based on ambient temperature with other charts based on the temperature of the component part. Since this temperature can be directly measured, we can ignore the true value of electrical power dissipation and the true temperature (if any) of the surrounding ambient.

This sounds too good. Unfortunately, one of the statements formulated above is false. We cannot measure directly the temperature of operating component parts: we can only measure their *surface temperature,* which in many instances will be close enough for good approximation, but which ignores the thermal gradient between the component's core and its outside surface.

The magnitude of this gradient is proportional to the power dissipation level; thus a distinct curve will be needed for every value of the ratio operating wattage/rated wattage. However, for low values of this ratio, the difference between curves is very small and often negligible.

Figure 142 shows how the family of curves of Figure 141 has been transformed when the "component surface temperature" parameter has replaced the "ambient temperature" parameter. This chart yields the real failure rate index for any precision wirewound resistor of a certain family, solely as a function of its surface temperature, without regard to the temperature (if any) of the ambient around it. From the chart we can see how, for the derating values currently used in conservative design, the various curves are so close to each other that the ratio W/W_R can be ignored and failure ratio values can be derived by merely taking one measurement: component surface temperature.

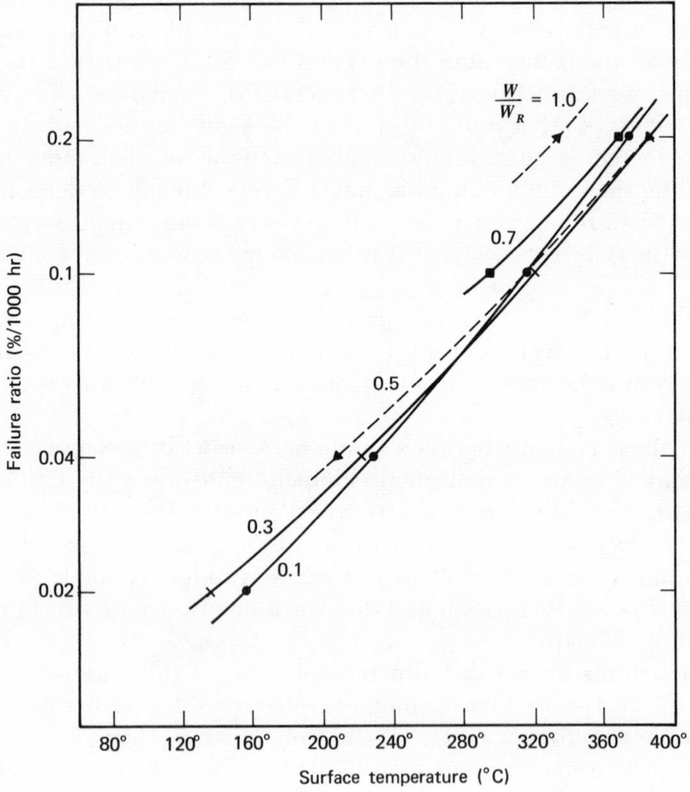

Figure 142. Failure rates for MIL-R-26C quality power wirewound resistors, surface temperature concept.

Since the chart of Figure 142 correlates failure ratio solely with component surface temperature, it eliminates the variable introduced by the higher or lower efficiency of heat sinking of the component. Good heat sinking, although of great importance, is too often neglected in the design of electronic assemblies, leading consequently to serious increase of the failure ratio.

This is dramatically shown in the chart of Figure 143, where line 1 gives the temperature above ambient for a 1-W carbon composition resistor connected to an "infinite" heat sink; in the same chart, line 2 gives the same data when the same resistor is connected to a heat sink of neglibible capacity. We can see that for operation at full rating, the temperature of the resistor jumps from 35 to 98°C above the ambient level, when the heat sinking is removed.

In terms of failure ratio, the corresponding jump is even greater: from a 0.01 figure it reaches the value of 0.50, as shown by the curves traversing the chart at an angle. Besides pointing out quite clearly the

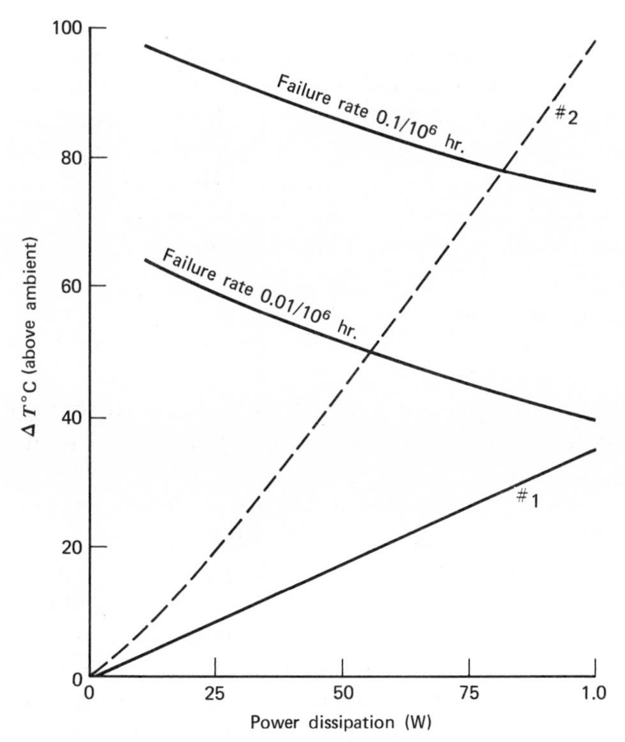

Figure 143. One-watt carbon composition resistors correlation chart.

importance of adequate heat sinking, the charts described in Figures 142 and 143 allow the evaluation of the quality of the engineering design of an electronic assembly and the realistic prediction of its overall failure ratio; that is, its reliability and its life expectancy.

The only indispensable requirement for such operation is the actual measurement of the surface temperature of each and every element operating in the electronic assembly under evaluation.

ENTER INFRARED

Of the many ways presently available for measuring such temperatures, by far the most convenient is the use of infrared scanning equipment. The major advantages of this approach are as follows:

1. It is a noncontact technique. Consequently, the thermal equilibrium of the target is not disturbed, and the measurement does not depend on the quality of the contact between target and probe.
2. It is a passive technique. No load or energy of any kind is imposed on the target.
3. It is all-encompassing. No matter how many elements are assembled in the system under evaluation, the infrared scanner will measure the temperature of all of them. The only requirement is that they be located within the field of view. And for viewing elements that are hidden behind opaque objects or out of the field of view, Chapter 3 describes ("optical fibers") ways and means to make it possible for the infrared detector to look around corners and through nontransparent envelopes.
4. It is fast. The millisecond or microsecond response of the infrared detectors makes it possible to scan a large field of view in only minutes or even just a few seconds.
5. It is accurate. Repeatability tests carried out on some of the best radiometers have shown a "standard deviation" of less than 1% of the measurement value.
6. It is safe. Since the measurements are made at a distance from the target, the danger of getting in contact with high-voltage or high-charge elements is eliminated.
7. It can achieve fine spatial resolution. Some scanning radiometers have a "spot size" as small as 0.5 mm in diameter. This allows mapping of thermal gradients on the surface of elements and components that are just a few millimeters in size. Infrared microscopes have a spot size that can be smaller than 0.001 in. in diameter, so that individual elements of integrated circuits can be resolved.

From the considerations, above we can draw the conclusion that infrared scanning equipment is at the present time the best tool for measuring the temperature of components of electronic assemblies.

We can also add that the knowledge of the operating temperature of the components of electronic assemblies will enable the reliability engineer to determine the true failure ratio figures and the consequent MTBF with an accuracy greatly superior to today's conventional methods based on the "ambient temperature" concept.

How this concept can be turned into practical implementation is the subject of the following chapters.

Chapter 7 Component Part Evaluation

RESISTORS

By its own nature, the resistor is a power dissipating element. It has been designed for this purpose, and in electrical circuitry it is used just for that. Consequently, in operation its temperature always increases above ambient and its infrared radiation stands out as a major source of power emitted by an electronic assembly.

As already discussed on p. 169, the temperature of a resistor depends on the electrical power dissipated through it, with due allowances for all other contributing elements such as physical mass, heat conduction, convection, surface emissivity, impinging radiation, and ambient temperature.

On the other hand, the DC resistance value is immaterial. As long as the power dissipation level is the same, and the other parameters do not change, a 1-Ω resistor emits the same amount of infrared radiation as a 1000-Ω or a 1-MΩ resistor. The infrared detector cannot discriminate these resistors from each other. This condition is not a drawback, but rather a great advantage, since all resistors belonging to a same family can be represented by a single curve, correlating power dissipation with infrared radiation.

We have already seen in Figure 143 one of these charts, drawn for 1-W carbon composition resistors with two different heat sink configurations. This chart is valid for the whole family of 1-W resistors, regardless of their ohmic resistance value. Similar curves are valid for resistors of different physical mass (which in resistor design is correlated with wattage rating) although they have different slopes. For instance, $\frac{1}{4}$-watt resistors, being smaller, have steeper curves, typical of faster temperature rise, and vice versa for higher wattage resistors.

DC and AC Operation of Resistors

When a resistor is DC energized at a steady current level, its temperature will gradually rise until thermal equilibrium is reached. At this point, its temperature will stabilize, since the amount of heat produced by power dissipation matches precisely the amount of heat transferred away by conduction, convection, and radiation.

An infrared reading taken at this point in time can be directly correlated to the amount of electrical power being dissipated through the resistor.

Things are different when AC or pulses are flowing through the resistor. Because of its nonnegligible physical mass, the resistor has thermal inertia, and its temperature variations cannot follow the variations of the electrical power being dissipated at every instant. Thus an infrared reading taken while the resistor is operating in AC regime will yield an rms measurement of the power being dissipated. When operating in pulse regime, the infrared measurement will be a function of the pulse amplitude and of its duty cycle.

Resistor Surface Mapping

However, more is possible than just measuring the amount of power dissipation by taking a single reading over a surface of relatively large size. By increasing the degree of area resolution, it is also possible to map the thermal profile of a resistor over its entire surface, or along any predetermined path.

In a normal configuration, the two terminal wires of a resistor must be connected to adequate heat sinks. As a matter of fact, power dissipation rating of resistors are established on the assumption that 20% of the heat is dissipated by the resistor's body, while the remaining 80% is carried away through conduction by the terminal wires (40% each). Thus the heat distribution along the resistor's body will show a peak at the center, gradually declining toward the two ends where the terminal wires are connected. Figure 144 is an example of such condition.

Profiles of this type can be obtained by scanning a resistor along an axial line. To avoid false readings, care must be taken to restrict the viewing area to the resistor's surface, avoiding even partial inclusion of background. As long as all the viewing area falls on the resistor surface, Lambert's cosine law tells us that the reading will be correct, independent of the angle at which the surface is observed.

Defects impossible to detect by conventional means can modify the "standard" infrared profile, thus pointing out the cause for the devia-

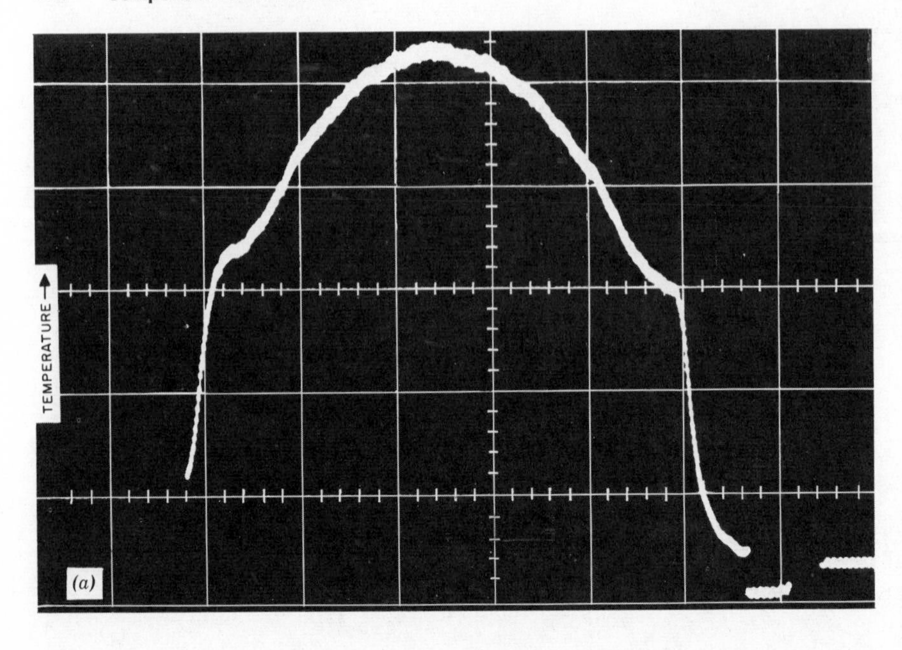

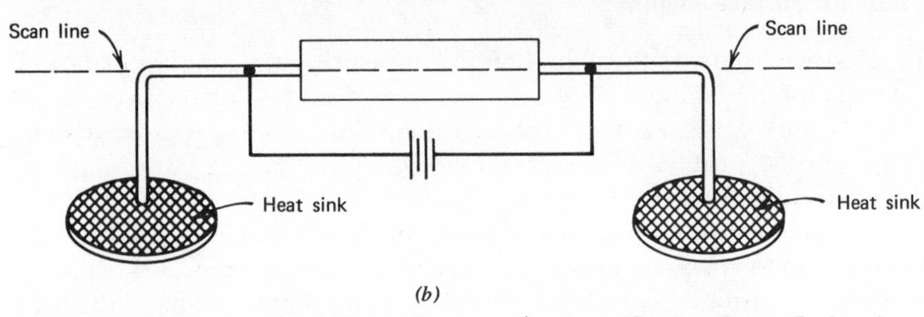

(b)

Figure 144. Thermal profile of electrically powered resistor. (Courtesy Barnes Engineering Company.)

tion. Figure 145 is an example. We can see how the infrared trace is not symmetrical, and the peak is located toward the right end. This indicates a high thermal resistance at the junction between resistor body and right terminal wire, so that the heat flow out of this end is severely restricted. As a result, less than the expected 40% heat is transferred to the right-side heat sink, and the operating temperature of the resistor is higher than specified.

Consequently, life expectancy is shorter and reliability is lower for this particular unit.

It goes without saying that a defective condition of this type cannot be

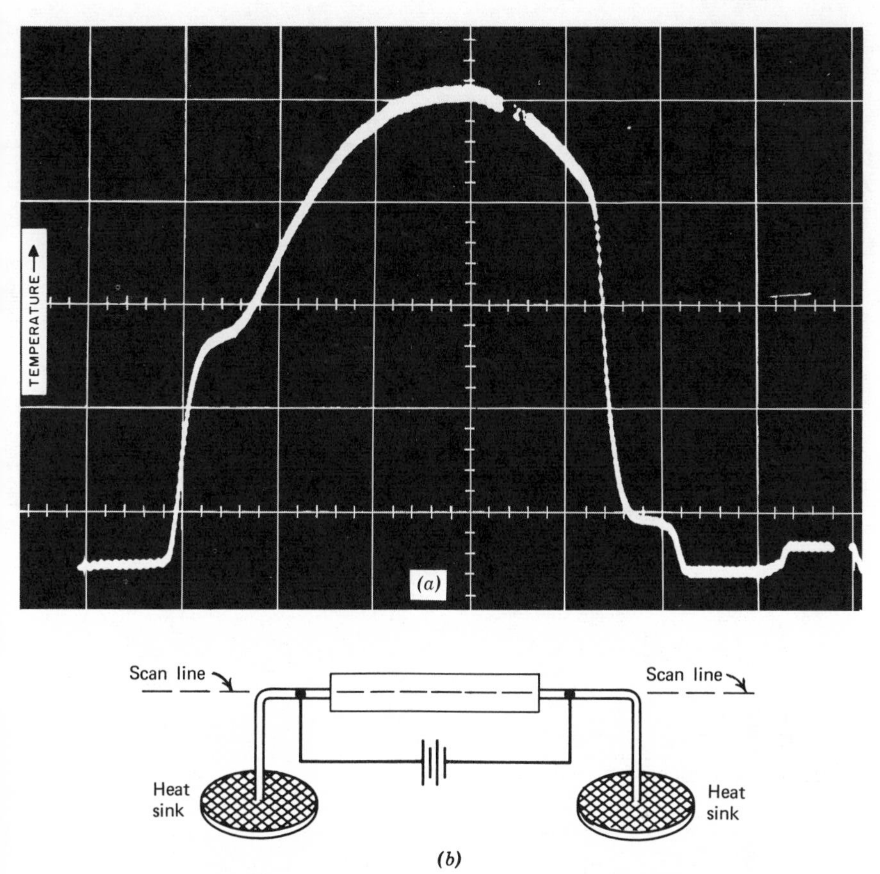

Figure 145. Poor bonding of terminal wire. (Courtesy Barnes Engineering Company.)

detected by electrical means. The minuscule DC resistance increment introduced by the poor mechanical connection of the terminal lead is several orders of magnitude below the value of the resistor, and its presence or its absence does not make any difference in the overall measurement.

Metal-Deposited Resistors

At the time of this writing, not much experience was available in regard to the extent of this condition for carbon composition resistors. However, a program run on a hundred 5-MΩ precision metal-deposited epoxy-encapsulated resistors showed a high incidence of units affected by poorly mounted end-caps.

The program was initiated to find out why some of these resistors

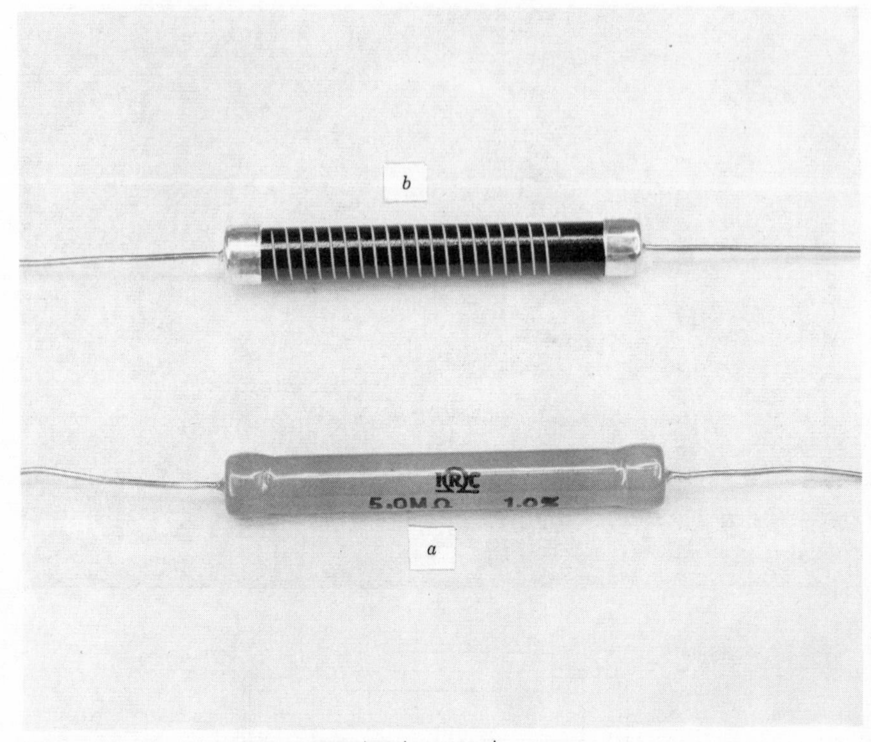

Figure 146. Precision resistor, coated and uncoated.

drifted toward higher resistance values as their operating time increased while others remained stable at the original level.

Figure 146*a* is the picture of one of these resistors, sealed in its epoxy encapsulation, and *b* is the picture of the same unit, stripped of said coating. It was noticed that the metal end-caps were clamped on the ceramic body through pressure exercised at four areas located 90° apart from each other. Axial scanning of a number of these resistors disclosed the fact that approximately 50% of them exhibited an asymmetrical infrared curve, such as the one shown in Figure 147*b*. In the illustration, curve *A* represents the average infrared profile of the "good" resistors of the lot. It is apparent that curve *a* has its peak temperature lower than the corresponding peak of curve *b*, for the same level of electrical power dissipation.

On the basis of these profiles, poor thermal conductivity through the body-to-cap connection located at the right end was diagnosed as the cause for the anomaly of resistors type *b*. Subsequent examination of the

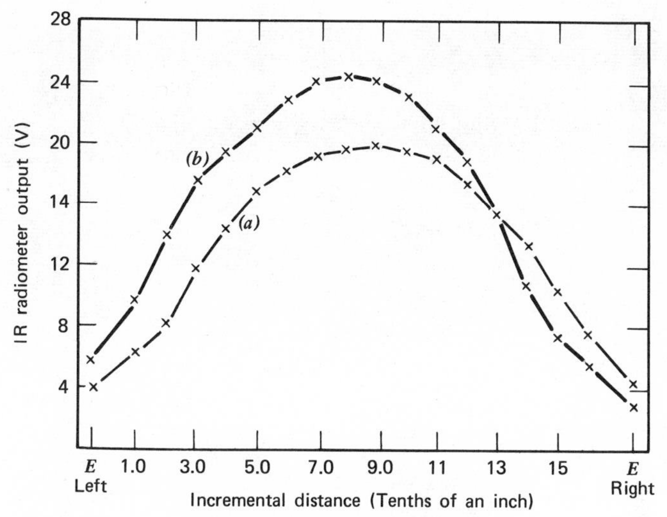

Figure 147. Infrared profiles of precision resistors.

suspected connection showed it to have high thermal resistance. As a consequence, the resistor's operating temperature was higher than normal, with predictable degradation of its reliability factor. The infrared test technique offered a quick, sure way to weed out the unreliable units from the good ones.

Wirewound Resistors

Besides finding hidden defects in the connection of the terminals, infrared scanning can be used to check the physical integrity of the resistance wire. A nick in it, a reduction in its cross-section, a slack of the tension of the turns and of their contact with the mounting core will show up in the infrared scan as a source of more intense radiation. Figure 148 illustrates how a nick in the wire shows up as a bulge of the infrared trace. Besides the nick, another defect is disclosed by the fast slope at the left end—poor mechanical connection of the metal cap. This condition has just been discussed in the last section. It is superfluous to mention that this resistor, whose life expectancy is severely reduced by these two defects, passes all conventional tests with flying colors.

Microminiature Resistors

These resistors are generally of the metal-deposited type, on glass or other insulating substrate. In spite of the tight process control, cross-

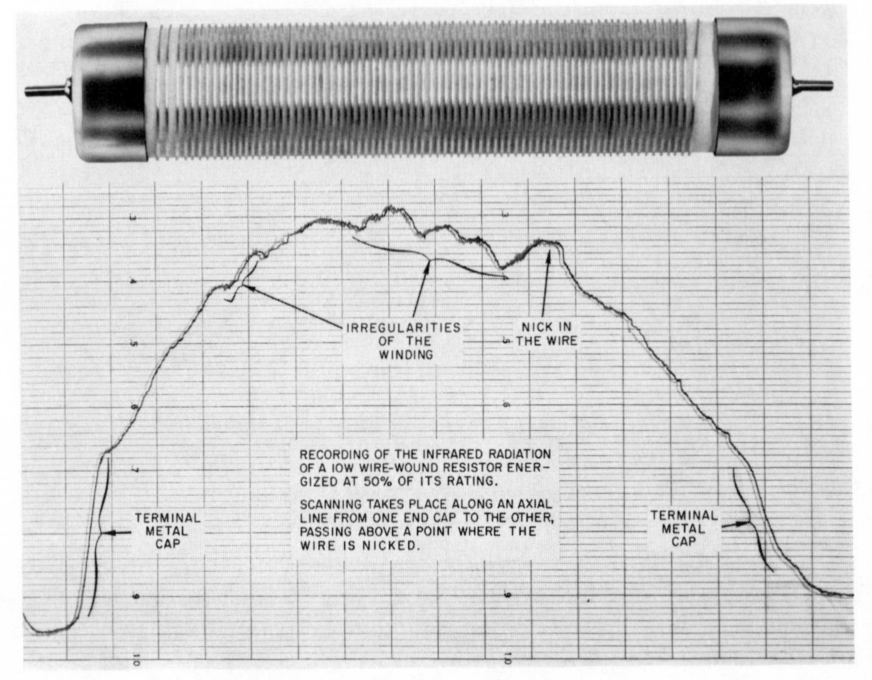

Figure 148. Infrared profile of damaged wire-wound resistor.

section variations can occur, to the extent that corrective action is needed to obtain the desired resistance value. This is usually done by carving a notch in the resistor metal until the correct resistance value is reached. Figure 149 shows a resistor of this type, with the corrective notch clearly visible.

The thermal map, hand-plotted on the basis of the data yielded by the infrared scan, gives information about the thermal regime of the unit at every point of the surface and can be used by the design engineer to optimize the size and shape of the resistor to obtain uniform, low operating temperature of the unit, so that reliability and life expectancy might be high.

Figure 150*a* is the temperature distribution of a 1-W resistor having rectangular shape and *b* is the temperature distribution of a 1-W resistor of diamond shape. In both instances the center is the hottest region, since the contact areas at the two ends act as heat sinks, but the diamond shape allows lower and more uniform heat distribution than the rectangular shape, as clearly shown by the thermal maps. This study was carried out by P. R. Young of Western Electric Company, Allentown,

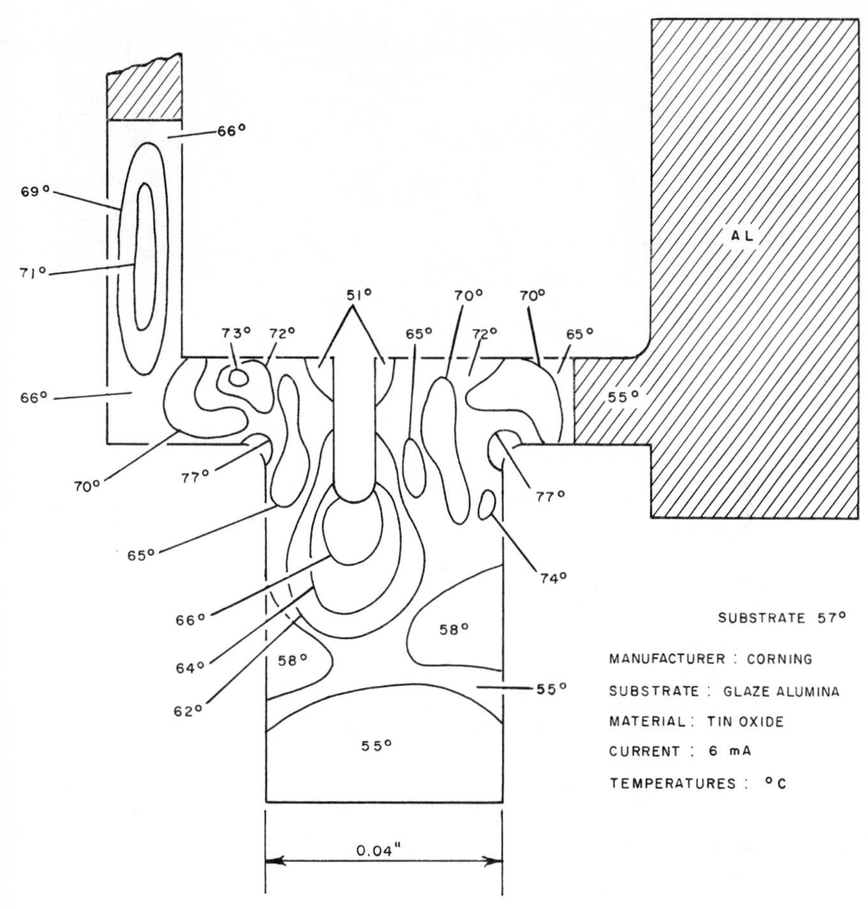

Figure 149. Isothermal map of energized thin film resistor. (Courtesy Barnes Engineering Company.)

Pa., and it was reported in a paper entitled "Application of Infrared Technology in Development of Thin Film Resistors."

The same thermal map can be used by the quality assurance engineer to make sure that the production unit matches exactly the accepted standard, and that no voids, cracks, discontinuities, or other anomalies might be present that would reduce the reliability of the manufactured unit.

Reliability of Resistors

It is clear how temperature variations can affect resistor reliability: physical expansion and contraction take place every time an increase or

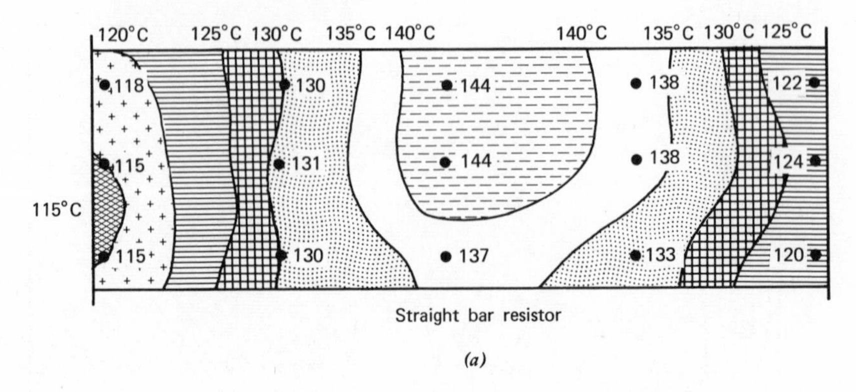

(a)

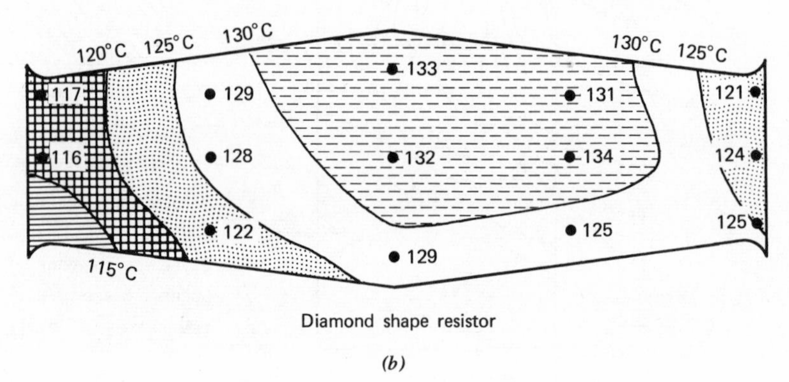

(b)

Figure 150. Design evaluation of metal deposited resistors. (Courtesy P. R. Young, Western Electric Company.)

a decrease in temperature occurs, and physical stresses are developed whose magnitude is directly dependent on the amount of temperature variation that takes place. Two elements make matters even worse: the first is the fact that throughout the resistor, there are thermal gradients causing different amounts of expansion (or contraction) in different areas; the second is the fact that the thermal expansion coefficients of the various elements of which a resistor is made are different, so that mechanical stress develops between adjacent elements even when they are subject to the same thermal gradient.

In any case, both these mechanisms concur in creating stresses which, if large enough, or if repeated often enough, will cause unbonding, cracks, discontinuities, and partial or total interruption of the electrical path along the resistor. Chemical changes in the composition of the resistive element or of its insulating coating can also be caused by overheating.

Most of these defective conditions will develop gradually, and their effect might become evident only after a long period of working time. Infrared testing, however, can disclose the existence of an anomalous condition that will develop into an eventual failure, or can detect a deteriorating trend that, in time, will bring about the same result.

DIODES

When we think of a diode, usually we think of a simple, discrete component that conducts current through a "junction" in one direction and not in the opposite. Things become quite different when we look at a diode from a molecular point of view. At this level, the junction appears made of an almost infinite number of diodes in parallel, all of them of molecular size, all of them identical to each other in the assumption that the molecular structure and the crystal lattice configuration are perfect.

Figure 151 shows the functional principle of how a diode works. *N* is a semiconductor material whose molecules can yield negatively charged current carriers (electrons), while *P* is a material whose molecules can provide positively charged current carriers (holes). When a source of electricity is connected to a diode with its negative end connected to *N*, and its positive end to *P*, current will flow through the junction since the electrons made available in the *N* region can move toward the *P* region under the influence of the electrovoltaic field created by the

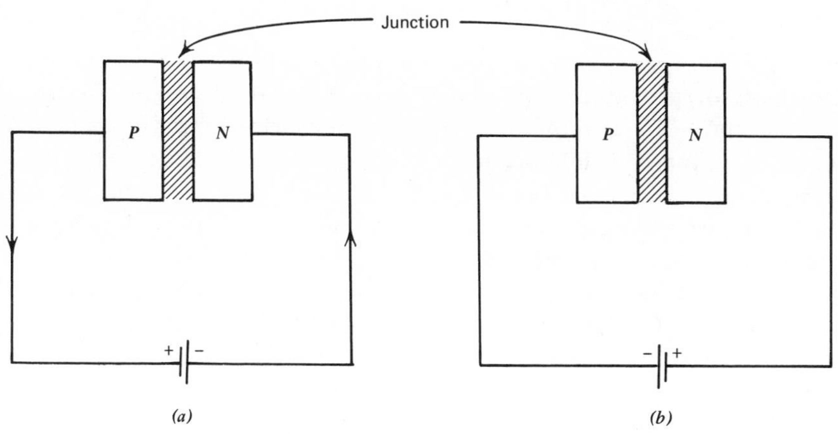

Figure 151. Functional principle of semiconductor diode.

power supply. As every electron passes into the *P* region, it is replaced in the *N* region by an electron supplied by the battery. Electrical current thus flows as shown in *A*, where the arrows indicate the movement of the electrons, according to the conventional way of showing the flow of electricity. Positive charges, or "holes" move in the opposite direction.

When the polarity of the battery is reversed, as shown in Figure 151*b*, current cannot flow, because the electrovoltaic field pulls the current carriers away from the junction, and there is no way to resupply the vacancies thus left in the *P* and *N* regions. This explains the rectifying action of the diode when connected to an AC source: only the half wave of the right polarity will go through, thus effecting a half-wave rectification of the input.

The *P* and *N* materials are generally crystals whose molecules contain positive or negative subatomic elements (respectively holes or electrons) that are relatively loosely connected, so that they can separate from the molecule under the influence of an electrovoltaic field. After separation, they become current carriers and can move through the *N* and *P* regions and along the conductors of the circuitry as elements of the electrical current flow.

These loosely connected subatomic elements belong to atoms that are part of the molecules forming the *N* and the *P* materials, and most of the time are the atoms of elements that have been introduced into their host molecules by artificial means. This artificial process is called "doping," and the concentration of the atoms so introduced is critically related to the electrical characteristics of the diode.

Many factors can affect the amount of energy required to "liberate" a subatomic particle, turning it into a current carrier. For instance, temperature has an effect, because heat is stored in the molecules as kinetic energy, so that the energy required to liberate a current carrier becomes less as the temperature increases. The presence of impurities, with their capability of providing intermediate energy levels different from the desired one (see Figure 38) will also affect the electrical performance of the diode. And crystal lattice imperfections, either from growth anomalies or from mechanical stresses of any origin, will also alter the expected energy level, because of the changes in the intramolecular distances and related intereffects.

We can conclude that the operation of a diode depends on the characteristics of each and every one of the elementary molecular diodes located in the junction area, whose performance can vary within very wide limits.

Conventional test equipment can only supply information related to the overall operation of the diode, but cannot tell, for instance, whether the electrical current flow is uniformly distributed through the junction area, or concentrated in one or more regions. An infrared microscope having adequate area and temperature resolution can supply the answer to the question above, along with the capability to detect other "hidden" anomalies that can make the diodes unreliable and likely to become an early failure.

When current flows through a diode, infrared radiation is emitted in two different spectral bands: incoherent radiation over a wide spectrum related to the blackbody emission curve corresponding to the temperature of the diode, and recombination radiation over a narrow spectrum

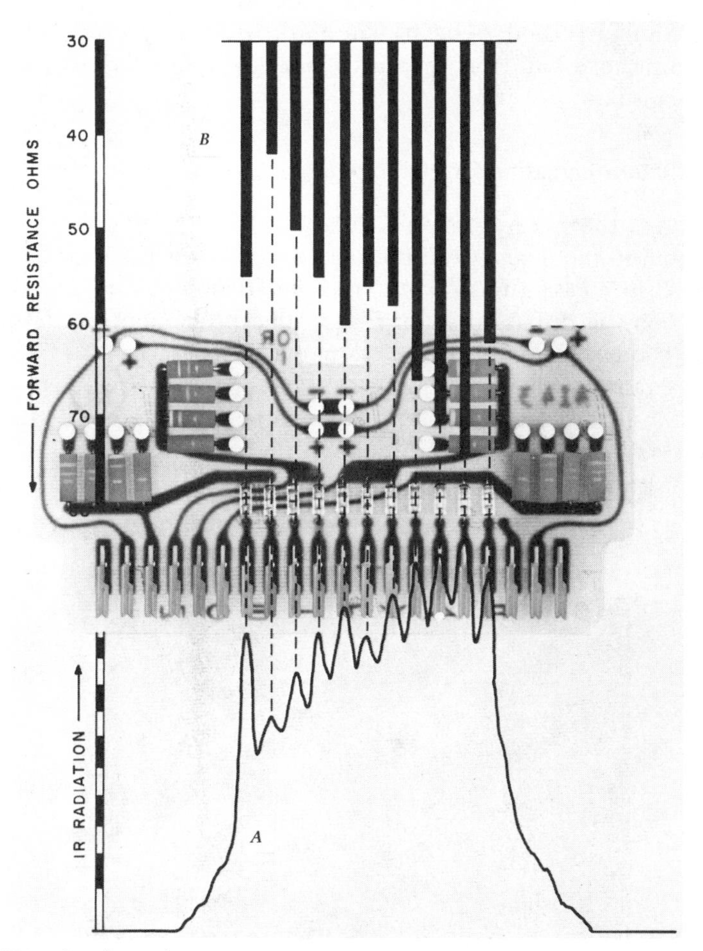

Figure 152. Correlation between infrared radiation and forward resistance of diodes.

whose wavelength is determined by the diode's material and its physical characteristics.

Measurement of infrared radiation of thermal origin emitted by diodes is done currently to measure their power dissipation. Figure 152 shows in trace *A* a line scan of an array of 11 diodes wired in parallel.

It can be seen that no two diodes are operating identically, since each of them emits infrared radiation at a different power level. Subsequent DC resistance measurements (Figure 152*b*) confirmed the existence of an inverse correlation between resistance and radiation. In this example, the diodes are glass-encapsulated, but in the infrared spectral range corresponding to temperatures slightly above room ambient the glass is opaque, so that the infrared detector reads the emission from the glass surface. Similarly, diodes encapsulated in plastic or epoxy or metal will radiate in accord with the emissivity index of the external surface of their encapsulant material.

Recombination Radiation Emitted by Diodes

Whenever the encapsulation is made of glass or of other material transparent in the near infrared range, current flow through the diode can be measured as a function of the recombination radiation emitted by the junction, as discussed in page 23. The instrument used for this purpose is the semiconductor junction analyzer described in page 134.

As an example of this technique and of the results obtained with it, Figure 153 shows a silicon diode, type 1N459 used in a program to verify

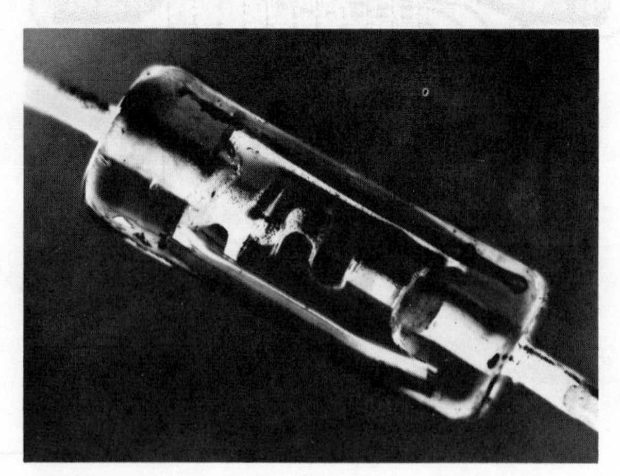

Figure 153. Silicon diode IN459.

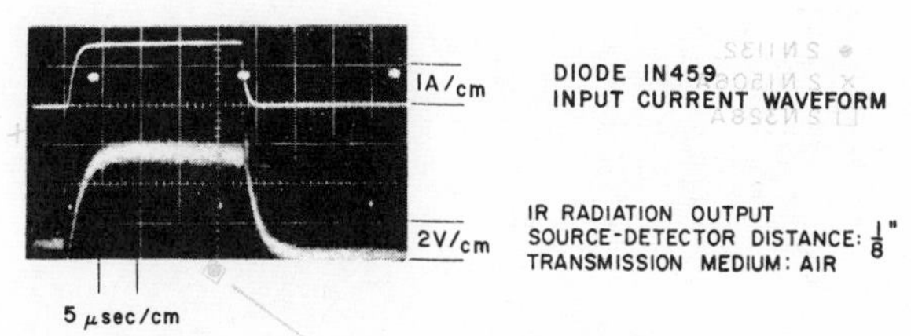

Figure 154. Silicon diode recombination radiation.

the feasibility of this approach. Figure 154 shows in the upper trace the oscilloscope display of the signal flowing through the diode's junction, and in the lower trace the recombination radiation signal appearing at the output of the infrared detector "looking" at the junction. The distortion appearing at the leading edge and at the trailing edge of the pulse read by the detector is due to the limited frequency response of the signal processing electronics. This point was discussed in page 24.

TRANSISTORS

The same considerations described for the diodes are true for the transistors; the only difference is that a transistor can be thought of as an assembly of two diodes joined back to back. However, the junction areas should still be considered as parallel arrays of an extremely large number of diodes whose individual characteristics are not necessarily identical, but rather depend on the molecular and lattice structure of the crystal.

Measurements of the infrared radiation due to the thermal condition of transistors have been carried out both on encapsulated units and on visually exposed semiconductor chips. Figure 155 shows the correlation between power dissipation and infrared radiation emitted by different types of transistors, all sealed in cans of the same type. It is apparent that the correlation is the same for all types of transistors, as long as they are enclosed in envelopes of the same type.

In all infrared measurements taken on energized transistors, the chip's temperature proved directly correlated with the power dissipation level, for units having identical geometry. However, differences in the length of the wires connecting the chip to the posts can affect the temperature of the unit, and so can the quality of the bond between the chip and the substrate. In the first case, the length of the wires will be an inverse function of the heat drain to the posts and this is shown in Figure 156; in the second case, a poor bond will reduce the amount of heat that can reach the substrate, so that the chip will run hotter.

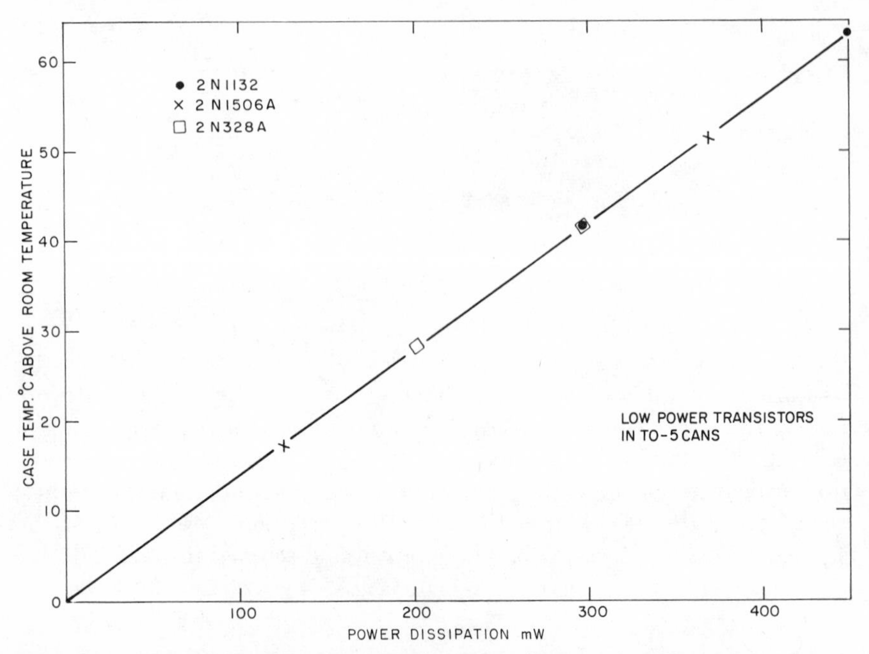

Figure 155. Transistors in TO-5 correlation between radiation and power.

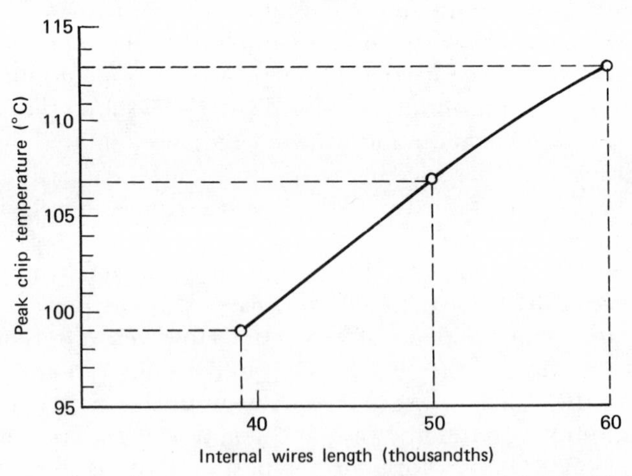

Figure 156. Correlation between transistor chip temperature and length of internal wires.

In practice, since it is possible to control by conventional means the exact amount of electrical power dissipation, and by visual means the consistency of the geometric configuration of transistors, infrared scanning of the chip can yield reliable information on the quality of the bond between chip and substrate, an item that is very difficult and laborious to evaluate by any other means.

A study carried out by the author on several lots of transistors of which infrared readings of the cap and of the chip were taken showed a remarkably good correlation between these readings. Figure 157 is an example of this finding on a lot of 2N930 transistors made by Fairchild.

The dispersion of the correlation points around the average is due to the presence of physical variations or geometry differences from unit to unit. In other words, the magnitude of the deviations is inversely correlated to the degree of process control of the operations carried out in the manufacturing cycle, which, in the example shown, appears good. This is an example of how infrared techniques can be used to assess how tight is the process control of semiconductor manufacture.

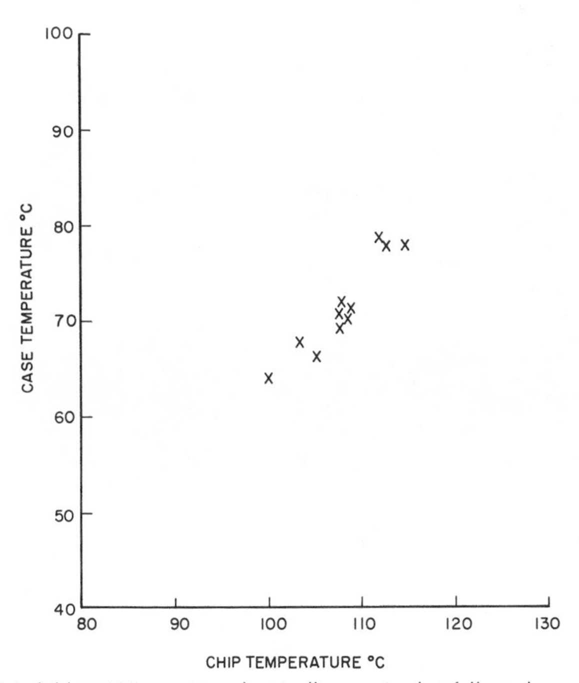

Figure 157. Fairchild 2N930 transistor electrically energized at full rated power: chip versus case temperature correlation.

Transistor Life Expectancy

Probably the major factor influencing the life expectancy of semiconductors is the junction temperature. For silicon transistors this correlation is clearly shown in Figure 158:[12] we can see that per every 50°C increase of the temperature of the junction, the life expectancy falls by one order of magnitude. This explains why so much importance is given to the measurement of transistor temperature.

In view of the difficulty in measuring junction temperature, it seems reasonable to expect that the case temperature of transistors will be a good indicator of their life expectancy.

An exploratory program designed to evaluate the merit of this approach was carried out on 20 power transistors, Type JAN2N174. These transistors, all meeting electrical acceptance specifications, were mounted on a $\frac{1}{8}$-in. sheet metal plate as illustrated in Figure 159, and wired in common base configuration with individual current control, as

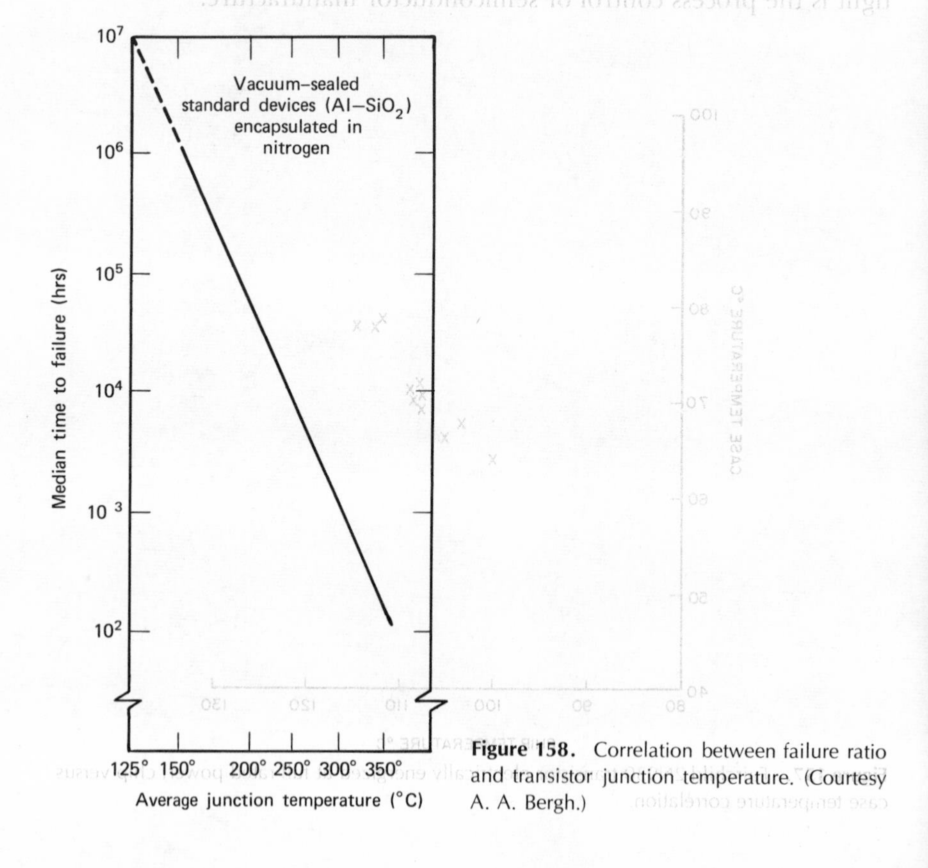

Figure 158. Correlation between failure ratio and transistor junction temperature. (Courtesy A. A. Bergh.)

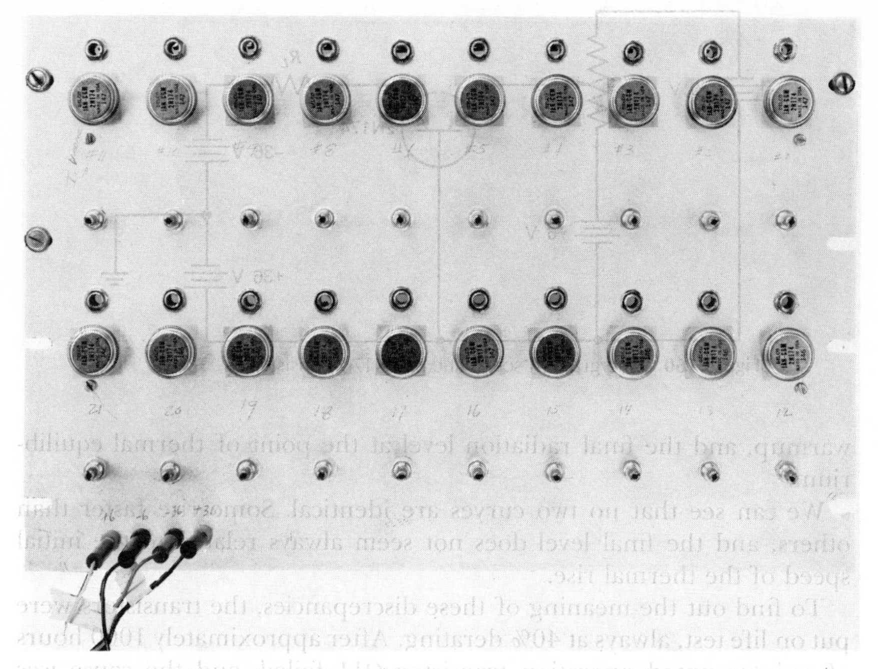

Figure 159. 2N174 transistors mounted for infrared and life correlation test.

visible in the same illustration. Figure 160 is the schematic of the electrical energization of each transistor.

During the setup phase of the experiment, the bias controls were adjusted to ensure that the identical amount of current flow through every transistor corresponded to a power dissipation level of 0.6 of their maximum rating.

Subsequently, an infrared radiometer was focused on transistor 1, which was then electrically energized. The detector output was recorded on a chart recorder from the instant the transistor was turned on until its temperature had reached thermal equilibrium.

At this point, transistor 1 was turned off. The same sequence of operations was carried out on transistor 2, and so on.

Figure 161 displays the infrared characteristic curves so recorded, grouped in such a way as to start from a common origin, which represents the transistors at ambient temperature at the instant when they are turned on.

Because it took almost half an hour for every transistor to reach thermal stability, in the illustration the corresponding curves have been shortened by showing only the portion related to the first 5 min of

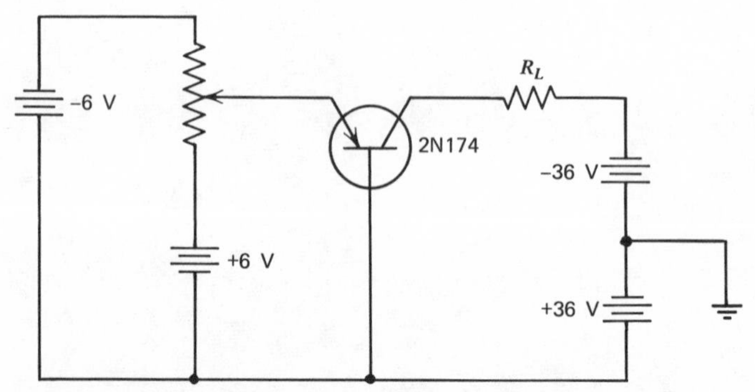

Figure 160. Energization schematic of 2N174 transistors.

warmup, and the final radiation level at the point of thermal equilibrium.

We can see that no two curves are identical. Some rise faster than others, and the final level does not seem always related to the initial speed of the thermal rise.

To find out the meaning of these discrepancies, the transistors were put on life test, always at 40% derating. After approximately 1000 hours of uninterrupted operation transistor *Q*11 failed, and the cause was diagnosed as thermal runaway due to increasing electrical leakage. From Figure 161, we can see that the infrared characteristic curve of *Q*11 runs well above all other curves, thus indicating a unit that is comparatively warmer.

A second failure took place after an additional 1000 hours of operation. This time *Q*18 failed. Its infrared characteristic curve lies at the bottom of the distribution, and subsequent autopsy seemed to indicate poor bonding of the chip to the substrate, so that a high thermal gradient might have existed between the chip and the case, which appeared cool at the infrared measurement.

After these two failures, which belong to the "early failure" class, no further failures developed during the subsequent 2 years of continuous life test, until the experiment was discontinued.

Similar work was performed by Air Force researchers at the Rome Air Development Center.[7] Many germanium power transistors, of type *PNP*, were processed by (*a*) energizing at rated power level, (*b*) allowing 5 min of warmup, (*c*) measuring the infrared radiation emitted by the top of the can, (*d*) classifying into three groups (low, medium, and high radiation level), and (*e*) step-stress life testing.

Figure 162 shows the results of the life test. At the end of the first 100 hours, 58% of the units in the high radiation group had failed, whereas

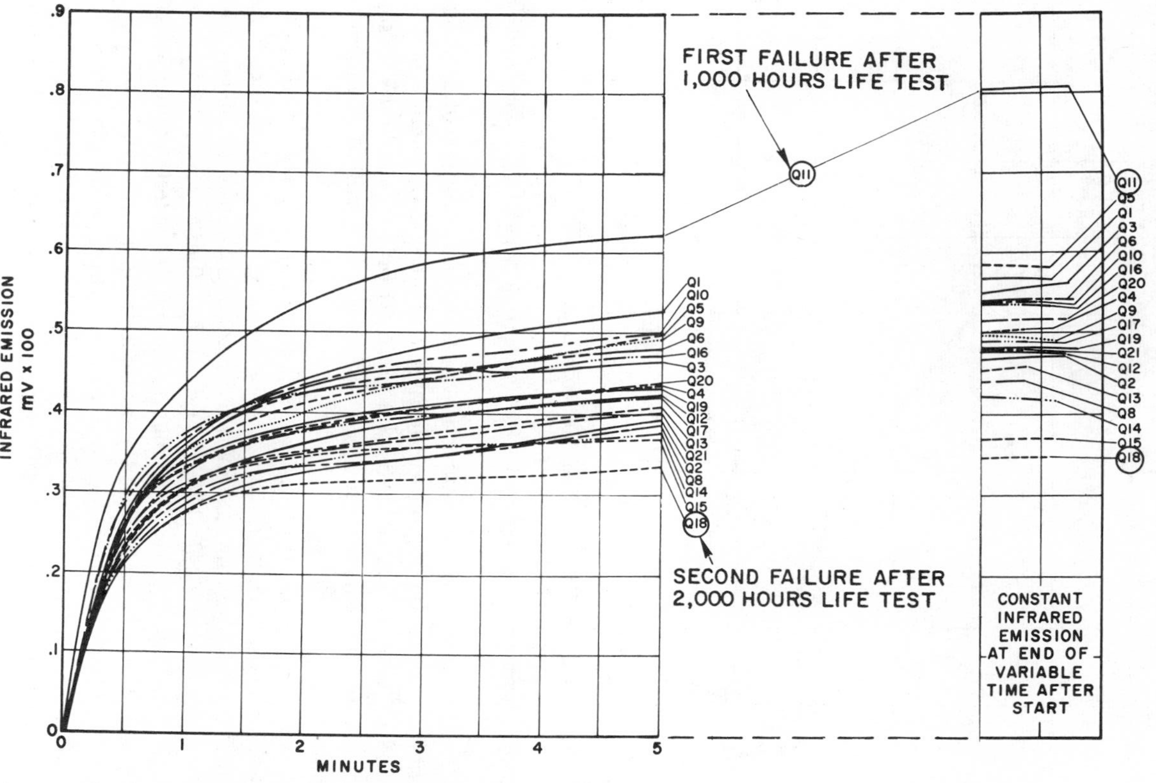

Figure 161. Transistors 2N174 warmup characteristics.

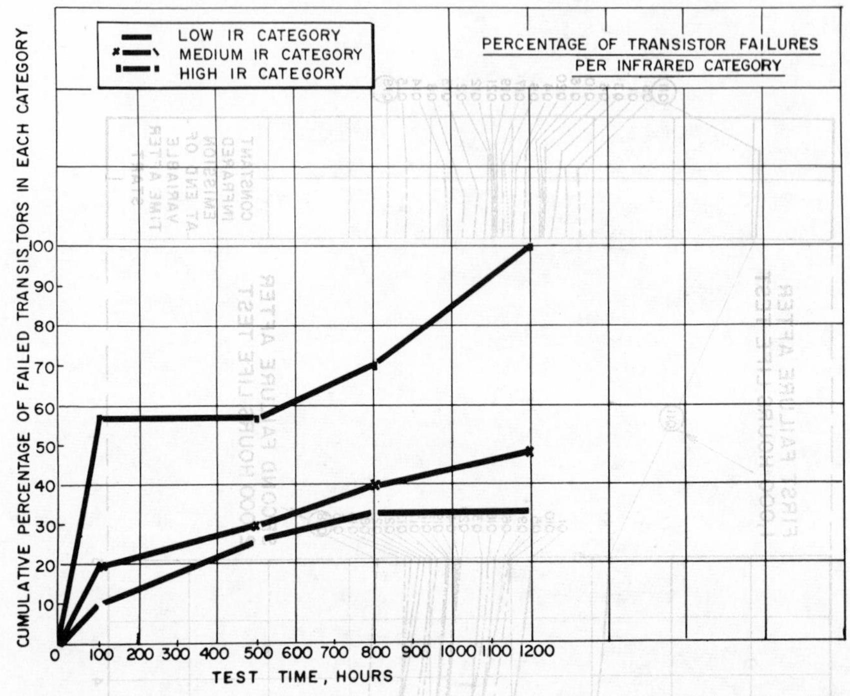

Figure 162. Transistor failures related to infrared emission. (Courtesy A. J. Feduccia, RADC.)

only 20 and 10% in the medium and low radiation groups had done so. All the transistors belonging to the high radiation group had failed at the end of 1200 hours of test, while 56 and 68% of the units in the medium and low radiation groups, respectively, were still in good working condition.

Statistical verification of these preliminary findings was carried out by Sylvania Electric Products, Inc., on a contract funded by the A.F.-RADC.[8] In this program, the infrared radiation emitted by 50,000 low power transistors, type 2N706, energized at 220-mW power dissipation level (75% of rated power) was measured and recorded. Tabulation of these data showed that almost 4% of the lot (approximately 1900 units) were radiating at abnormal levels (beyond the three σ limit). Subsequent step-stress life testing proved that the failure ratio of the anomalous units was almost ten times higher than the failure ratio of the normal ones.

Second Breakdown

During the last few years, infrared evaluation of transistors has been shifting from encapsulated units to direct scanning of the semiconductor chip. This approach can disclose minute anomalies that conventional equipment can not detect and that might eventually cause early failures.

An interesting example is the second breakdown phenomenon, which so far had prompted more questions than answers. In the course of a study carried out for NASA-MSFC by Raytheon Company,[9] it was found that infrared methods can exactly locate the point where second breakdown will occur and that the magnitude of the infrared radiation emitted by such spot is indicative of the approximate voltage level at which Second Breakdown will take place.

According to the definition, second breakdown is a self-sustaining incremental negative-resistance process. This means that at a certain spot of the semiconductor, the temperature increase generated by the localized current flow makes available more current carriers, so that a greater current can flow through that point. This in turn will raise the localized temperature, thus making available more carriers for a further current increase, and so on. This process will bring about a very fast temperature rise in the breakdown point, so that semiconductor melting will take place in very short time, usually a fraction of a second. At the completion of this process, the device undergoes a catastrophic, irreversible failure, and a small crater surrounded by molten crystal appears at the point of second breakdown. It is usually called the "punch-through point."

Pulsed energization of the transistor avoids completion of the breakdown process, because it cuts the current flow before semiconductor melting temperature is reached.

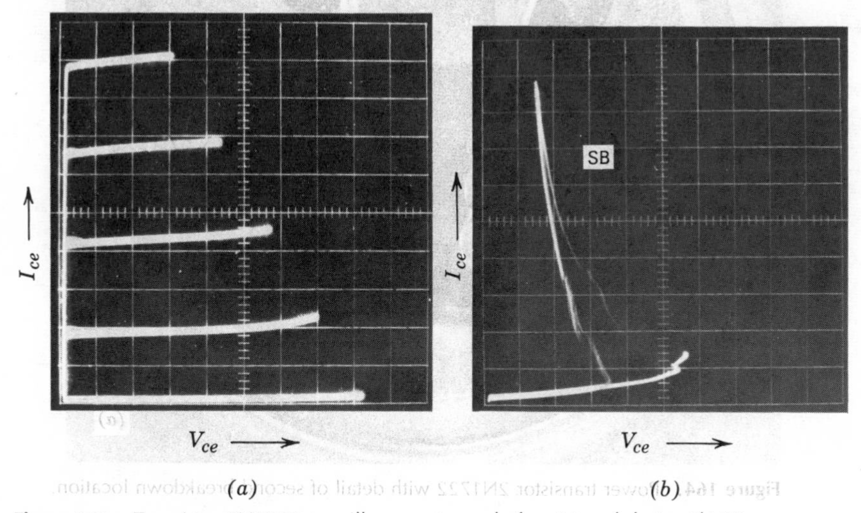

Figure 163. Transistor 2N1722—oscilloscope traces before (*a*) and during (*b*) SB.

Energization of this type can be carried out, for instance, with the use of a Tektronik 575 curve tracer in the "oscillator sweep" mode of operation. In that mode, the transistor is intermittently energized at a rate of 120 pulses/sec. The magnitude of these pulses is gradually increased until second breakdown initiates in correspondence of the peak of every pulse. This condition is displayed on the oscilloscope as shown in Figure 163*b*: the voltage drop and the corresponding large increase of the current flowing through the device are readily apparent in the trace marked SB. It might be worth emphasizing that almost all the current flows through the SB point, while the rest of the junction is almost totally depleted of current due to the large difference in ohmic resistance. For comparison purposes, the transistor characteristic curves in normal operation are shown in Figure 163*a*.

During the feasibility study mentioned above, the surface of power transistors type 2N1722 were scanned with the fast scan infrared microscope, while operating in pulsed SB condition. Figure 164 is the visible picture of one of these transistors, whose area measures $\frac{1}{4} \times \frac{1}{4}$ in. The point of second breakdown was precisely located at the center of the scan line from *X* to *Y*, where the infrared detector disclosed a very high peak of radiation, emitted in synchronism with the pulses energizing the transistor.

Figure 164. Power transistor 2N1722 with detail of second breakdown location.

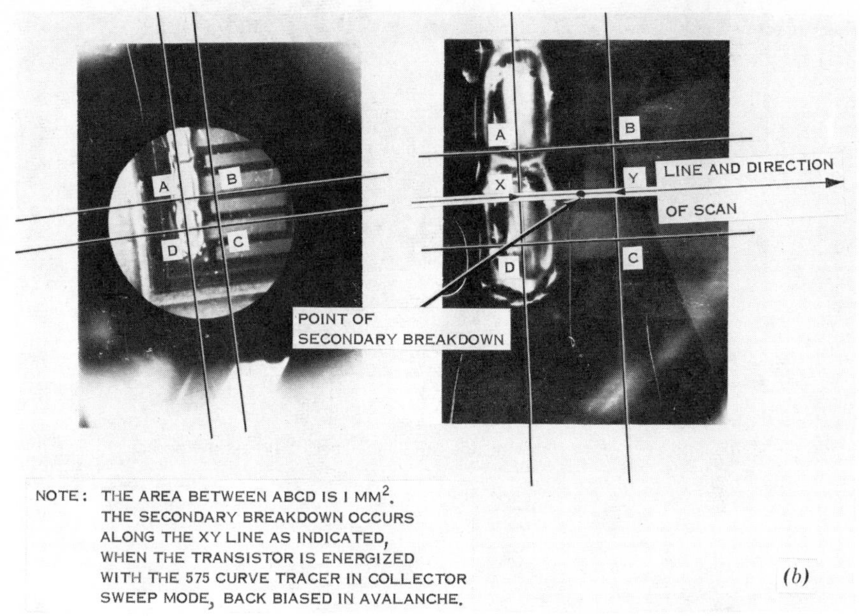
A
B
D
C
A
B
X
Y
LINE AND DIRECTION
OF SCAN
D
C
POINT OF
SECONDARY BREAKDOWN
NOTE: THE AREA BETWEEN ABCD IS 1 MM^2.
THE SECONDARY BREAKDOWN OCCURS
ALONG THE XY LINE AS INDICATED,
WHEN THE TRANSISTOR IS ENERGIZED
WITH THE 575 CURVE TRACER IN COLLECTOR
SWEEP MODE, BACK BIASED IN AVALANCHE.
(b)

Figure 164. (Continued).

Figure 165 shows the oscilloscope traces depicting the detector output scanning from *X* to *Y*. The upper trace is taken during SB, while the lower trace is taken in the intervals between energizing pulses. The temperature of the crystal at the SB point was estimated around 800°C. Of course, shorter pulses reduce this value, while longer pulses increase it, until melting temperature is reached.

Most of the time, a "hot spot" is noticed at the point where SB will occur at higher energization level, and this anomaly can be detected long before the curve tracer will show that SB is taking place. Figure 166 shows how the thermal profile builds up at different power levels: the four pictures in the upper row are the oscilloscope display of the infrared scan lines, while the four pictures in the lower row are the oscilloscope display of the transistor's characteristics. As the V_{CE} increases, a thermal peak appears at the location where second breakdown will occur. The important feature of this technique is that the thermal peak becomes detectable *before* the actual occurrence of second breakdown, which in this way can be predicted, both in location and in electrical energization level without reaching overstress conditions. As evident from the illustration, the thermal peak becomes conspicuous while the transistor is energized within rated level, and no indication of impending trouble is given by the oscilloscope display of its electrical character-

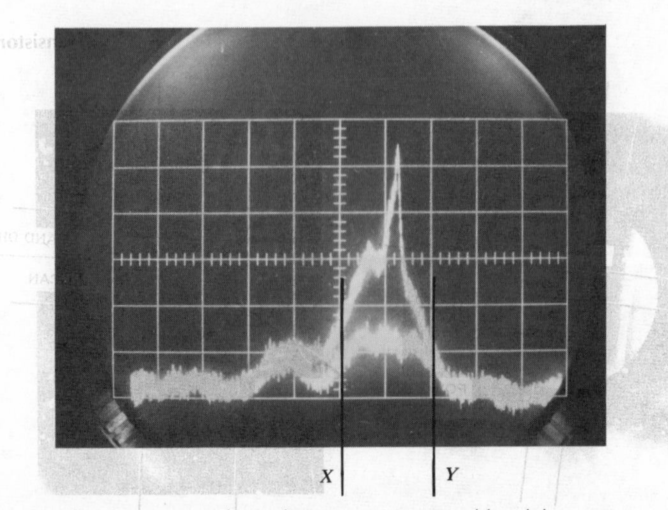

Figure 165. Oscilloscope traces of scan line traversing second breakdown area.

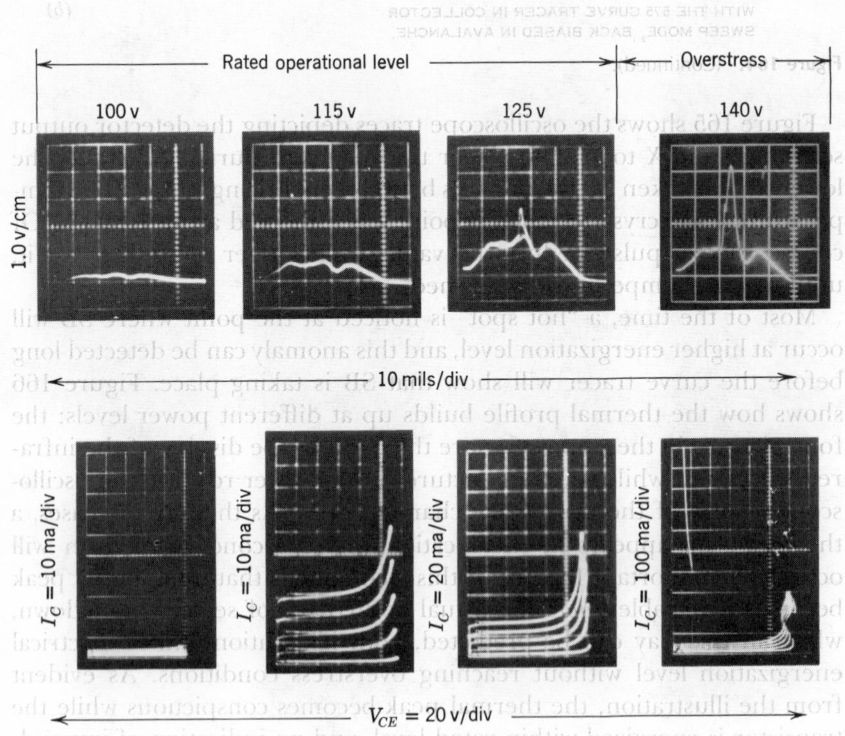

Figure 166. Predicting second breakdown in power transistors.

istics (first three pictures, bottom line). With this technique, grading of power transistors according to their resistance to second breakdown is feasible, and it allows the weeding out of second breakdown prone units without exceeding the rated operating levels, therefore avoiding any chance of damaging the units under test.

RECOMBINATION RADIATION

The availability of infrared test equipment capable of measuring the recombination radiation emitted by semiconductor junctions (see page 23 and 134, Junction Analyzer) can be useful in the evaluation of transistors as follows:

1. Measuring, without contact, the current pulses flowing through the junction, both in amplitude and waveform.
2. Scanning the junction along its length to check whether it is uniformly active, or whether points of current crowding or current voids are present.

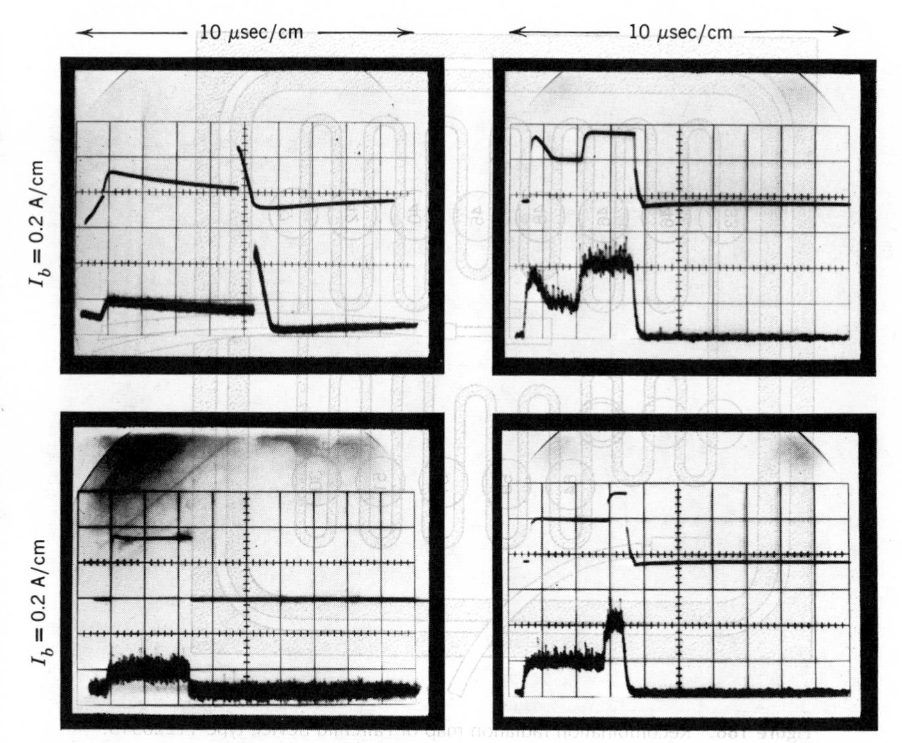

Figure 167. Recombination radiation emitted by a 2N930 transistor.

Figure 167 shows examples of waveform measurements taken as mentioned in (1) above. The oscilloscope trace shown in solid line is the base current of the transistor, while the "fuzzy" trace is the output of the detector measuring the recombination radiation emitted. These illustrations are taken from the reports submitted by Raytheon to NASA-ERC during the work performed under Contract NAS 12-585, with a breadboard instrument utilizing a 0.005-in. optical fiber and a germanium avalanche photodiode as the sensing element.

The high level of noise present in the detector output was reduced after the completion of the NASA program, through work performed after the development of the semiconductor junction analyzer.

Scanning the junction along its length, as mentioned under (2) above, was carried out, by the author, under contract to Raytheon Company, in a study entitled "Correlation between Second Breakdown and Infrared Radiation in Power Transistors."

During this work, recombination radiation maps were drawn of transistors type 11220513 and SES2196B, respectively manufactured by Fairchild and by Solitron. Figures 168 and 169 show two of these maps,

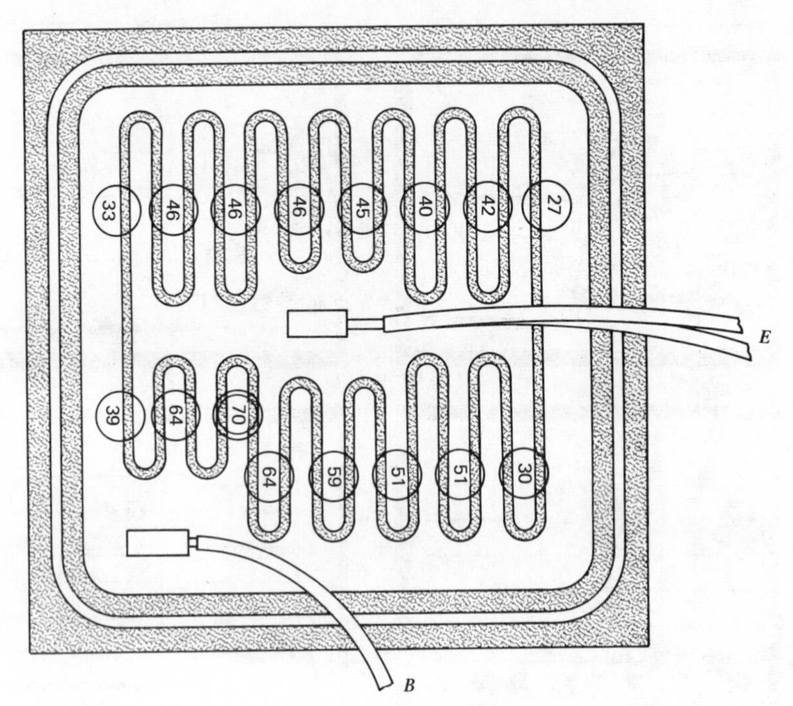

Figure 168. Recombination radiation map of Fairchild device type 11220513.

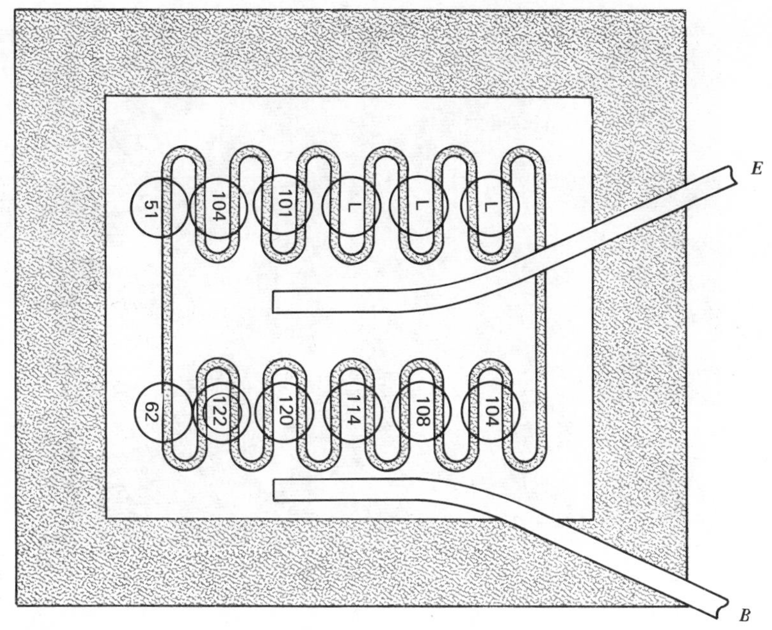

Figure 169. Recombination radiation map of Solitron device type SES 2196B.

where the intensity of the radiation is indicated in plain numbers inside circles representing the area located in center of the field of view of the optical fiber.

It is of interest to note that the intensity of the radiation varies from point to point of the transistor junction, and that a peak was found, consistently located in the region between the points where the emitter and the base leads are bonded to the semiconductor.

It is also of interest to note that the radiation values of the Solitron device are higher than those of the Fairchild units. This might be due to the larger physical size of the latter, which allows for a lower flux density of the electrical current flowing between base and emitter.

INTEGRATED CIRCUITS, MSI, AND LSI DEVICES

The modern trend toward miniaturization is rapidly pushing ahead the development of integrated circuits of ever increasing complexity. From the very simple circuit shown in Figure 170, first made by Philco a few years ago, we have now progressed to circuits containing several thousand active elements, such as the LSI (for large scale integration) cir-

Figure 170. Inverter circuit, integrated. (Courtesy Philco Company.)

cuit of Figure 171, made by Fairchild. At the time of this writing, complete digital systems made of hundreds of gates (and every gate is composed of ten or so elements) had been etched on Silicon "wafers" about $1\frac{1}{2}$ in. in diameter, with a component density that can exceed 100,000 units/in.^{2}.

The more complex the circuitry, the larger the number of active elements whose performance cannot be checked with conventional test equipment, because of the impossibility of reaching them through the outside connections.

Probably out of desperation, microprobes sometimes are used to check the electrical performance of discrete elements or of circuits otherwise inaccessible. Microprobes are tiny needles that must pierce through the insulating layer of silicon oxide to make contact with the electrical connections of the elements to be measured. However, besides the physical damage to the insulation, the mechanical pressure needed to assure good electrical contact can alter the performance characteristics of the semiconductor, so that the measurements are hardly reliable.

Infrared does not know such limitations. A point-detector infrared microscope can be focused upon each element of the energized circuit, or a scanning infrared microscope can sequentially measure the infrared radiation emitted by every point of the device, thus supplying us with thermal and stress information about each and every element. Further-

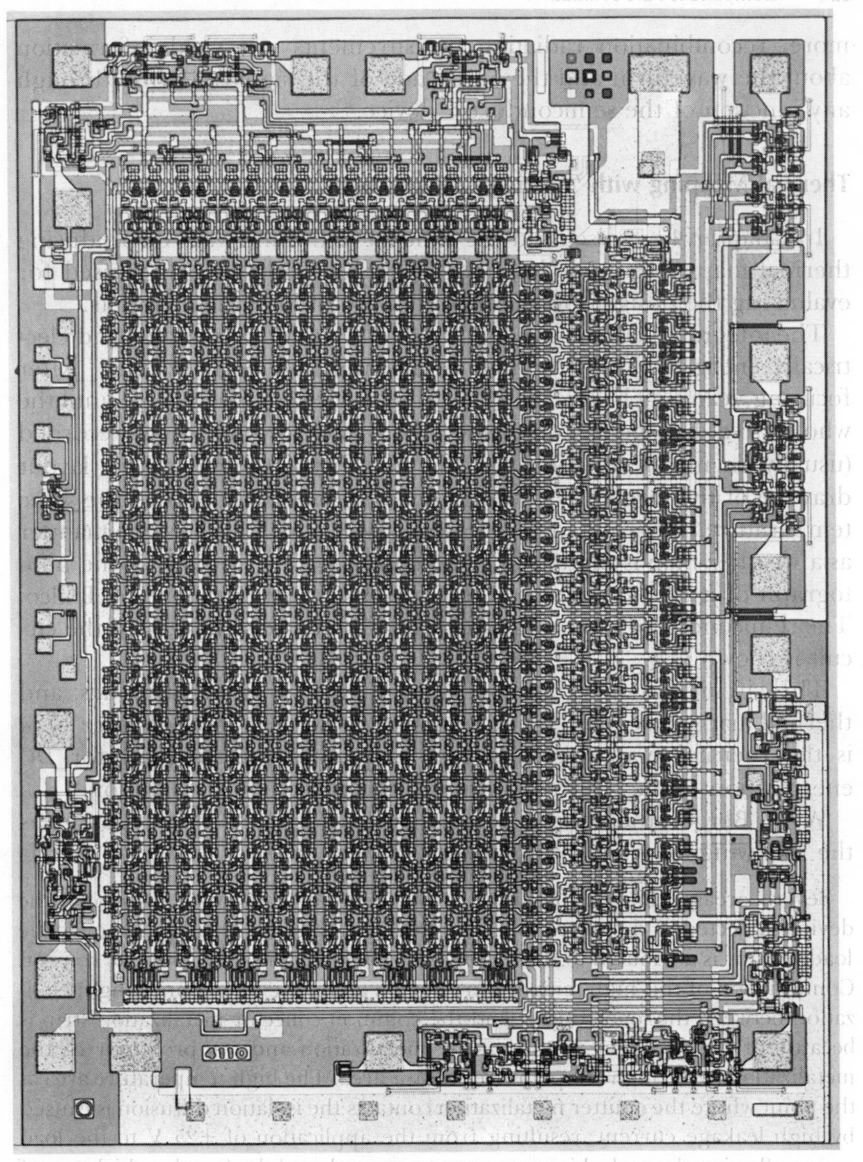

Figure 171. LSI memory circuit. (Courtesy Fairchild Semiconductor Company.)

more, recombination radiation measurements can yield information about the waveform and the amplitude of the signals flowing through any junction of the semiconductor device.

Thermal Mapping with "Staring" Microscopes

In Chapter 4 we discussed the concept and some ways of producing thermal maps. Here we describe how this type of display is used for evaluating the design and the manufacture of integrated circuits.

The integrated circuits are mounted on a scanning substage, and electrically energized until thermal equilibrium has been reached. After focusing, scanning begins, and the target is viewed, line by line, until the whole area has been completed. The detector output is recorded (usually on magnetic tape), and this data is transferred onto a layout drawing of the target. By connecting all the points that are at the same temperature, a thermal map is obtained, very much in the same manner as a weather map is plotted by a meteorologist. Figure 172*a* is the photograph of a silicon epitaxial three input gate circuit made by Philco. The semiconductor chip measures 1 mm^2, and the schematic of the circuit is shown in *b*.

This circuit was energized, successively, in two different modes, and thermal maps were plotted in the manner described above. Figure 173*a* is the thermal map depicting the temperature distribution for *DC* energization, while *b* is the map related to pulse power condition.

W. M. Berger, the Philco researcher who developed these maps, gives the following analysis of the findings:[10]

Several areas of high temperature are observed on the isothermal plot of the device subjected to d-c operation. The high temperature area located over the load resistor is attributed to the high power (808 mW) dissipated in this resistor. Continuation of this high temperature region along the interconnecting metalization between the load resistor and the common collector metalization strip is because of high current density in the metalization and the proximity of the metalization to the high temperature resistor area. The high temperature area at the point where the emitter metalization contacts the isolation diffusion is caused by high leakage current, resulting from the application of +25 V to the load resistor flowing through this contact cut to ground, and the fact that thickness of the metalization is somewhat reduced where it passes over the edge of the contact cut. The high temperature region at both contact cuts in the metalization connecting base resistor number 1 to the base of transistor number 1 is due to a reduction in the cross sectional area of the metalization where it crosses the edge of the contact cut.

For the device subjected to pulsed power dissipation, the area of high

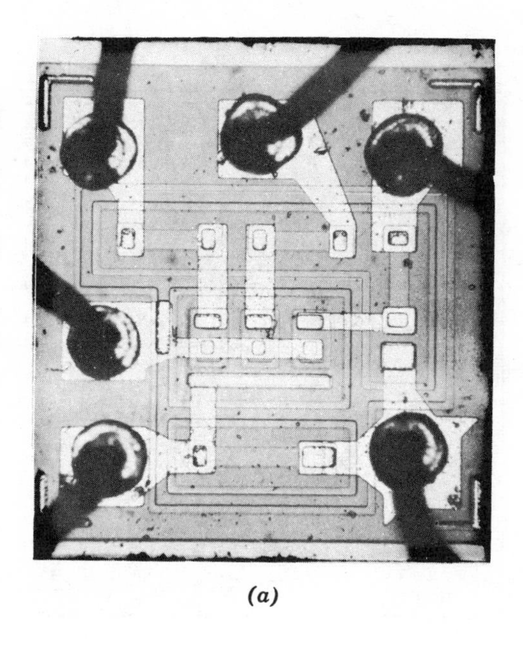

(a)

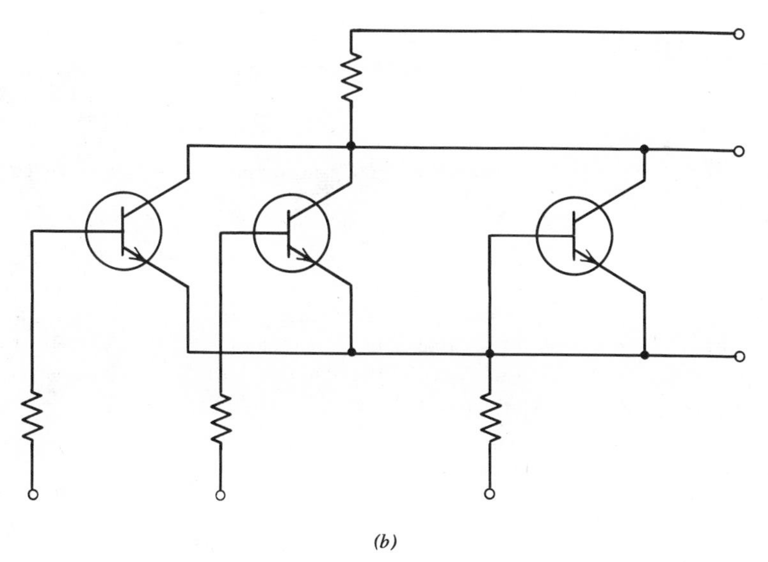

(b)

Figure 172. Philco epitaxial three-input gate. (*a*) Photomicrograph. (*b*) Schematic. (Courtesy W. M. Berger, Philco Company.)

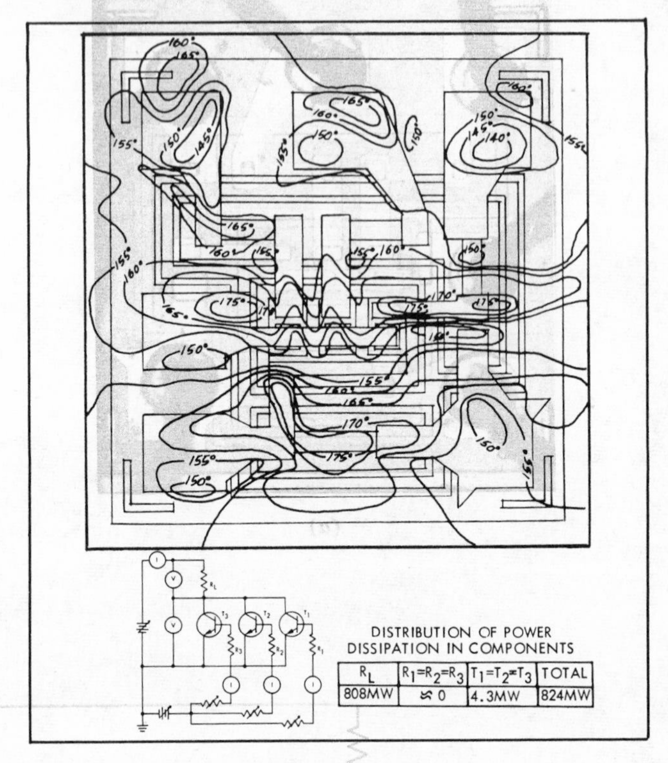

(a)

Figure 173. Thermal maps of Philco three-input gate. (Courtesy W. M. Berger, Philco Company.)

operating temperature over the load resistor is because of the large amount of power being dissipated. The high temperature region located over the base resistors is due to the power dissipated in these elements; the fact that this region extends between the bonding pads is because this portion of the substrate was not bonded well to the header, preventing efficient conduction of heat from this region. The reduction of the temperature of the metalized areas is because the high thermal conductivity of the aluminum permits efficient removal of heat during the periods when power is not being dissipated. The hot spots in the transistor region are due to dissipation of power in these elements.

According to the same author, the dominant failure mechanism detected during operating life tests has been open interconnecting metalization, located at the crossing of an oxide step. Infrared analysis, with its capability of detecting reduced crossections of the deposited metaliza-

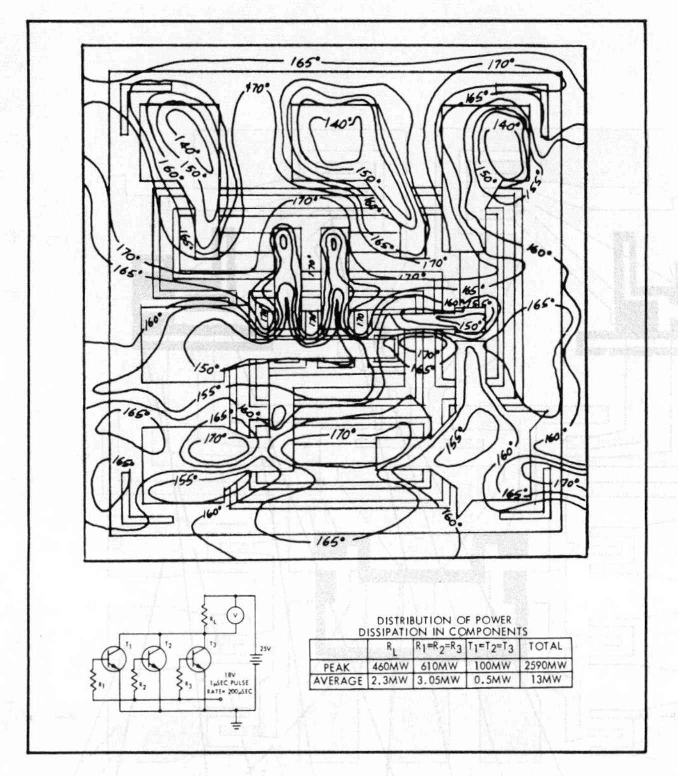

	R_L	$R_1=R_2=R_3$	$T_1=T_2=T_3$	TOTAL
PEAK	460MW	610MW	100MW	2590MW
AVERAGE	2.3MW	3.05MW	0.5MW	13MW

PULSED POWER ISOTHERMAL PLOT

(b)

Figure 173. (Continued).

tion, appears to be one technique capable of pinpointing such a defect and to assess its severity.

Thermal Mapping with the Fast Scan Microscope[9]

The use of this instrument allows the plotting of a thermal map of a semiconductor chip at any point in time during warmup or in steady-state condition. As previously mentioned, infrared scanning while the target is in transient condition might be useful to detect thermal details before they are masked by lateral heat transfer.

Figure 174 shows the schematic and the physical layout of two identical circuits, integrated onto the same semiconductor chip, that measures 0.050 × 0.050 in. Scanning took place along the raster outlined in

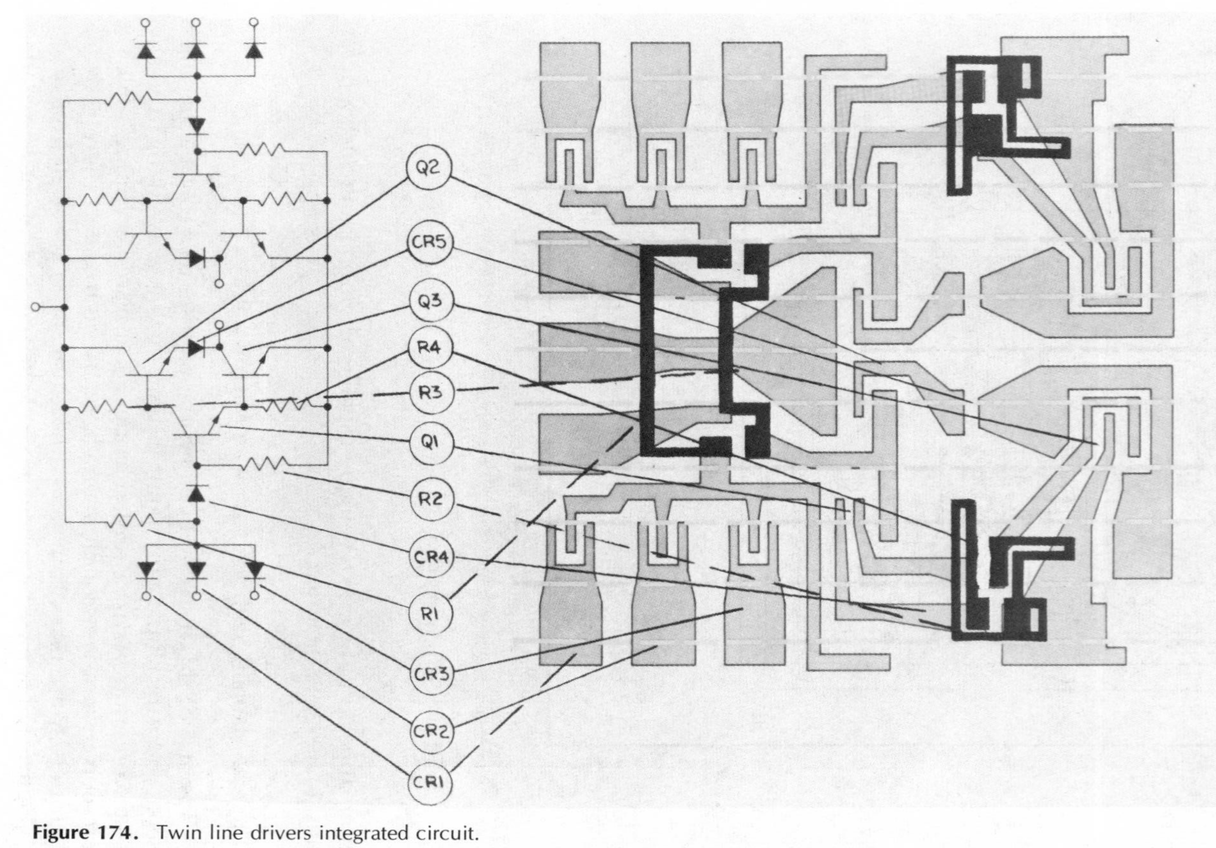

Figure 174. Twin line drivers integrated circuit.

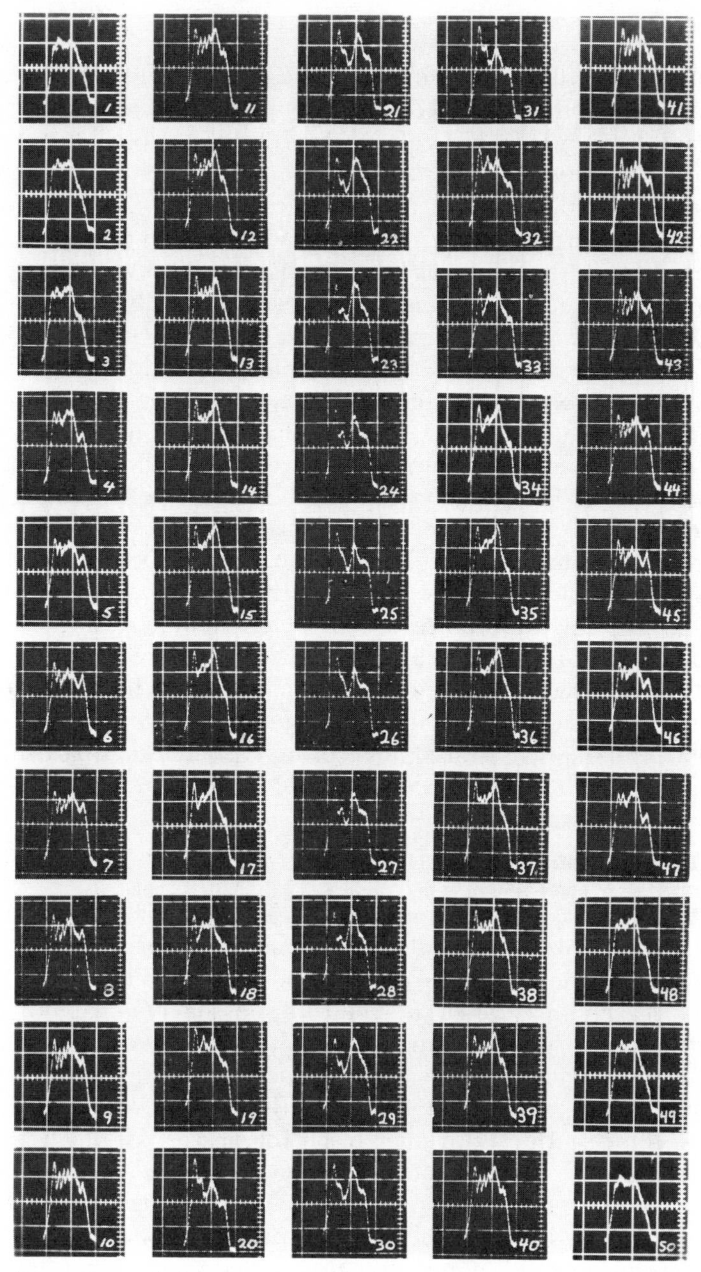

Figure 175. Fifty-line scan traces of twin line drivers integrated circuit.

the illustration, where every fifth scan line is shown and numbered. The analog output of the infrared detector, for each of the 50 lines forming a full frame is shown in Figure 175. Analysis of the oscilloscope traces shows that the highest temperature is located in correspondence to transistor Q_2, which dissipates by far the largest amount of power: 75 mW. Next to it, the output diode *CR*5 dissipates 15 mW and contributes to the centrally located thermal peak. The R_1 dissipates 8 mW, while all other components are at a power dissipation level practically negligible.

Therefore, the temperature of the semiconductor chip is highest in the center and decreases toward the periphery. The slope is not smooth, however, because of the different emissivity coefficient of the various materials of which the surface is made: silicon, metal, and oxide. Consequently, every scan line has a jagged profile, which is typical and consistent for all integrated circuits of the same type, operating in the same electrical mode. Any deviation from this profile is caused by some anomaly in the physical structure or in the electrical performance of the corresponding element. The study of these anomalies and of their causes are of key importance for the better understanding of failure phenomena and for more reliable life expectancy prediction.

Figure 176 shows a three-dimensional model of the infrared radiation emitted by the integrated circuit that has just been described. In a way, it can be considered its infrared fingerprint, typical of that particular circuit, operating in a given electrical mode.

Bond Quality Evaluation

Fast infrared scan of integrated circuits and of semiconductors in general after they have been bonded to their substrate (or header) can yield reliable information about the quality of such bonds. The technique calls for electrical energization of the unit, and subsequent infrared mapping of the surface of the semiconductor "chip." Whenever a discontinuity exists in the bond area, an infrared anomaly will appear, namely a thermal "bulge" due to the fact that the heat developed in correspondence of said defect cannot reach the heat sink located below it.

An example of such application is shown in Figure 177 where, for simplicity, only one scan line is shown; it is line 10 of Figure 174. The oscilloscope traces depicting the warmup process of the chip's elements along this line are shown in the picture located at the upper left corner of Figure 177: starting with the bottom trace, taken before electrical energization, we can watch how the temperature increases as time goes by: after 1 sec from energization, the temperature in the center has increased approximately by 8°C, and by successive increments has

Figure 176. Three-dimensional model of 50-line scan of integrated circuit.

reached 73°C after 46 sec from the instant when power was applied. This is the normal warmup pattern of a properly bonded chip, as verified by the x-ray picture located directly under the oscilloscope traces: no discontinuities or voids are visible under the chip.

The right-hand side of Figure 177, instead, shows a much faster warmup of the elements of scan line 10. After 1 sec from application of

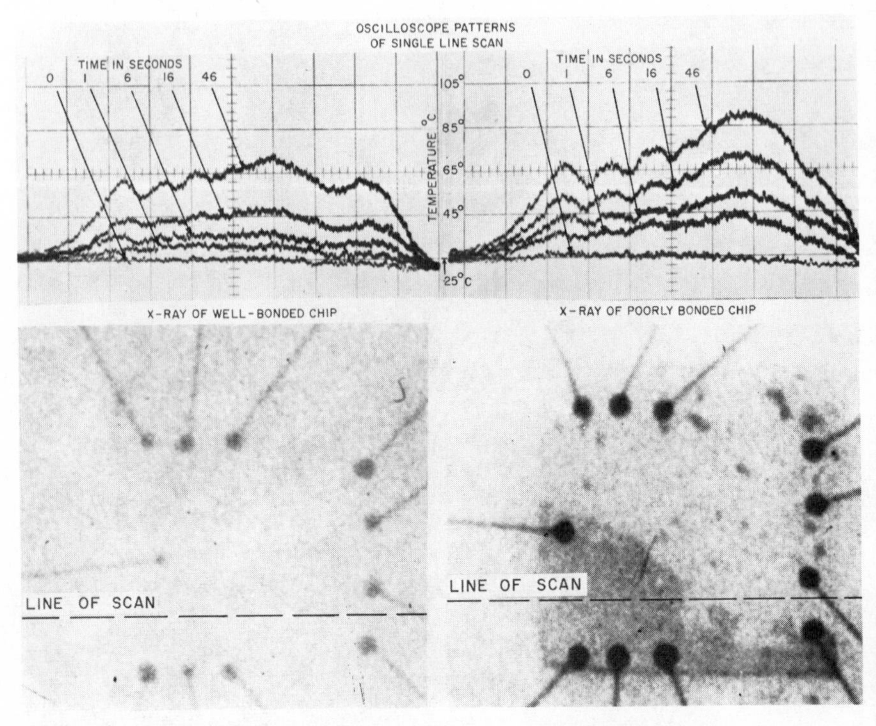

Figure 177. Infrared and X-ray check of chip-to-header bond quality.

electrical power, the temperature in the area has increased approximately 20°C, and at the end of 46 sec, the peak has reached 92°C. Besides showing the presence of a poor bond, the shape of the oscilloscope traces indicates also its location: the right half of the trace shows the greater deformation, and the x-ray picture shows that it is exactly in this area where the void is located. In effect, the chip is bonded only at its left lower corner and along its bottom edge: all the rest is separated from the substrate, except for small droplets of eutectic near the upper right corner. This condition, of course, shows quite conspicuously in the complete raster scan of the semiconductor chip. The advantage of using a fast scan infrared microscope lies in the capability to inspect a large number of units in a short time. When time is of no concern the same bond quality evaluation can be carried out with the use of a point-detecting infrared microscope, and with the target mounted on a scanning substage.

The approach described above could be called "a posteriori," since the bond evaluation is carried out after the fact. Its implementation can only tell us whether a chip is properly bonded to the substrate or if the bond is defective.

This situation might change in the near future, in view of the recent availability of the thermal bond monitors (see page 133 for their description). These instruments carry out a semiconductor bond evaluation *in real time,* and through automated feedback can control the bonding operation so that no defective bonds can be produced during the manufacturing process. Figure 178 shows, in schematic form, the average thermal behavior of a semiconductor chip during the die-attach cycle, when using a thermocompression bonder equipped with a hot plate on which the substrate is placed. As the temperature of the substrate increases, the chip's temperature also increases, at a slower rate because of the collet's retarding action. Since the chip is heated primarily by the heat conducted through the small areas of the points of contact with the substrate, there is a significant temperature differential between these two elements. The gold-silicon eutectic flows when the temperature of 375°C has been reached, and the "scrubbing" has succeeded in achieving intimate contact at the silicon-gold interface. During the scrubbing phase, the chip's temperature increases quite fast, due to the large heat conduction through the increased area of contact, so that at the completion of the bond the chip's temperature approaches that of the substrate. At this point heating must stop to avoid chip spoilage by overheating.

Figure 179 is a good summary of this capability. In the illustration, the three pictures at right show three double-transistor chips bonded to the substrate at different degrees of eutectic wetting: the chip *A* located at top right shows a very poor bond, as evidenced by the missing eutectic

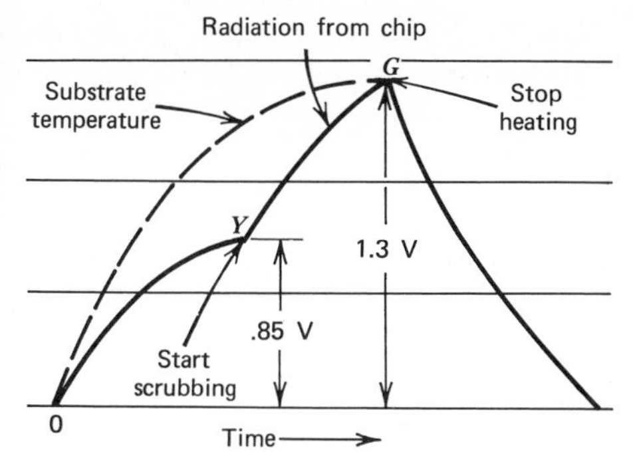

Figure 178. Chip bonding temperature analysis. (Courtesy A. S. Dostoomian, Vanzetti Infrared and Computer Systems, Inc.)

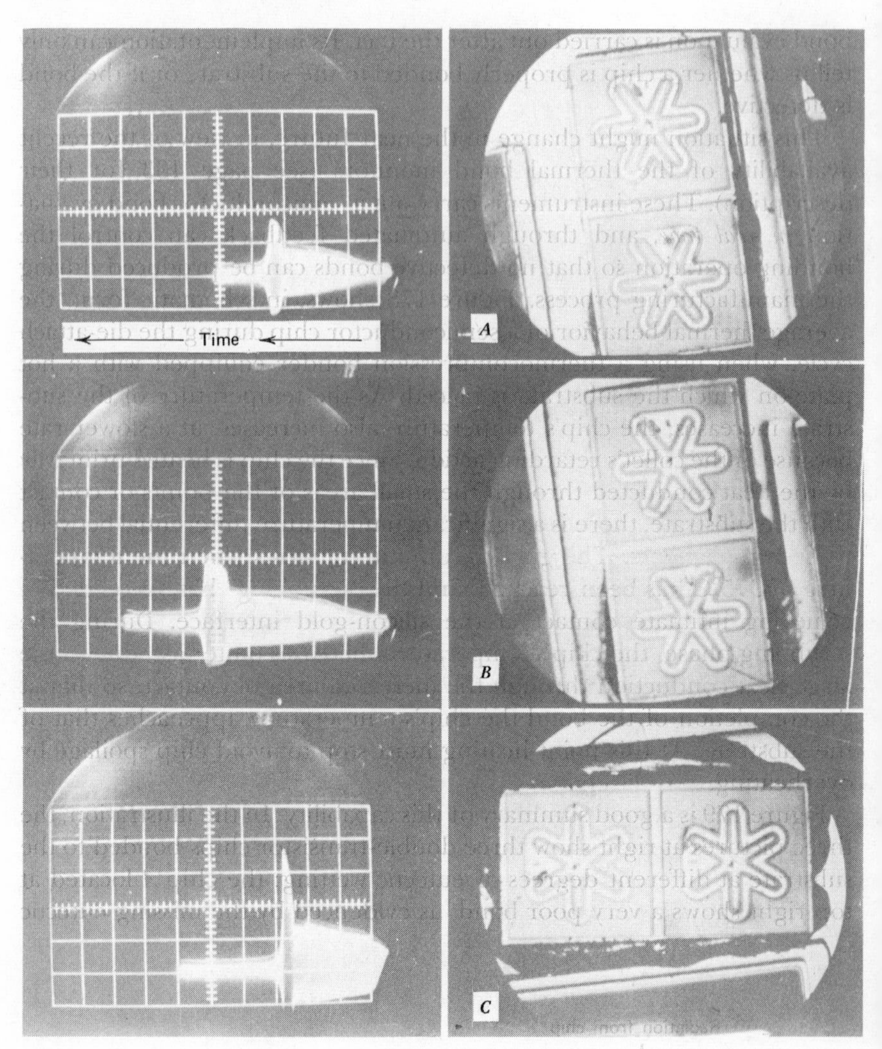

Figure 179. In-process control of semiconductor chip eutectic bonding.

fillet at top and right sides of the chip; the chip *B* located at middle right shows a somewhat improved bond, since the eutectic fillet is a little more pronounced around the two sides of the lower area of the chip; finally, the chip *C* at bottom right shows good eutectic fillet all around its sides. The three oscilloscope traces at left are the corresponding signals obtained at the output of the thermal bond monitor, using a slow horizontal sweep (5 sec/cm) to allow the recording of the whole operation in one single scan picture. The temperature above ambient is shown in the vertical direction, at an approximate scale of 100°C/cm.

In the top picture, the peak of the trace shows a Δt of approximately 170°C; in the center picture it is barely above 200°C; and in the bottom picture it is 360°C. Adding to these Δts the 25°C of ambient temperature, we see that only in the last instance we have reached the full eutectic temperature and at the same time a good bond between chip and substrate.

To ensure a perfect bond each and every time, the feedback loop of the thermal bond monitor must be set in such a way as to automatically stop heating as soon as the eutectic temperature has been reached at the upper surface of the chip.

Recombination Radiation Measurements

The electrical signals flowing through the transistors and diodes incorporated in integrated circuits can be monitored by measuring the recombination radiation emitted by their junctions, very much in the manner that was described in the preceding sections under "Diodes" and "Transistors," on pp. 208 and 221. The differences are mainly due to the smaller dimensions of these elements, and also to the smaller magnitude of the signals flowing through them.

Fortunately in this instance smaller multiplied by smaller yields equal electrical flux per unit area of junction, which is not surprising, since design engineers use the average figure of 12 mA/mil of emitter periphery when designing integrated circuits.

Because of this condition, the sensitivity required by the measuring instrument is more or less the same whether large or small geometry integrated circuits are used. The measuring instrument is again the semiconductor junction analyzer described in page 134.

At the time of this writing, recombination radiation measurements had been carried out on several ICs during the work of two NASA contracts.[11] The highlights of this work can be summarized in the following section.

Recombination Radiation from Diode Junctions in Integrated Circuits

Measurement of recombination radiation emitted by diodes in integrated circuits were conducted on Fairchild Micro-DTL, type 933. This particular circuit was chosen because the diodes could be energized individually or in parallel without other components interfering with the measurements. Figure 180 is the circuit diagram of the type 933 integrated circuit, which is deposited on a 1 mm^2 silicon chip. Figure 181 is

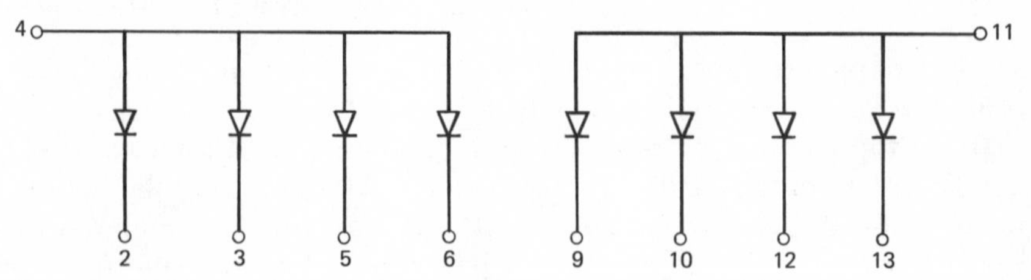

Figure 180. Schematic of Fairchild Micro DTL-933. (Courtesy Fairchild Semiconductor Company.)

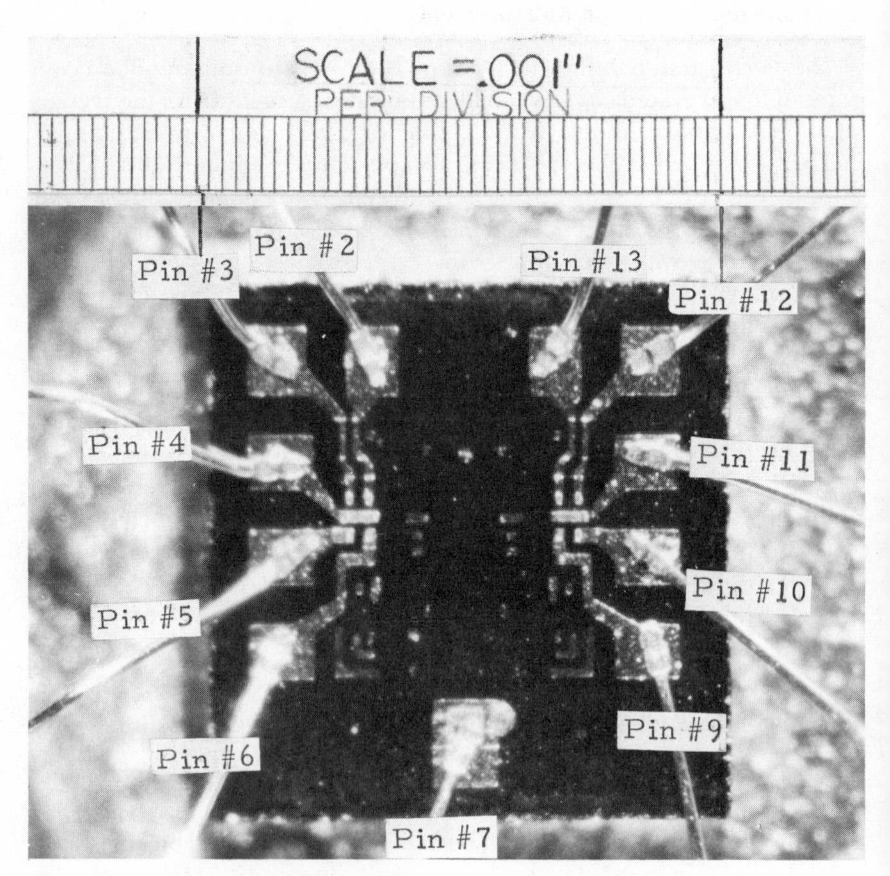

Figure 181. Fairchild Micro DTL-933 Chip.

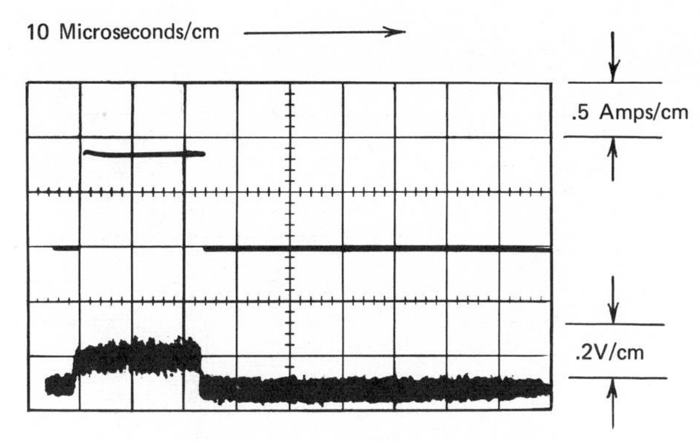

Figure 182. Recombination radiation emitted by diodes of integrated circuit μDTL-933.

the visible picture of the same unit with a dimensional scale added for clarity.

Electrical energization of one diode at a time was carried out with a 22-μsec pulse of 0.8A, at a repetition rate of 1000 pulses/sec. This low duty cycle was chosen in order to avoid overheating. Figure 182 shows in the top trace the current pulse flowing through the diode junction, and in the bottom trace the detector output, produced by the recombination radiation emitted by the junction during the process.

The proof that the recombination radiation signal is directly proportional to the current flow and independent from the physical size of the junction was given by the following finding:

> When using an optical fiber whose field of view would cover the entire surface of the semiconductor chip, the detector output resulted exactly the same whether the energizing current pulse would flow through a single diode, or through four diodes wired in parallel.

Recombination Radiation from Transistor Junctions in Integrated Circuits

Experiments to observe recombination radiation from transistors in integrated circuits were conducted on Motorola MC-305G units. This particular integrated circuit was chosen because the transistors could be energized individually or in parallel without other components in series. Figure 183 is a circuit diagram of an MG-305G integrated circuit which is deposited on a 1-mm^2 silicon chip.

The visible picture of this chip is shown in Figure 184, with a dimensional scale added for clarity.

The first experiments were run on only one transistor. Electrical

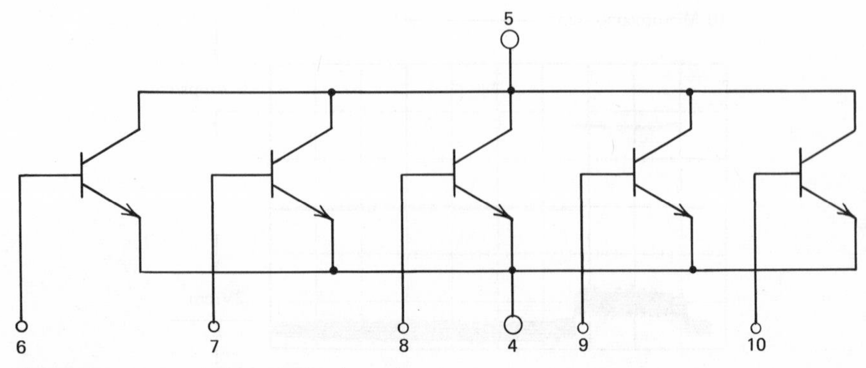

Figure 183. Motorola MC-305G schematic.

energization consisted of a $V_{CE} = 5.0$ V and a 20-μsec pulse into the base. The base drive was increased gradually with the results shown in Figure 185. As the base drive is increased, the radiation also increases, so that the correlation between current flowing through the junction and the power of the radiation emitted appears to be essentially linear, as already found in the case of diodes.

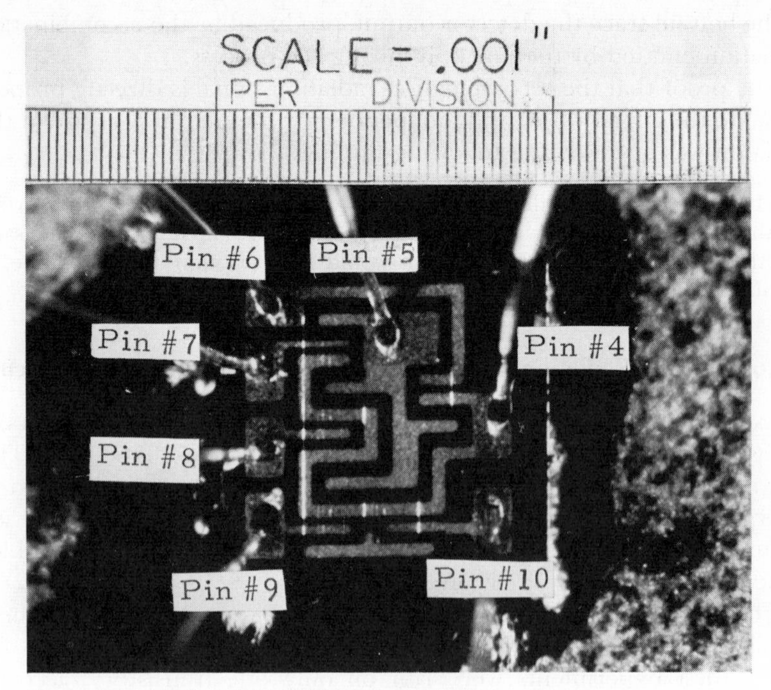

Figure 184. Motorola MC-305G integrated circuit.

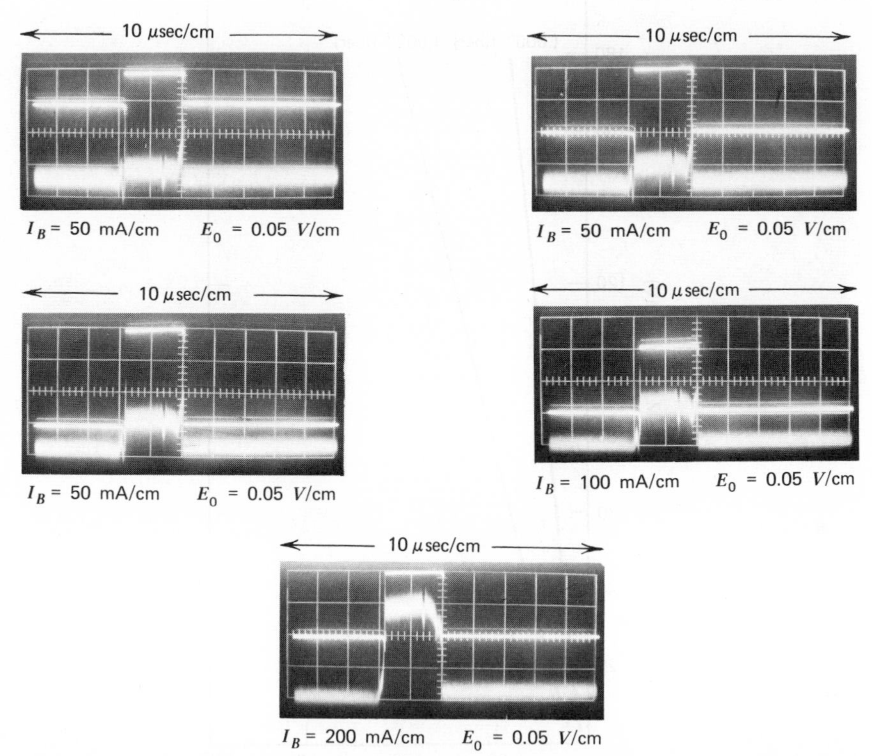

Figure 185. Recombination radiation emitted by a transistor of MC-305G integrated circuit.

However, these measurements were taken with the transistors operating in saturation. Further work performed with lower base current values, below the saturation level, yielded a different ratio between the base current and the recombination radiation emitted. Measurements were taken with optical fibers of different diameters (namely 0.006 and 0.002 in.) and the results are plotted in the correlation chart of Figure 186, for a transistor of Motorola Integrated Circuit type Mc355. The slope of the curves is related to the different amounts of radiation picked up by fibers of different diameter.

Further work was carried out on another Motorola integrated circuit type Mc-304G. This circuit was tested with all components energized, as shown in the schematic of Figure 187. Figure 188 is the visible picture of this device, which measures 1 mm². The field of view of the optical fiber used in this instance was covering the whole surface of the unit, so that the recombination radiation emitted by the transistor and by the two diodes was simultaneously picked up by the detector. Figure 189 shows

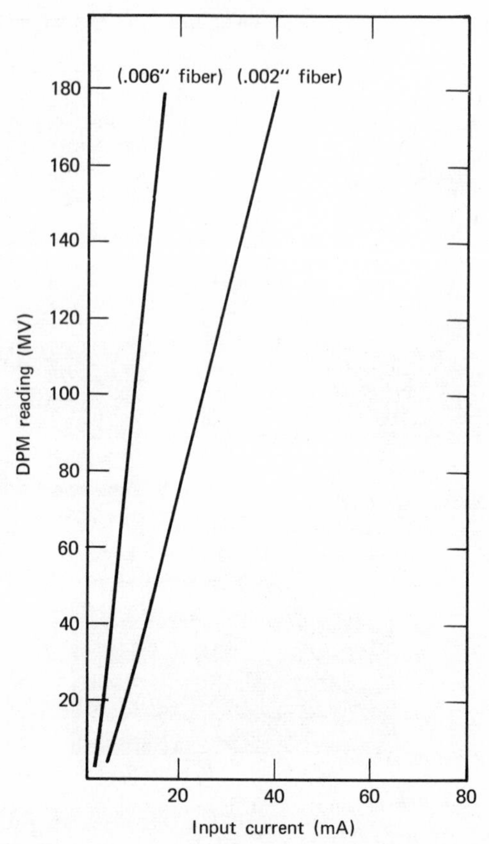

Figure 186. Recombination radiation versus base input current in a transistor of MC-355 integrated circuit.

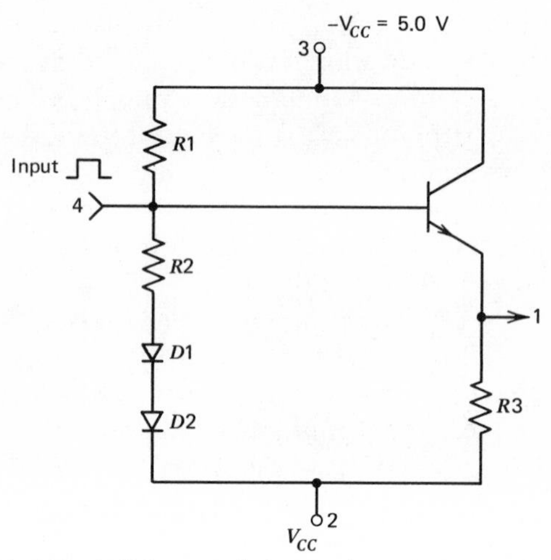

Figure 187. Motorola MC-304G integrated circuit schematic.

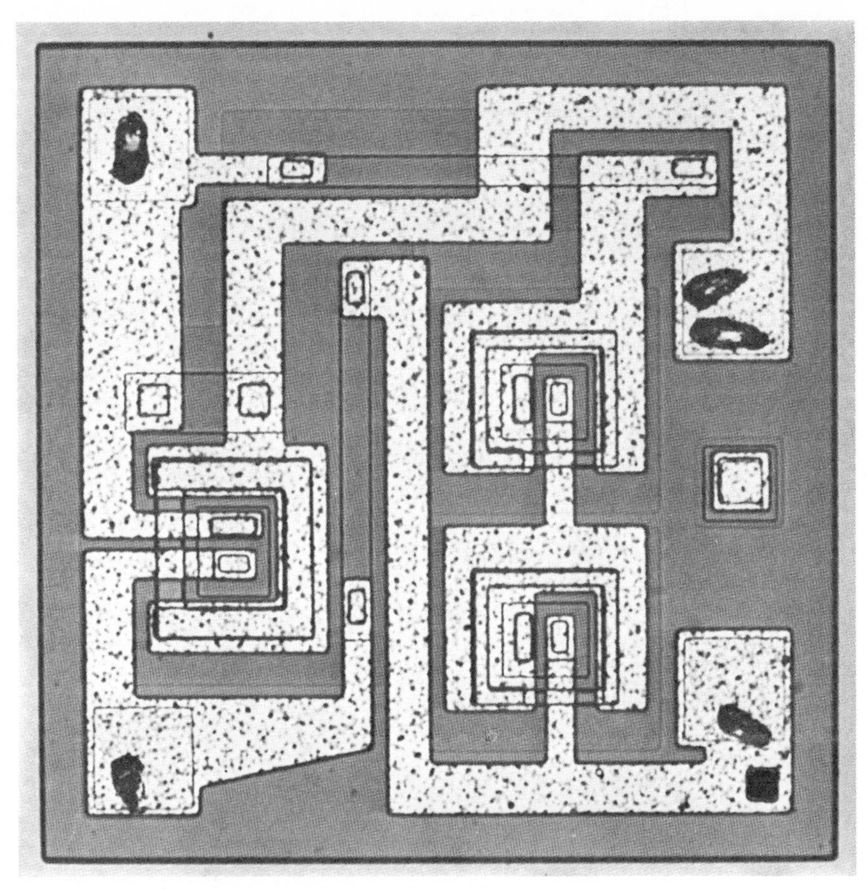

Figure 188. Visible picture of MC-304G integrated circuit.

this signal in the trace at the bottom, and the energizing input pulse in the top traces, respectively for collector, base and emitter.

Recombination radiation from discrete junctions of more complex integrated circuits was measured in another phase of the NASA-sponsored work. The integrated circuits used were transistor-transistor logic (TTL) devices type SN5400J, made by Texas Instruments Incorporated. They are quadruple two-input NAND gates, whose symbolic diagram is shown in Figure 190, while Figure 191 shows the schematic of one of the four gates of which the whole integrated circuit is made. The visible picture of the unit is shown in Figure 192, with a dimensional scale for clarity. The individual diodes and transistors of each gate are pinpointed in the physical layout of Figure 193, which is derived from an actual photomicrograph of one of these devices.

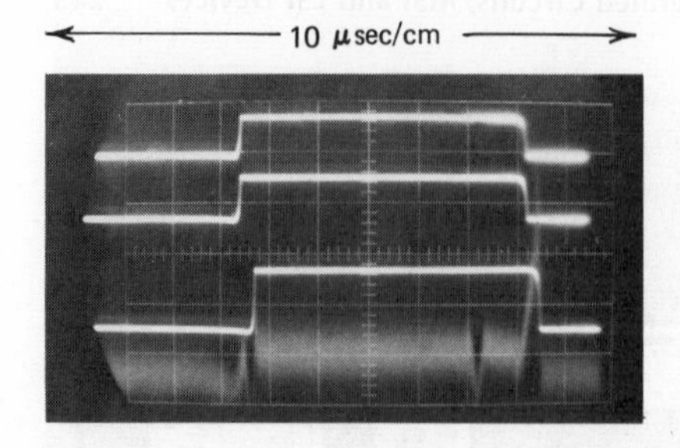

Figure 189. Recombination radiation from MC-304G integrated circuit.

$I_C = 0.5$ A/cm

$I_B = 0.5$ A/cm

$I_E = 0.5$ A/cm

$E_O = 0.05$ V/cm

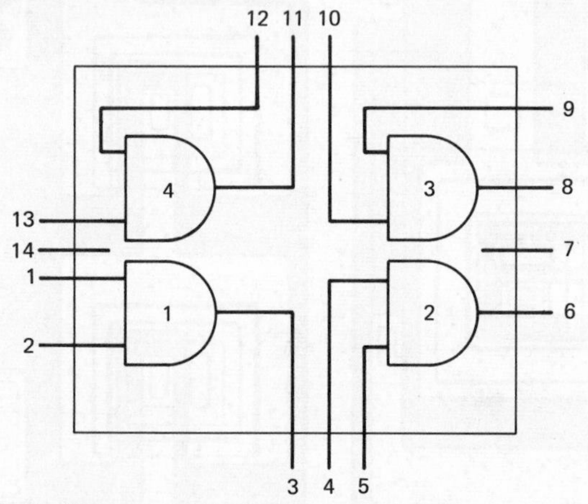

Figure 190. Symbolic diagram of TI SN5400J integrated circuit.

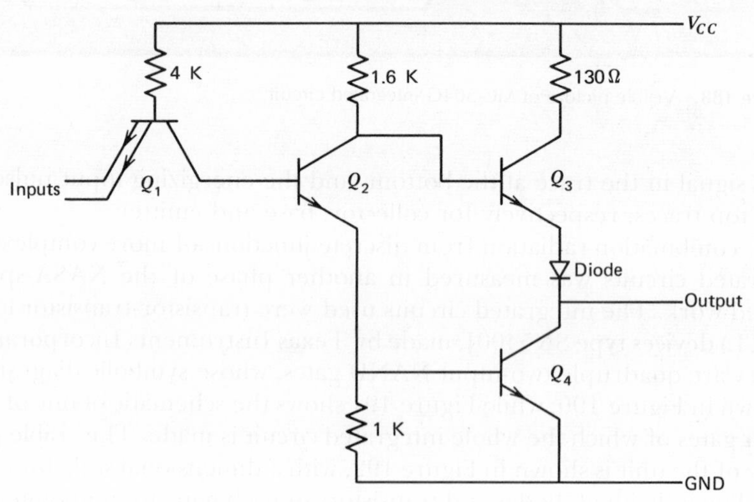

Figure 191. Schematic of one gate of TI SN5400J integrated circuit.

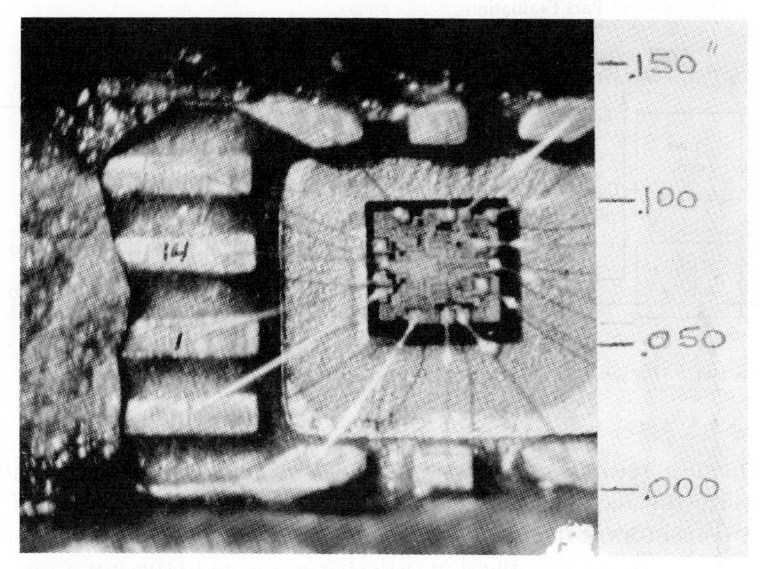

Figure 192. Visible picture of TI SN5400J integrated circuit.

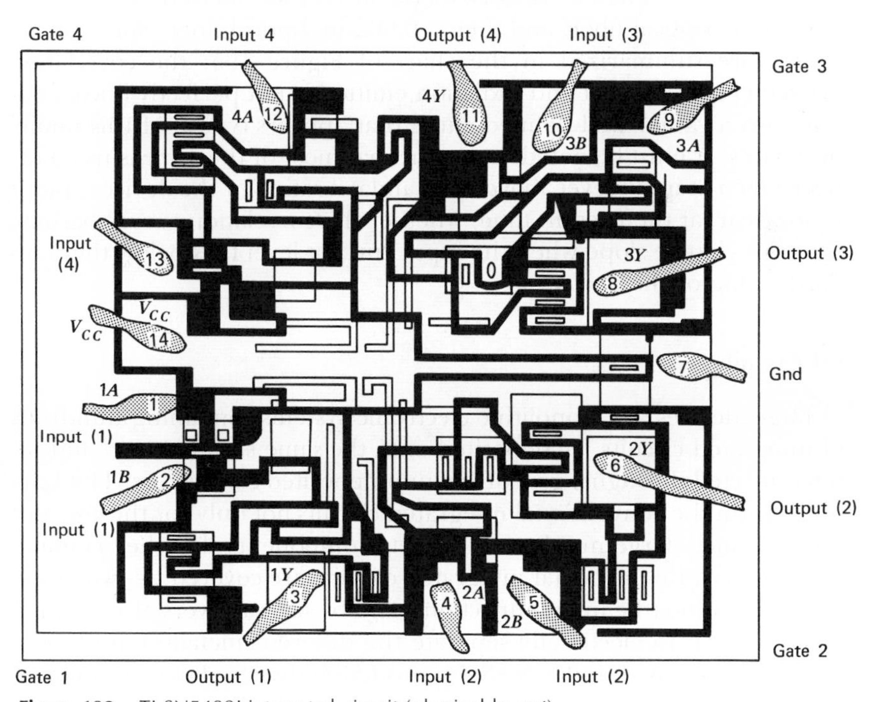

Figure 193. TI SN5400J integrated circuit (physical layout).

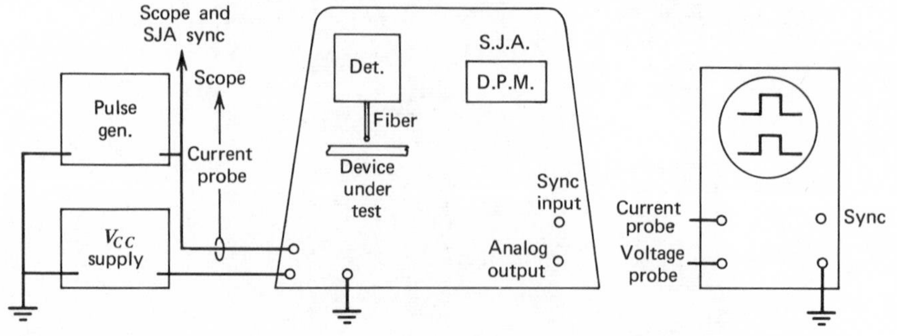

Figure 194. Test setup of TI SN5400J integrated circuit.

The test setup used to energize electrically these devices and to measure the recombination radiation emitted by the junctions of discrete transistors of the circuitry is shown in Figure 194: a positive voltage pulse of 20 μsec was applied in orderly succession to the input terminal of each gate at a repetition rate of 1 kHz. During operation, recombination radiation emitted by transistors *Q*2 and *Q*4 was measured using first a 0.006-in. optical fiber, and then a 0.002-in. optical fiber. Some of the results are summarized in the chart of Figure 195: the correlation between input current and radiation emitted is not perfectly linear, but it follows a rather well-defined curve that appears typical of this family of devices. It must be noted, however, that most of these measurements were taken at input level above the standard rating of the devices, and it is not clear, at the present time, whether the correlation curves become linear or change slope when the input current level falls below the standard rating of the device.

LSI Circuits

These devices are monolithic electronic systems containing hundreds of integrated circuits, all deposited onto the same silicon wafer, and interconnected through a grid of vacuum-deposited conductors. The LSIs differ from the conventional integrated circuits not only for the size and the extremely large number of discrete components that they contain, but also because the actual integrated circuits are covered by two layers of metallization (the interconnections) and by several layers of insulating material used to electrically separate the different metallization layers.

Work performed under NASA sponsorship disclosed that recombination radiation measurements are possible, notwithstanding some absorption by the layers covering the discrete junctions of the devices. Verifica-

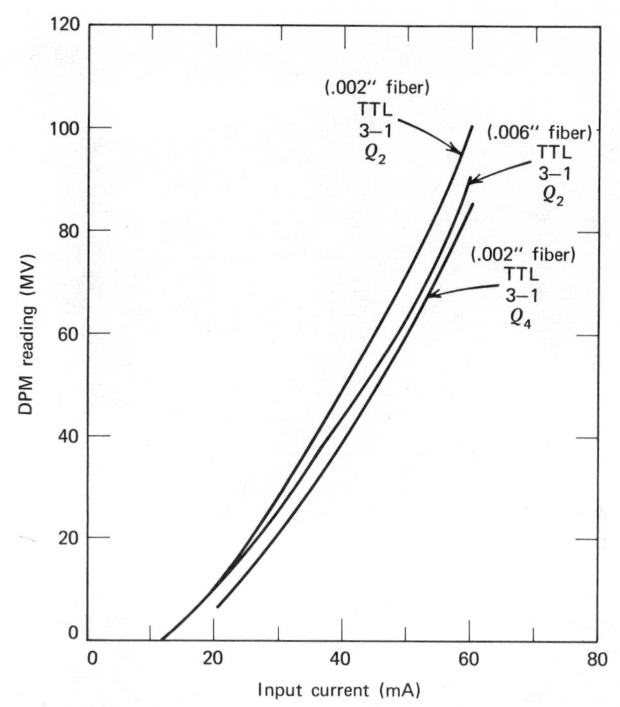

Figure 195. Recombination radiation versus gate input current for TI SN5400J integrated circuit.

tion of the above was carried out using fragments cut off real LSI units, to isolate TTL circuits electrically and physically identical to those described in the former section. Figure 196 shows one of these units, designated as LSI Special Unit 5, in two micrographs taken respectively at 30 and 100×. The custom-made connections of the circuit to the terminals are clearly visible as thin wires "plugged" to the circuit's pads and running diagonally toward the outside terminal connectors.

For electrical energization and test the same setup already shown in Figure 194 was used. The readings obtained from the junction of discrete transistors of five "special units" of this type were plotted onto the chart shown in Figure 197. It is of interest to note that the response curves of four units (2*a*, 2*b*, 5 and 6*b*) show a similar trend, while the curves of unit 6*a* show an opposite curvature. One of the hypotheses formulated at the time of this finding was that the discrepancy might be caused by anomalies in the forbidden gap of the semiconductor material where the junctions are located.

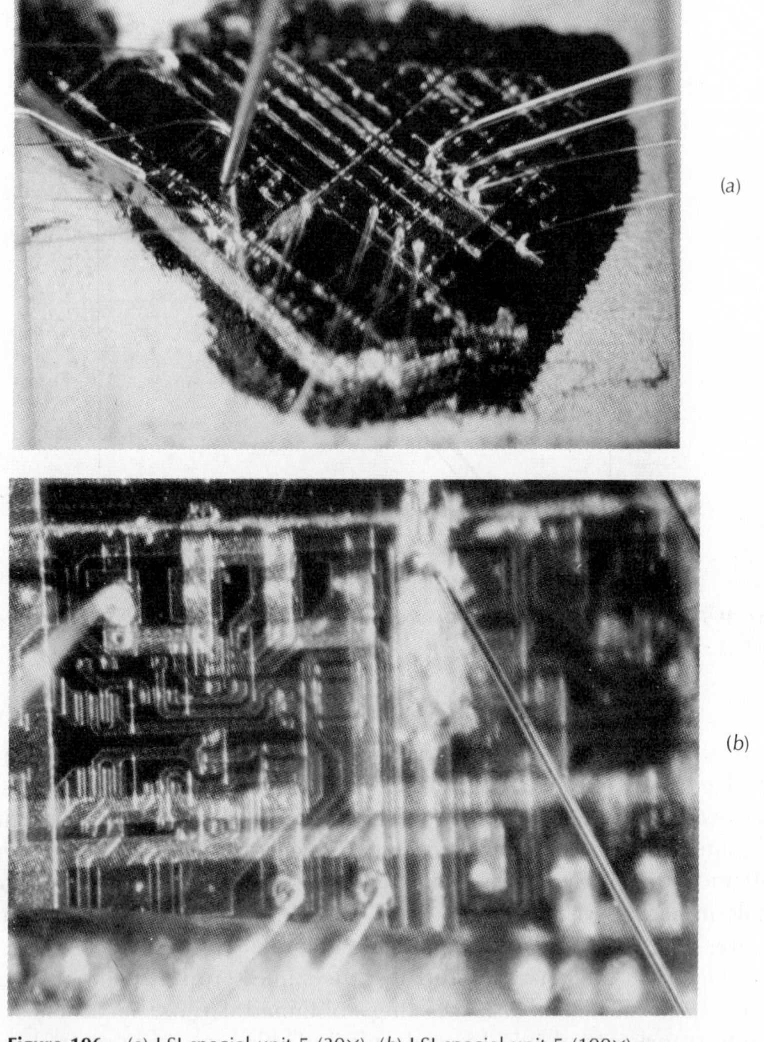

(a)

(b)

Figure 196. (*a*) LSI special unit 5 (30×). (*b*) LSI special unit 5 (100×).

To find out what effect the metallization has on the shape of the emitted radiation, the base-emitter junction of *Q4* of LSI unit 6*b* was selected for area and contour measurements.

The data so generated were plotted first in a map configuration, and then a three-dimensional model was derived, as illustrated in Figures 198 and 199. Some asymmetry is apparent, evidently due to the dif-

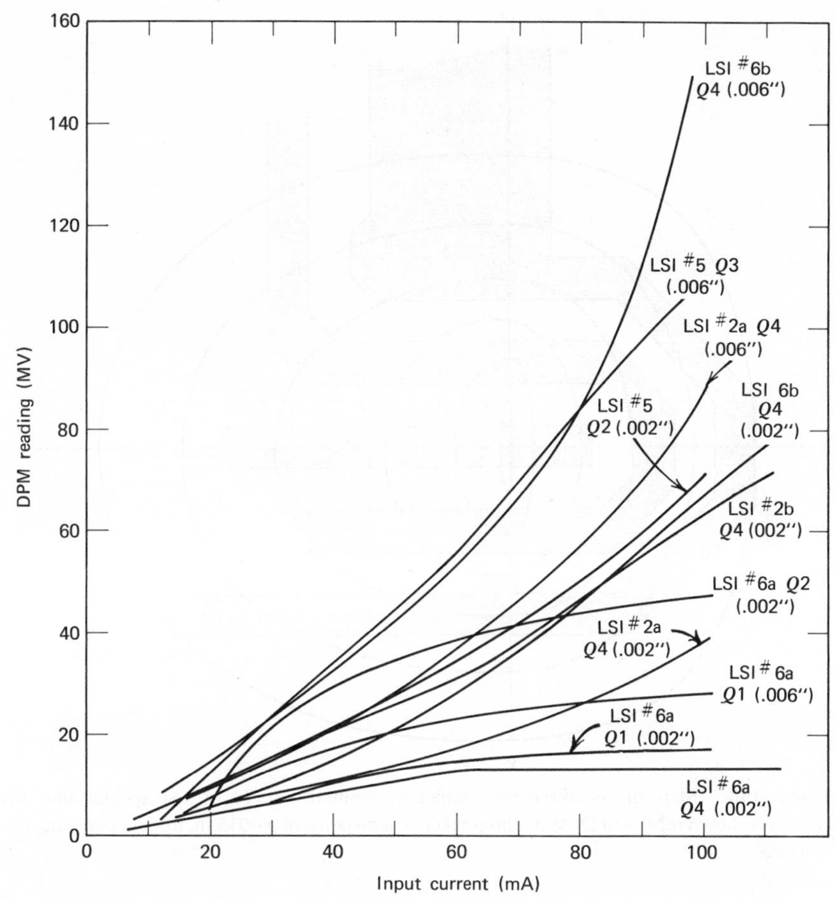

Figure 197. LSI special units: correlation between transistor base current and recombination radiation from junction.

ferent shape of the metallization in the various areas around the junction.

Summary of Applications

When applied to semiconductor devices, infrared techniques must be divided in two areas:

1. Information related to the radiation of thermal origin.
2. Information related to the recombination radiation

The data obtained under Step 1 carries information about electrical power dissipation and about mechanical characteristics related to the

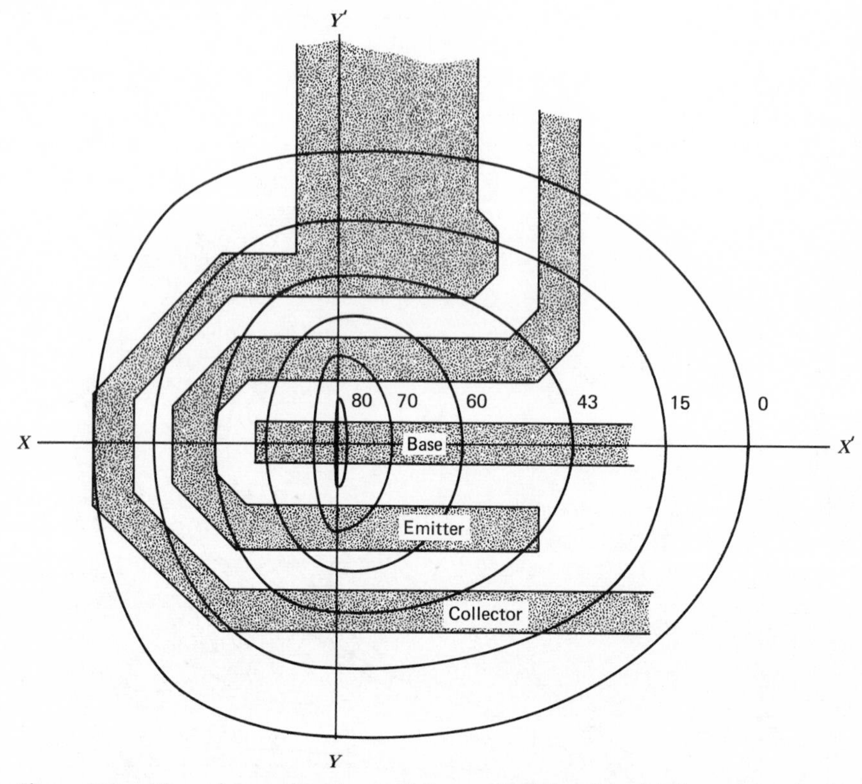

Figure 198. Map of recombination radiation emitted by Q4 of LSI special unit 6b. Numbers shown represent DPM readings taken at a distance of 0.0015 in. from upper surface of device.

performance of the semiconductor devices. Its usefulness can be broken down in the following areas:

1. Engineering design evaluation. The ability to measure the true operating temperature of each and every element of the circuit enables the design engineer to determine the electrical operating characteristics of transistors and diodes, which are heavily temperature dependent. With this information, actual stress levels and derating factors can be established and unsound design features can be disclosed and corrected.

2. Reliability calculation. Knowledge of the actual operating temperature of the components enables the reliability engineer to obtain the true failure ratios for all components, so that the reliability calculations can be trusted as being really representative of the unit under evaluation.

3. Manufacturing quality control. So far the following defects have

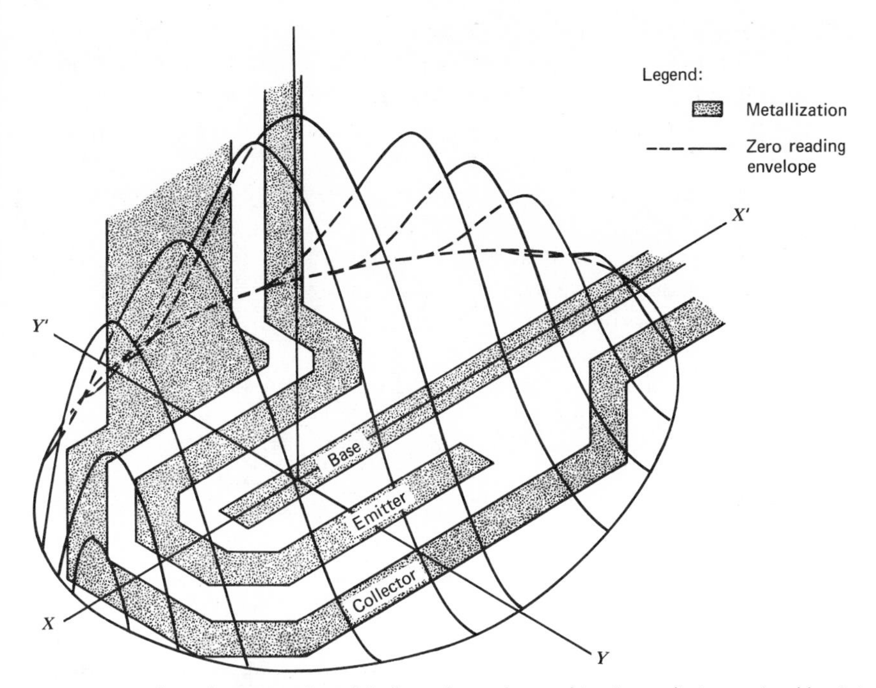

Figure 199. Three-dimensional model of envelope of recombination radiation emitted by Q4 of LSI special unit 6b.

been detected through infrared measurements: poor chip-to-header bonding; defective metalization; reduced cross-section of conducting elements; hot-spots capable of inducing breakdown conditions; excessive or insufficient power dissipation by junctions or by resistors. Investigation is still proceeding in other areas of interest, such as semiconductor cracks and quality of the bond between the chip and elements attached to it (wire bond) or deposited onto it (resistors and metal interconnect).

The data obtained under step 2 carries information about electrical characteristics (other than power dissipation) such as waveform and current magnitude of the signals flowing through the junction of transistors and diodes. It also contains information about carrier lifetime, crystal doping, and degree of uniformity of current density along the junction.

Such data is especially useful to:

1. The design engineer who can find how the signal travels through the circuit and how it gets modified in waveshape and amplitude as it proceeds from point to point.

2. The quality control engineer who can verify how well the process keeps within the upper and lower control limits established as a result of the recombination radiation measurements. Of course, this refers not only to the electrical characteristics of the signals, but also to the uniformity of current flow density along each junction.

3. The failure analyst who will be able to investigate the circumstances and the effects related to the faulty condition under examination.

4. The physicist who will be able to actually "see" on the oscilloscope display the carrier lifetime characteristics and other physical qualities of the semiconductors that are the object of his study.

5. The purchasing specialist who will have at his disposal a new tool to evaluate the product of competing vendors to a much finer and deeper degree than was possible thus far.

CAPACITORS

As opposed to resistors, diodes, and transistors, capacitors are not normally power-dissipating elements. Consequently, their temperature is not bound to change with different conditions of electrical energization. As a result, the infrared emission of a capacitor is not indicative of its electrical operation.

However, the failure ratio and the life expectancy of a capacitor are definitely related to its operating temperature, so that infrared measurements are important for correct reliability calculations.

Furthermore, faulty conditions of a capacitor might be reflected in a temperature increase: electrical leakage will produce a directly proportional increase in temperature, and electrolytic capacitors that underwent some "deforming" during an extended storage period will heat up when first energized, because the "reforming" process is an exothermal reaction.

Incidentally, these "deforming" and "reforming" processes, whose nature is essentially electrochemical, could stand some investigation on their correlation to reliability. The speed of these processes is not the same for all capacitors, and it should be of interest to the reliability engineer to find out whether fast "deforming" electrolytics are more or less reliable than the slow "deforming" capacitors of the same type.

Finally, electrolytic capacitors mistakenly mounted with reverse polarity in electronic assemblies will be readily detected as defective during infrared scanning of the electrically energized assembly. This is a condition that conventional test equipment is likely to miss most of the time.

In concluding this brief section, it can be stated that infrared measurements of capacitors are useful for reliability purposes and can point out some defective operative conditions.

COILS AND TRANSFORMERS

When provided with a magnetic core, these units are bound to show a temperature increase related to the characteristics of the flux variations of the electrical current flowing through them.

Besides this, failure ratios are related to their operating temperature; so that reliability calculations and life expectancy predictions based on actual infrared measurements are likely to yield true results.

Unfortunately, surface infrared readings are incapable of giving the measurement of the temperature of hot spots located inside the windings. It is usually in correspondence of these hot spots that failures will occur. This is basically a design problem, and it must be investigated and solved by conventional means.

MECHANICAL BONDS

The most common types of mechanical bond used in today's space-age hardware are as follows:

1. Fusion bonds (solder, brazed, welds).
2. Adhesion bonds obtained through the use of chemical or mechanical bonding agents.
3. Bonds obtained through electrical deposition of metallic materials (plating, metalizing).
4. Bonds obtained through vacuum deposition of elements or compounds (metalization in semiconductor devices).
5. Compression bonds, obtained through application of pressure at the interface of the metals to be joined. The pressure can be steady or intermittent (hammering, ultrasonic) and the temperature of the metals can be ambient or raised to a convenient thermal level.

In many instances, infrared thermal mapping can be of help in the evaluation of the quality of the bond. Normally, a temperature differential is generated between two points located across the area to be investigated, and the surface temperature distribution gives an adequate description of the magnitude of the thermal resistance at each point of the bond area.

According to the application, one of the following systems is used:

1. Flooding the surface of the target with radiant heat and mapping the temperature at every point of it. This discloses how the heat diffuses through the material located under the surface, thus yielding information about the presence of hidden anomalies along the diffusion path.

2. Soaking the target in an oven-like environment at a higher-than-ambient temperature, and mapping the thermal distribution on the surface, after removal from the oven. Physical characteristics along the heat-flow path from the core to the outside surface will be reflected in a corresponding thermal distribution that can reveal hidden anomalies.

3. Introducing a thermal differential between two points of the target chosen in such a way that the heat flow between them will be affected by the characteristics of the element under evaluation. Thermal mapping of the surface between the two points will depict how the heat flow takes place, thus disclosing the presence of any anomaly hidden under the surface. The temperature differential can be introduced by heat injection at one point, or along a line conveniently chosen.

4. Developing heat due to friction generated in the unbonded region. Thermal mapping of the outside surface of the target will show a warm area in correspondence to the location where the friction takes place.

Examples

1. *Heat flood system.* Figure 200 shows a typical setup for this approach. Radiant energy from one or more sources heats one surface of the target *T* in a uniform, even manner, while an infrared detector scans the opposite surface. A physical discontinuity between these surfaces (such as an unbonded area, a void, an inclusion) will reduce the heat

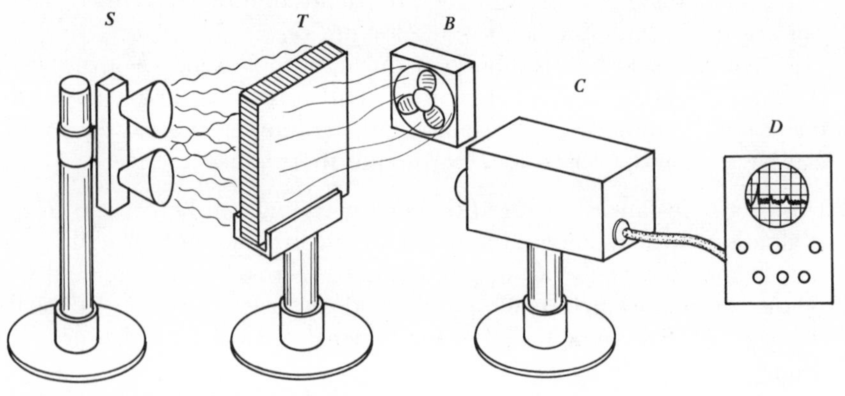

Figure 200. Heat flood system setup.

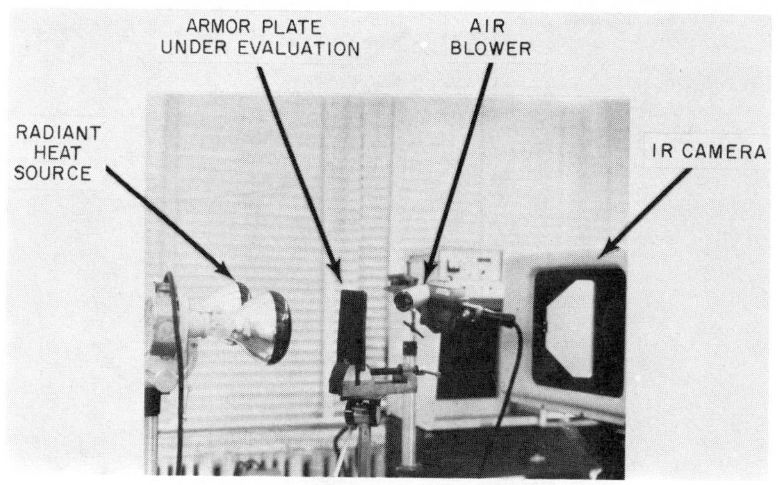

Figure 201. Armor plate delamination test. (Courtesy P. Vogel, U.S. Army Materials Research Agency.)

transfer to the surface being scanned. Consequently, the temperature in this region will be lower than in the area where the bonding is good. Cooling the heat-receiving surface with an air blower, *B*, increases the ΔT between bonded and unbonded areas, thus enhancing the detection of defects of this type that will show as negative peaks in the oscilloscope display, *D*. Applications of this system are numerous and cover the whole gamut from multilayer boards to honeycomb structures to spot-welds. Programs already completed or in progress at the present time include evaluation of rocket motor cases, honeycomb panels, helicopter propeller blades, and jet rotor blades. Figure 201 shows how armor plate delamination caused by armor piercing bullet impacts is being measured with this system at the Army's Material Research Agency, Watertown, Mass. The setup for this study is illustrated in Figure 200, and the areas where delamination took place appear clearly in Figure 202 where the intensity-modulated Polaroid picture shows dark regions, which agrees with the fact that the heat from the warmer surface cannot readily transfer to these areas.

2. *Heat soak system.* The evaluation of targets having considerable thickness can best be carried out with this system, which in effect places the heat source at the core of the target. The schematic of Figure 203 shows how the bonding between the solid propellant, the "liner" enveloping it, and the outside "skin" is evaluated with this system, in a test of a solid propellant missile motor. A practical application of this system is illustrated in Figure 204, where the target is a Polaris rocket motor case

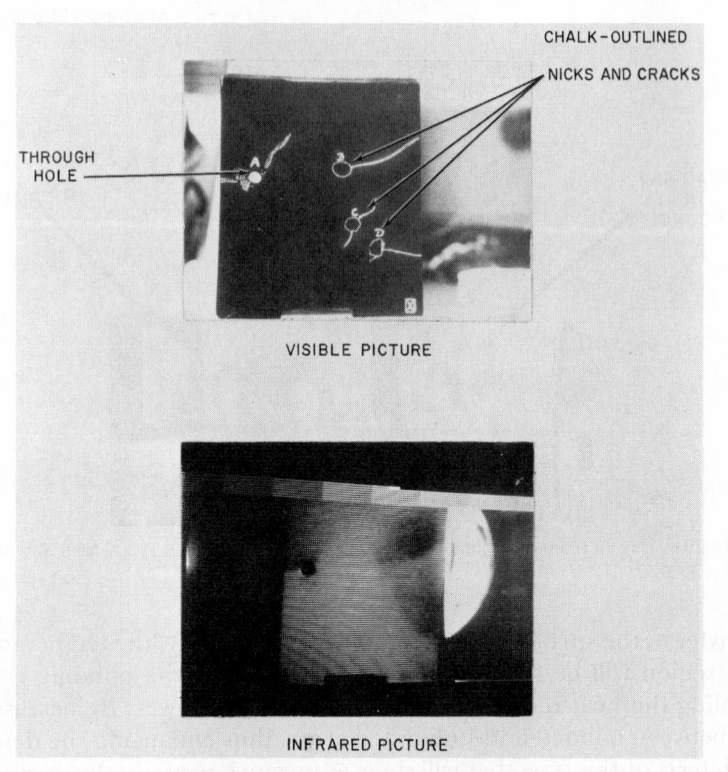

Figure 202. Armor plate delamination. (Courtesy P. Vogel, U.S. Army Materials Research Agency.)

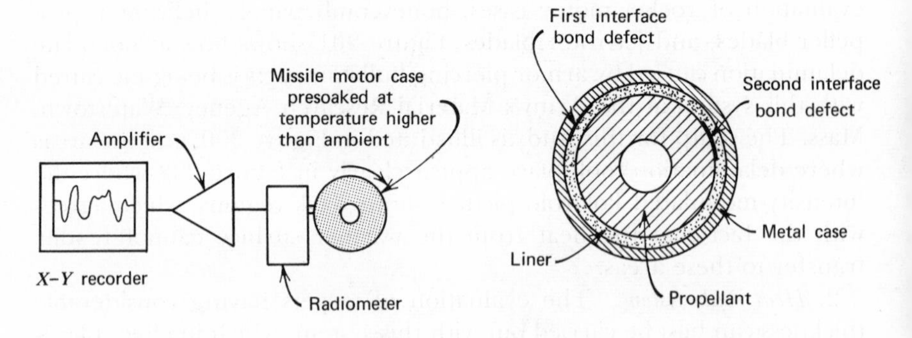

Figure 203. Infrared bond defect detection with heat soak system.

Figure 204. Infrared scan of Polaris motor for hidden bond defect detection. (Official photo, U.S. Navy MHL-13/9209/4060/538.)

and the infrared scanner is a Barnes T-4 infrared camera. The core temperature was 105°F and the surface temperature 80°F when the scanning took place. Ambient temperature was 75°F. The motor case was rotating at 1.5 rpm while the infrared camera was slowly moved in the vertical direction to allow complete coverage of the surface. Areas affected by poor bonding between the core, the liner, and the skin are cooler than the surrounding areas and appear as dark shadows in the photographic recordings. This work was successfully carried out at the U.S. Naval Weapons Station, Concord, Calif.

3. *Heat injection system.* Figure 205 illustrates the physical principle upon which the system described in (3) is based. A good soldered or welded joint is characterized by a large area of metal fusion, and an absence of narrow cross-section which could act as a bottleneck for the heat flow. This corresponds to a very low value of the thermal resistance, and this condition is reflected in an almost linear temperature variation between the points P_1 and P_2. Line A depicts this case. On the other hand, a poor bond will present a narrow cross-section that will restrict the heat flow through it. The thermal resistance value at this point is

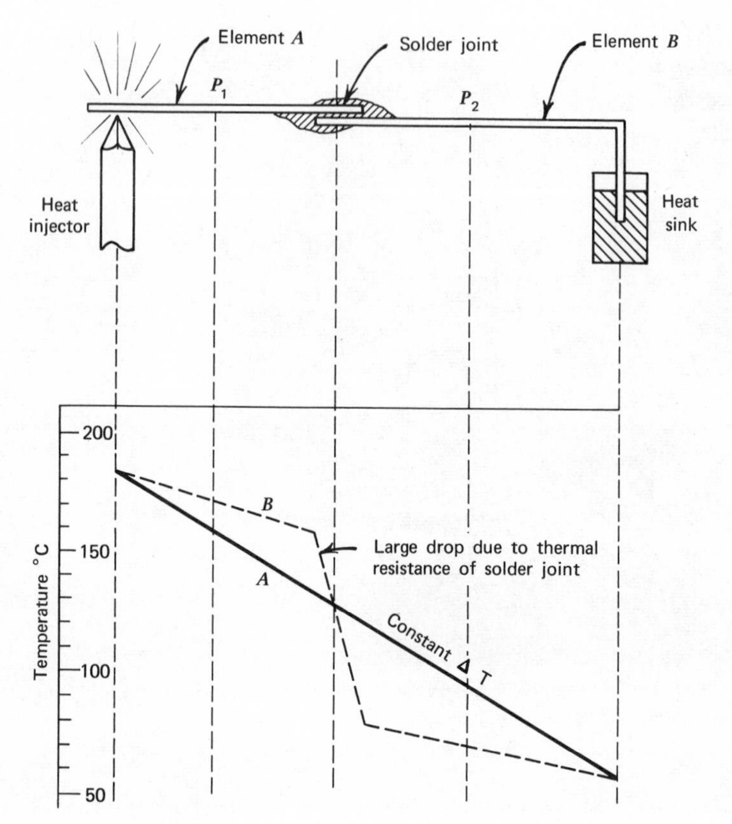

Figure 205. Heat injection system: temperature distribution of *A* and *B* elements between point of heat injection and heat sink.
——— Good solder joint
– – – Poor solder joint

high, and the temperature will show a large drop, as shown in line *B* of Figure 205. Implementation of this approach does not require full thermal plotting along the surface between points P_1 and P_2. Just two measurements, taken at two points on either side of the bonded areas, are sufficient to determine whether a temperature drop exists at the joint. Figure 206 shows an application of this type.

In this instance, instead of heat injection from an outside source, thermal energization from within is used. This approach is based on the fact that the flow of electrical current produces an increase in the temperature of the element carrying the current. This phenomenon, known as power dissipation, turns a fraction of the electrical current into heat and takes place in every conductive element presenting a finite resis-

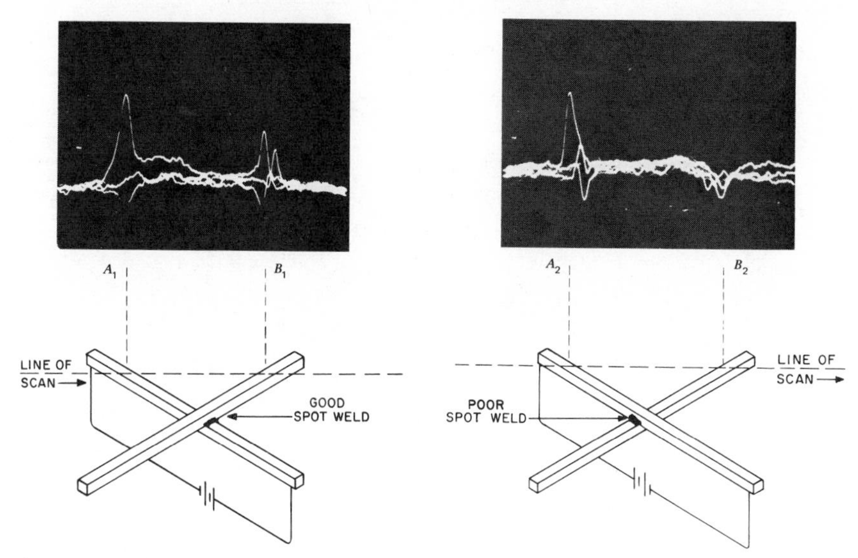

Figure 206. Thermal heat transfer in spot-welds.

tance to the current flow. The consequent thermal rise produces a proportional increase in the power emitted by radiation from the surface of the element, and this parameter is representative of the temperature of the component.

The example shown in Figure 206 is based on the fact that a good spot weld allows good heat transfer from the electrically heated wire onto the wire welded to it. When such an assembly is scanned with an infrared detector along the dotted line shown in the illustration, the scope display of the detector output shows in A_1 and B_1 positive peaks corresponding to the temperature of the two points crossed by the scan line.

Conversely, when the spot weld is poor (right part of Figure 204) the thermal resistance existing at the weld introduces a large thermal gradient, and this prevents any sizable amount of heat from transferring from the heated wire onto the welded wire. This condition is depicted by the oscilloscope display, where point B_2 remains at ambient temperature in spite of the increasing temperature of point A_2.

The same considerations are valid for brazed or soldered connections and for multiple joints either in series or in parallel, and can be extended to the evaluation of seam welds along lines of almost any configuration. For this application, heat injection could be achieved by focusing radiant energy, by means of optics mounted on a carriage supporting also the

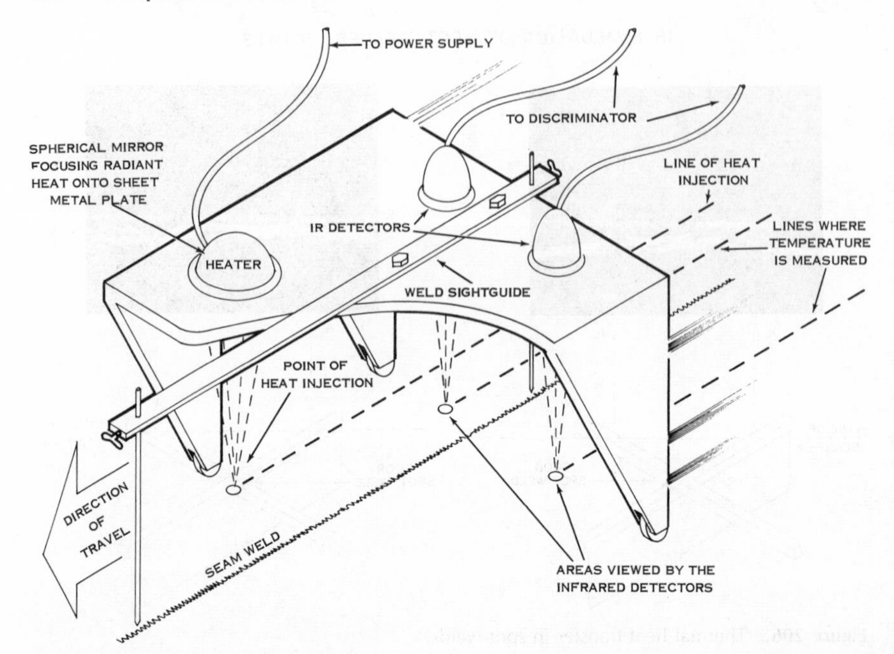

Figure 207. Seam weld evaluation.

two infrared detectors, whose output is being fed into a discriminator for display of the gradient between the two points.

The schematic drawing of Figure 207 shows how these elements could be mounted in a setup designed for this operation. Surface emissivity variations could be equalized by using one of the several coatings available for this purpose (see last section of Chapter 5).

4. *Friction-generated heat.* A typical application of this system can be found in the infrared inspection of tires. During the test, the tire is rotated under a static load. At the same time a fast scan infrared camera (Model Barnes T-101) develops a thermogram of the tire on a CRT screen. Simultaneous viewing of the front of the tire and of both sides of it is obtained with the help of two side mirrors, mounted at 45°, as shown in Figure 208.

In the illustration, an unbonded area between the carcass and the rubber tread is visible as a hot spot in the center of the frontal projection of the tire. The apparent stroboscopic effect is due to the different speed between the rotation of the tire and the camera's scan system. Therefore the hot spot appears repeated in the CRT presentation. When the tire is stopped, the hot spot can become a cold spot, if the heat developed in the layers of the carcass can not easily reach the outside surface through the unbonded area.

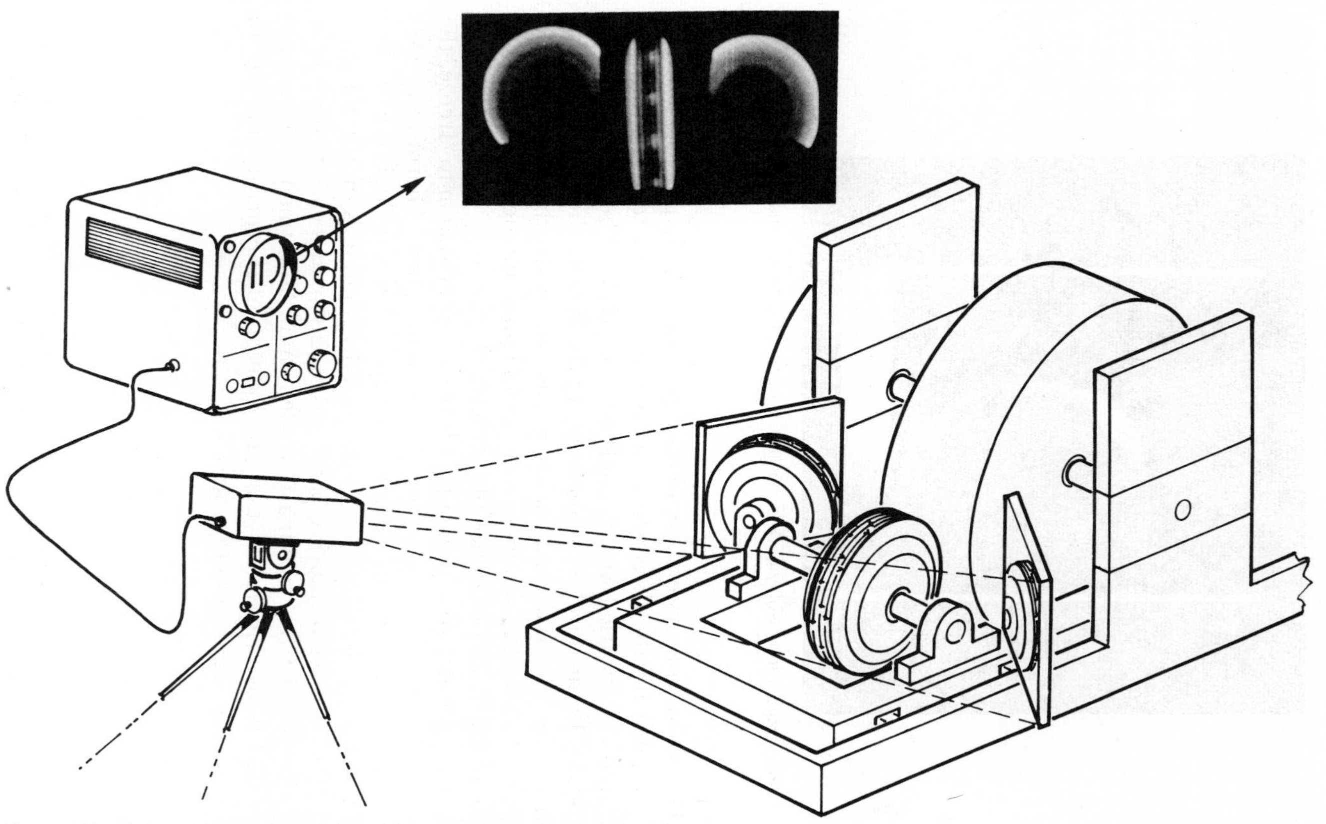

Figure 208. Infrared inspection of tires. (Courtesy Barnes Engineering Company.)

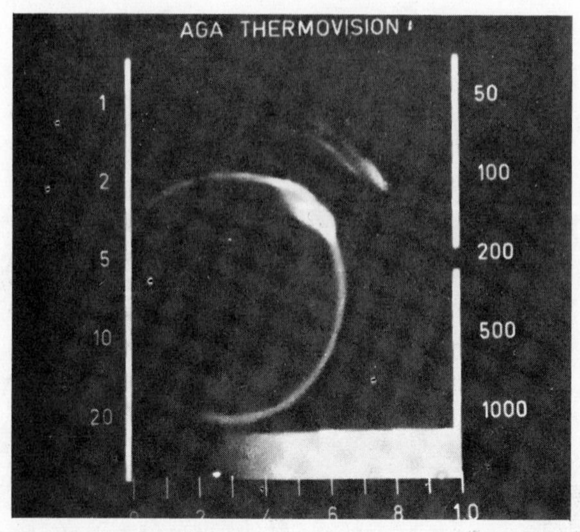

Figure 209. Spinvision display of defect in tire. (Courtesy Comstock and Wescott Company.)

The spinvision system mentioned previously (Chapter 3, p. 81) is also applied for the detection of hidden defects in tires. The fast scan infrared camera used in this application is AGA's Thermovision, and, according to the developers and users of the system, the static presentation of the infrared data makes it easy to locate the hidden fault. Figure 54 shows the system's setup, and Figure 209 the CRT display of a rotating tire with evidence of a defect.

Lately, Novatek, Inc. came up with an electronic system to "freeze" the infrared image of a rotating tire. The system, called "Thermscan," uses an AGA camera, and its operating setup is shown in Figure 210. The CRT display of the tire's surface thermal distribution is shown in Figure 211. The absence of moving parts makes this system easy to operate at any tire speed from stationary to 200 mph. Adjustment controls are provided to determine precisely the location of thermal events and to alter the tire segment being viewed.

LASER INJECTION SYSTEM

A novel approach for thermal target energization has recently become possible because of the availability of lasers of various types. It could be called "pulsed point heating," and it eliminates the difficulty due to lateral heat transfer. This difficulty is common to all four heat energization

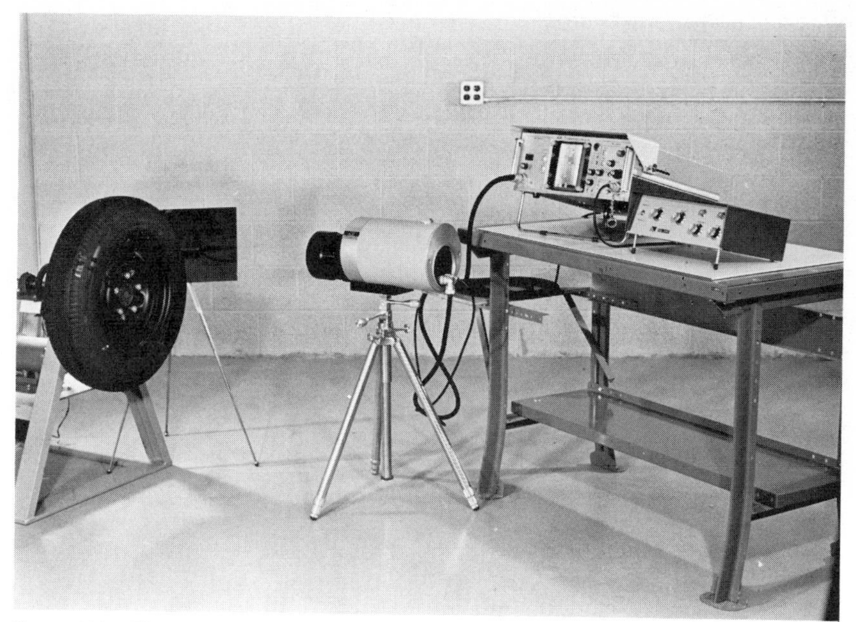

Figure 210. Thermoscan system setup. (Courtesy Novatek, Inc.)

systems just described, and it greatly reduces spatial resolution, often masking indications of anomalies that consequently remain undetected.

The use of a laser for thermal energization reduces to a small point the area where heat is injected. This point can be either stationary or moving along a preestablished trajectory on the target's surface. The laser beam is pulsed during the blanking of the detector to avoid reflected signal pickup by the latter, whose field of view can be either coincident with the point of heat injection, or following the heat injection point along its trajectory, with a time delay whose magnitude is dictated by the physical characteristics of the target and of the defects to be detected. In operation, the detector will monitor the speed at which the injected heat diffuses into the surrounding area. For instance, a void under the surface will so slow down the heat transfer toward the substrate that the temperature in that area will be higher than expected for a good bond condition. Comparing the detector pattern output with the "standard" pattern established by scanning "good" units of the same type will allow detection of physical anomalies of the material at or near the heat injection point.

The configuration of an infrared scanner incorporating a laser for implementation of the pulsed point injection system is shown in Figure 212. Its operation is as follows:

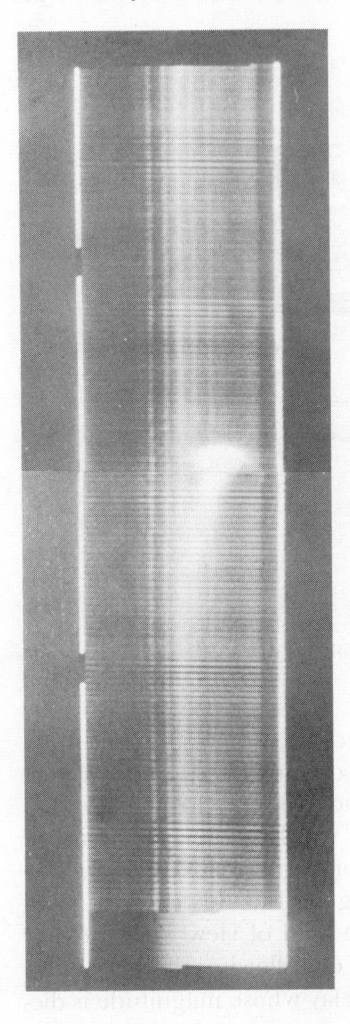

Figure 211. Thermoscan system CRT display. (Courtesy Novatek, Inc.)

To the left is the laser, shown firing a burst of energy during that portion of the spinning mirror's rotation that reflects the heating pulse onto a scan line of the target; during this period of time the infrared detector is blanked out, since its field of view is physically cut by the interposition of the back side of the spinning mirror. After adequate rotation of the mirror, the detector located to the right scans the same line on the target while the laser is turned off. Variation of the precession interval between heat injection and infrared scanning is made easy by the position adjust control that is partially shown in the illustration.

For special applications it might be desirable to reduce the laser precession to zero, that is, to have the point of heat injection coincident with the detector's field of view. The setup for this configuration is

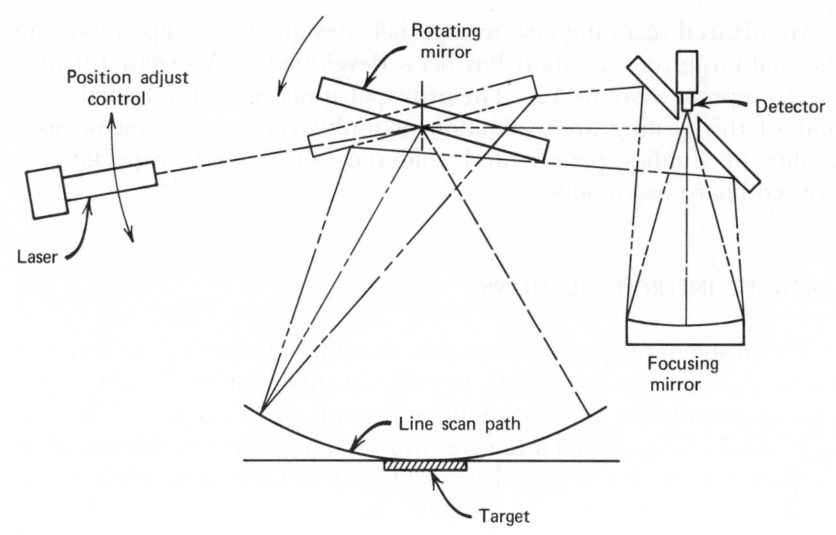

Figure 212. Scanning point laser injection system.

shown in Figure 213. To avoid laser radiation, reflected by the target, being picked up by the detector, out-of-phase operation is necessary. Thus the detector is blanked out when the laser fires, and it "looks" when the laser is off. Another solution is offered by the use of optical filters, to "blind" the detector at the wavelength at which the laser emits radiation.

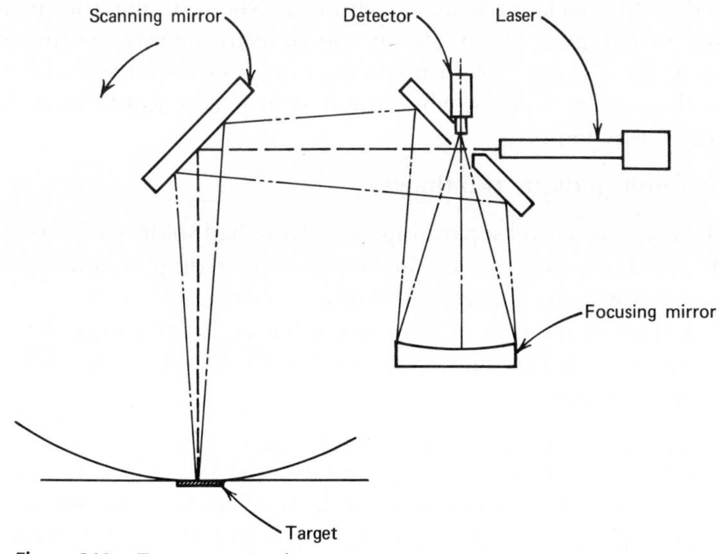

Figure 213. Zero precession laser injection system.

An infrared scanning system especially designed to accept a laser for thermal target energization has been developed by Vanzetti Infrared and Computer Systems, Inc. The principal applications forecasted at the time of this writing are evaluation of multilayer boards, test of bond quality of fluidics panels, and soldered connections inspection on printed board assemblies.

SOLDERED INTERCONNECTIONS

The problem of quality evaluation of soldered electrical connections has been solved thus far mainly by visual examination by trained inspectors. As usual when human judgment is involved, this type of evaluation lacks consistency and quantitative information. A survey conducted by the author in his former capacity of Quality Control Manager at Fisher Radio Corporation revealed the following:

1. When production runs especially good, the inspectors feel bound to reject anyway a modest percentage of the soldered joints coming off the line.
2. When line production runs especially poor, the inspectors feel bound to accept a limited percentage of the soldered joints coming off the line. Examination of the soldered joints accepted as good during such time shows their quality to be much lower than the quality of the soldered joints rejected under the first case.
3. A double check on inspector findings showed that the average inspector would miss about 4% of the defective joints existing in a "prepared" lot of units. A poor inspector would miss up to 8% of the existing faulty soldered joints, while the best inspector would miss about 2% of the faulty joints.

These results indicate two things:

1. The threshold level separating good from bad soldered joints is not a fixed value but is adjusted by the inspector according to, among other factors, the prevalent quality of the day's production.
2. A significant number of defective soldered joints escape detection by the inspectors and are allowed to reach the field where they will eventually cause failures.

This situation is due to the fact that human judgment is highly personal and lacks the element for a quantitative instead of merely a qualitative measurement. Furthermore, even the best visual inspection can only detect the defects extending onto the surface of the soldered connec-

Figure 214. "Hidden" defect in soldered connection.

tions, but certainly not those hidden under a surface of normal appearance, such as the one shown in Figure 214. This condition will be corrected only when the quality of soldered connections is measured throughout their whole thickness by test equipment and reduced to quantitative information and data.

The pulsed point heating system shows promise to solve this problem, with its capability to measure the speed of thermal decay at each point of the target.

Figure 215 shows in a schematic way how this system will work for a subassembly where the electronic components are held to the printed board by "folding" their leads. As the infrared detector follows the laser's heating spot moving along the scan raster pattern, it measures the amount of heat remaining at each and every point of the surface. When a soldered interconnection is good, the added heat sink action of the component will reduce the temperature of the lands immediately surrounding the interconnection, thus a drop will appear in the trace of the detector output. When the soldered interconnection is insufficient or missing, this drop will not be there, or it will not be as pronounced, as shown in the illustration.

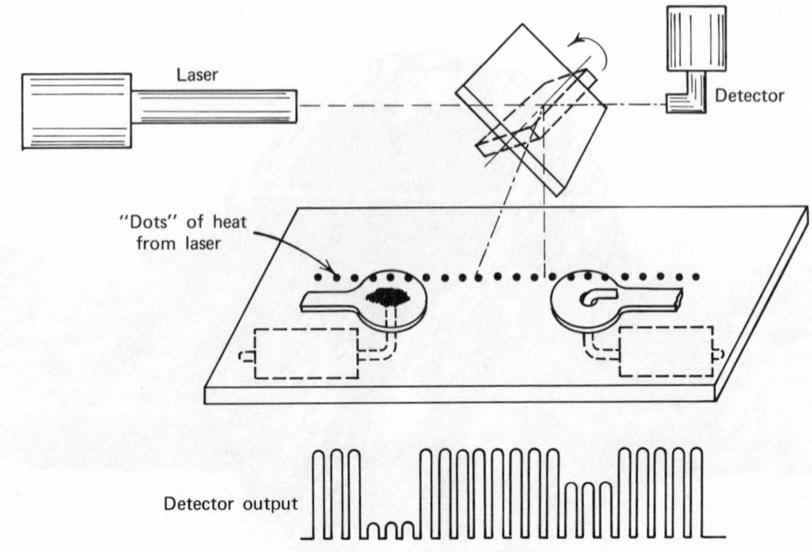

Figure 215. Unsoldered joint detection with laser injection system.

In this way, the quantitative measurement of the thermal gradient will indicate the quality of the soldered joint, and the detector output, duly processed and recorded, could be kept as the quality record of each solder joint.

Spot welds

Spot welds are usually made one at a time, either by the "opposite electrodes" configuration, or by the "parallel gap" systems (see Figure 216).

In both systems, the current flowing across the elements to be bonded must bring their temperature above their melting point, and for every type of spot weld, there is an optimum length of time during which this temperature must be kept constant to obtain a perfect weld. Conventional spot-welding equipment has adjustable settings for current intensity, length of application, and electrode pressure and shape; thus the best combination can be arranged for every type of weld. Unfortunately, tolerance variations of the various elements of the weld (metal dimensions and conductivity, surface condition, etc.) and of the equipment settings will affect the weld temperature and the time length at which the optimum temperature is kept. Thus the weld quality will vary within limits that depend on the magnitude of the tolerance variations of all the above-mentioned elements and their combinations.

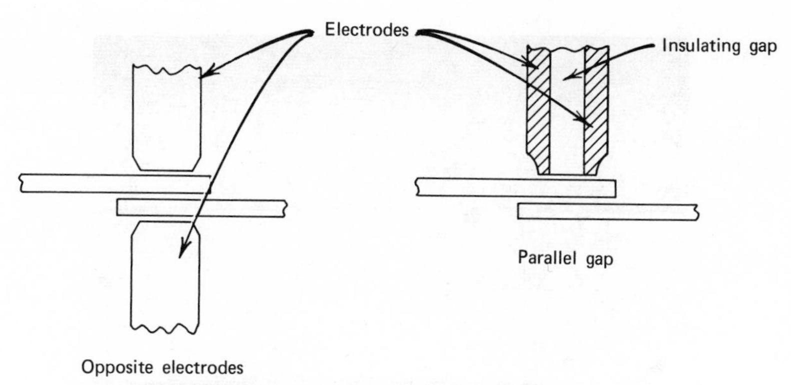

Figure 216. Spot welding systems.

Of course, there are tests and inspection. Tests are of a destructive type and can only be performed on production samples that can be disposed of. The so-called pull-test is the most common, and it is carried out by ripping apart the two welded elements and measuring the force (usually in pounds) needed to break the weld. Another destructive test is the micrographic examination of the welded area. This is done by cross-sectioning this area, polishing the surface of the cut, and examining or photographing it under adequate magnification. Figure 217 shows the photomicrographs of two spotwelds; *a* shows no defects, while *b* shows poor bonding at the interface.

Since destructive tests can only establish that the machine settings are adequate for the average weld, visual inspection must weed out whatever poor welds are produced by random variations of the several elements involved. Because it is a subjective judgment process, visual inspection cannot be 100% foolproof. In a large manufacturing program carried out by the Raytheon Company, it was estimated that the reliability index of the microwelds accepted was as high as 0.997.

This figure is quite impressive. Only three out of every 1000 welds are unreliable. This is almost perfection.

Or is it? These microwelds are made in large numbers. A single electronics system might have as many as 200,000 of them. At a 0.3% rate, it contains 600 unreliable welds. It appears that we must get closer to perfection if we want our electronic system to work trouble-free for a long time.

Infrared can help. The approach is to monitor the metal temperature while the weld is being made. If this can be achieved, we have the choice between rejecting those welds whose temperature falls outside the op-

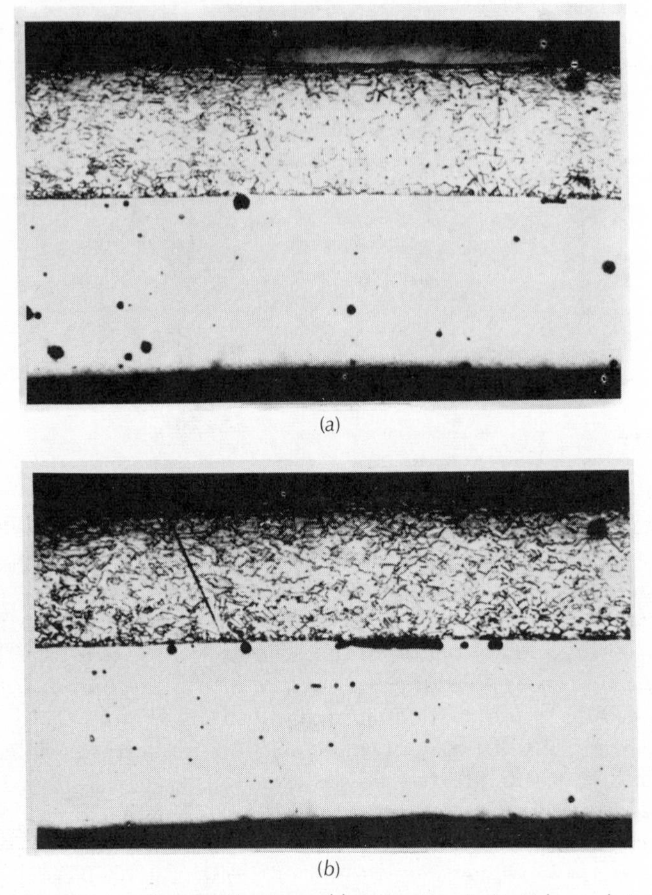

(a)

(b)

Figure 217. Good (*a*) and poor (*b*) microwelds. (Courtesy S. N. Bobo and A. Crowley.)

timum limits, or using a feedback loop to control the weld current to obtain a conforming weld each and every time.

The difficulty lies in the problem of viewing the weld while it is being made. Alas! The electrodes are in the way; they cover the whole weld area, so that no detector can observe the welding process until the electrodes are removed. And then it is too late.

The problem was solved with the use of optical fibers. These are light waveguides and inside them the light travels by total internal reflection, following their circonvolutions. Chapter 3, page 84 describes the principle and the applications of fiber optics to adequate extent. The equipment used for this purpose is described in the same Chapter.

To monitor the welding process, a thin optical fiber bundle is inserted in the small gap located between the two electrodes of a "parallel gap" welder, or is brought down along side one electrode of an "opposite electrodes" welder. In the first case, the front end of the fiber touches the center of the area where the weld is going to be located. In the second case, the front end of the fiber is aimed at a point immediately adjacent to the weld area. In both instances the output end of the fiber faces an infrared detector that in this manner "looks" at the target without having to be physically near it. Figure 218 shows the assembly setup for the two conditions mentioned above.

In view of the fact that the temperature of molten steel is around 1200°C, the peak of the radiation emitted is located in the near infrared, its energy content is quite large, and it can be detected by an uncooled photodetector such as a Silicon or a PbS cell. The detector output is processed through a preamplifier and then displayed onto an oscilloscope, along with a second trace depicting the pulse of weld current. Figure 219 shows these traces in an application[14] to a parallel gap welding process: we can see that for the same parameters of the weld pulse (12 msec and 0.66 V) the peak of the detector output reaches full scale for the good weld (*a*), but hardly rises from the ground level set as the lower sensitivity threshold of the system for the poor weld (*b*).

The two welds shown here are the ones whose photomicrographs appear in Figure 217. It is clear that the infrared system allows a degree of quality resolution that is at least one order of magnitude better than the level attainable by conventional techniques. For instance, in this case weld *b* had been judged of borderline quality by the inspector. However, its infrared pattern is definitely below the lower control limit level, and no doubt is left that it is a defective weld, as subsequently verified by micrographic examination.

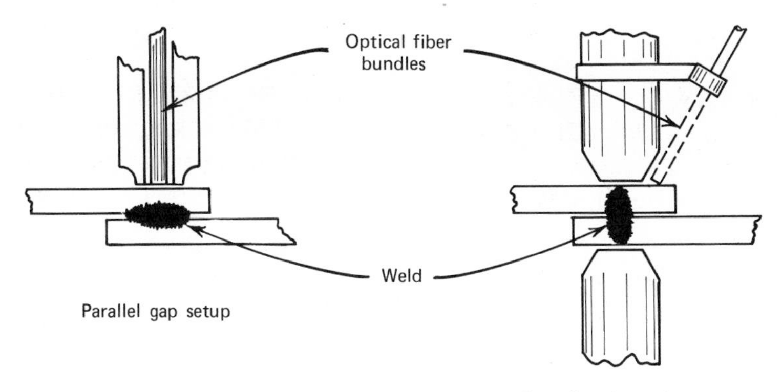

Figure 218. Infrared fiber setup for weld monitoring.

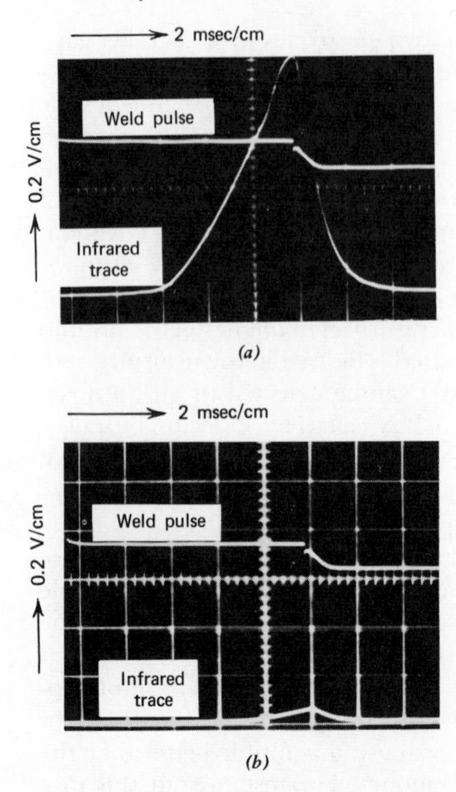

Figure 219. Oscilloscope traces of good (*a*) and poor (*b*) welds. (Courtesy S. N. Bobo and A. Crowley.)

In the case of opposite electrodes operation, the consistent location of the spot at which the optical fiber is aimed is of critical importance. Once this is assured, upper and lower control limits for the temperature of such spot can be established, as a function of the desired temperature of the weld area. In this way the quality of spot welds can be monitored during the welding process.

However, the capability to measure the weld temperature in real time offers the greatest advantage: namely, the ability to control, through a feedback loop, the heating of the weld region to the optimum level each and every time. In this way, no defective welds can be produced.

Electrically Deposited Metals

The same approach described for the evaluation of adhesion bonds is valid for testing the quality of the bonding of electrically deposited metals. Besides metal plating on metal, electroplating is widely used in the production of printed boards for electronic assemblies. A serious problem, however, is the detection of "delaminations," that is, the separation of portions of the deposited metal interconnections (usually called "lands") from the isolating substrate.

Conventional techniques call for destructive "pull tests" and visual inspection. The first give only a statistical rating of the board's quality, while the latter is faced with the almost impossible task of detecting delaminations where the layer separation is in the order of 0.001 in. or less.

With infrared, it is easier. Figure 220 shows how a delaminated land was detected by Lockheed Q. C. engineers on a printed board that had been brought back for routine maintenance check-up. After the unit had been electrically energized long enough to bring it to thermal equilibrium, a thermograph was taken with a Barnes T-4 Infrared Camera. This thermograph shows the land surface of the board, that is, the surface opposite to that where the components are mounted. The infrared radiation of these components is still visible, since the board is translucent to it, but this is not all. Robert E. Horne, the Lockheed-Georgia Company engineer who performed the infrared evaluation, so describes his findings as follows:[15]

In analysing the thermograph it can be seen that in some cases the printed circuit lines change from light to dark. A visual inspection showed that the circuit

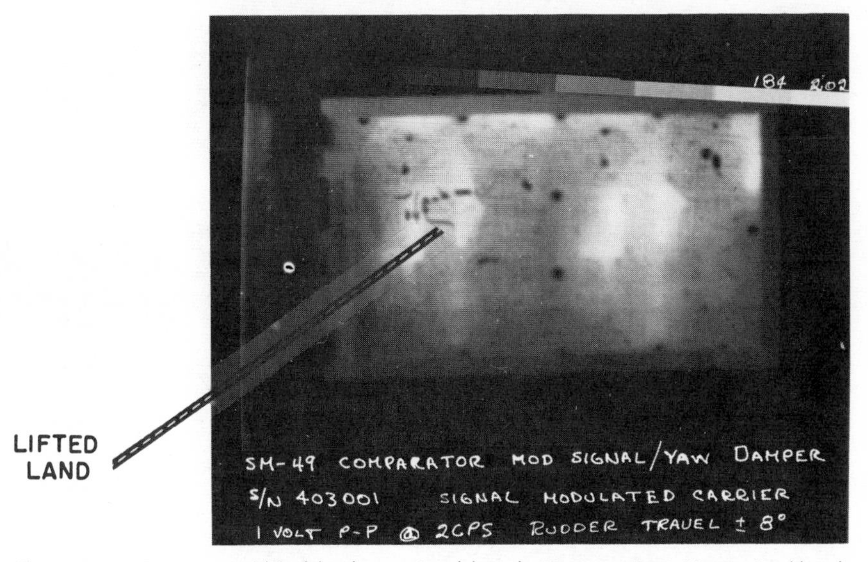

Figure 220. Detection of lifted land on printed board. (Courtesy R. S. Horne, Lockheed-Georgia Company.)

had separated from its glass laminate supporting backing. The cold flow solder circuit actually separated from this backing as a result of heat dissipated by components on the front of the board. Mismatch in the expansion coefficient of the two materials probably accounted for this separation. In any event, the connection is now unsupported and as such is susceptible to the customary aircraft shock and vibration which could result in an open circuit. The flow lines are not hot due to current carried but due to heat conduction from adjacent electronic components.

Multilayer Printed Boards

Multilayer printed boards, imaginatively described as "two-dimensional cable harnesses," are the interconnecting elements for the increasing complexity of ever-shrinking microminiaturized and microelectronic assemblies and systems.

Multilayer boards, as their name tells us, are laminar assemblies of printed circuits stacked and cemented together into compact, planar shape. Electrical connections between the "lands" of the different layers are made by means of plated-through holes, sometimes called "barrels," whose number, always large, is dictated by the number of interconnections and outside connections required.

Figure 221 shows a multilayer board, made by Hazeltine, having three

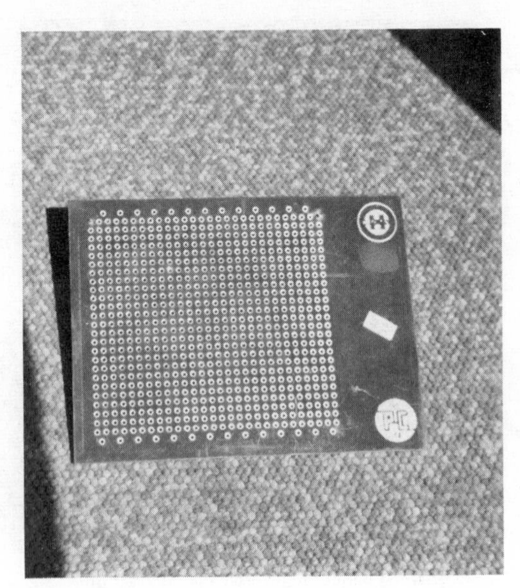

Figure 221. Multi-layer printed board.

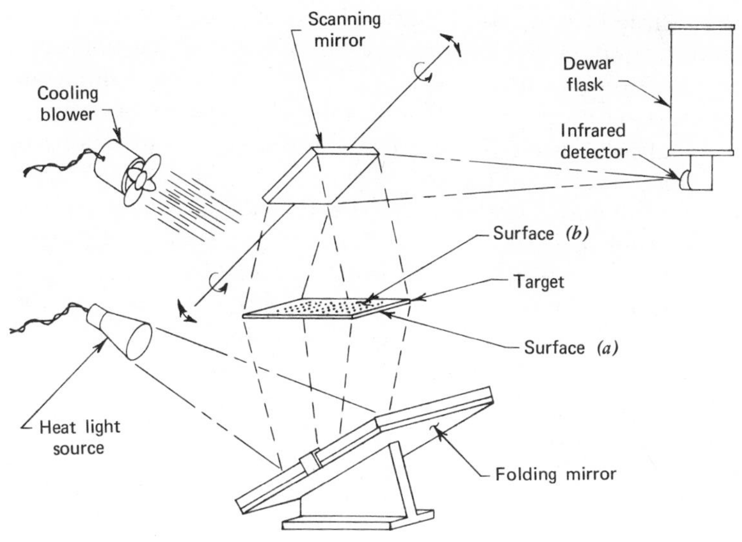

Figure 222. Multi-layer board quality evaluation.

layers of printed circuitry, a ground layer, and a total of 727 plated-through holes.

It also has problems—more problems than layers or holes, since each of these must make good electrical contact with the desired lands and avoid contact with the other lands or with the all-covering ground or shield layer.

Here is a list of the problem areas where infrared techniques have shown promise of bringing help:

1. Quality of the bond between land and substrates.
2. Quality of the bond between layers.
3. Reduced cross-section of conductive paths.
4. Reduced cross-section of plated-through holes.
5. Points of unwanted electrical leakage.

Bonding between layers is best checked with the heat flood method (see Figure 222), using a separate heat source to flood one side of the board (*a*), and a blower to cool the opposite side (*b*). A bond discontinuity will reduce the heat flow from surface *a* to surface *b* in the affected area, and will appear cooler than its surroundings to an infrared detector scanning surface *b*. There is a limit, however, to the resolving power of this method, and this limit is a function of the size of the discontinuity, its depth below the surface, and the thermal conductivity of the interposed material.

The localized heat injection method has been used to check the quality of plated-through holes. The principle used in this application (shown in Figure 223) is that the heat-flow rate between two points is limited by the size of the narrowest cross section along the heat propagation path.

As already discussed for the case of soldered connections in the thermal diagram of Figure 205 in page 258, the temperature reading, taken at point P_2, is a function of the wall thickness of the element connecting element A with element B, in our case a plated-through hole; as such, it will yield information about its quality. In the case of inadequate wall thickness of the metal, the temperature reading at point P_2 will be lower than expected, and the warm-up rate will be slower. Dynamic testing can then be used instead of static testing.

Electrical energization from within has been used to detect reduced cross-section of conductors, the location of unwanted shorts, and leakage paths. By feeding high enough levels of current, these defects will appear as hot spots that can be identified in location and magnitude by infrared scanning.

For instance, an unwanted electrical leakage path can be disclosed by running a current between the two points that should be isolated from each other, while the surface of the board is being scanned. Soon the point where the short is located will stand out as a hot spot, thus permit-

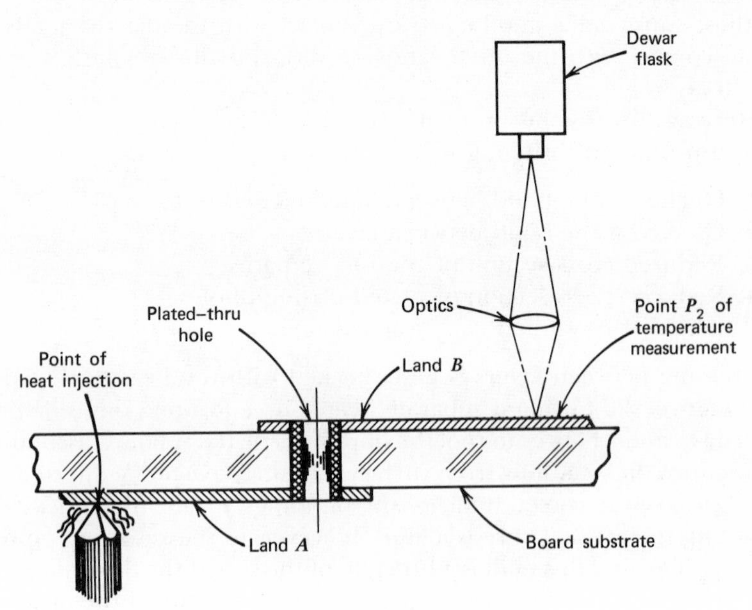

Figure 223. Quality evaluation of plated-through holes.

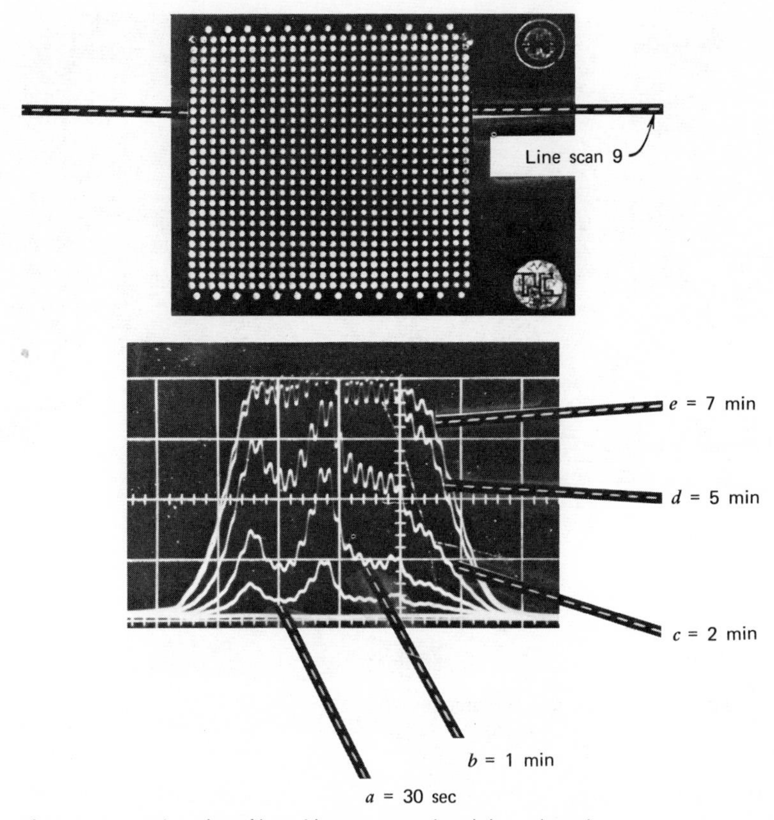

Figure 224. Infrared profiles of line scan 9 of multilayer board.

ting repair and salvage of the board. Figure 224 shows the infrared profiles of line 9 of the board shown in Figure 221. As indicated by the increasing height of the lines, these profiles represent the warmup process of the board surface along line 9. Profile *a* was taken $\frac{1}{2}$-min after electrical energization of the leakage path; profile *b* after 1-min, (*c*) after 2 min, (*d*) after 5 min, and (*e*) after 7 min. It is clear that there is an optimum point in time, when the leakage points stand out conspicuously. Before this time, most of the heat is still below the surface, while beyond this time lateral heat transfer fills the thermal voids and reduces contrast until the picture becomes flat and loses significance.

Of course, this optimum time length depends on the depth at which the leakage points are located, and therefore varies from defect to defect. Fast, repetitive scanning of the boards, and close monitoring of

Figure 225. Infrared evaluation of large multilayer board.

the infrared profiles generated on the oscilloscope is required for best results.

In the case of Figure 224 it appears that the best contrast is available approximately 2 min after energization. The infrared profile also indicates that there are three distinct points of leakage, the major one in correspondence of hole 12, a smaller one at hole 5, and a third one, even smaller, at hole 19.

Figure 225 shows how a large multilayer board (19 layers) was scanned to find the location of a short between a ground layer and a shield layer. This short, which could have developed at any one of the 2800 feedthrough barrels, was quickly pinpointed by the hot spot appearing soon after electrical energization.

CASTINGS

Recent work performed on jet turbine blades and vanes[16] has disclosed the practicality of using fast scan infrared techiques to detect such

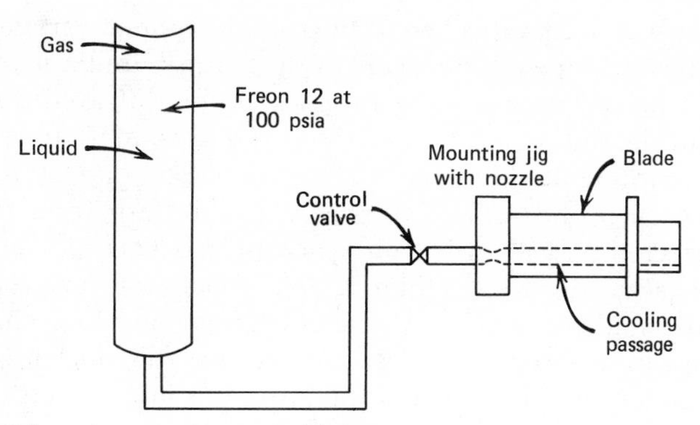

Figure 226. Schematic setup of jet engine blade test. (Courtesy W. T. Lawrence, Novatek, Inc.)

defects as occlusions of internal cooling passages and out of specs wall thickness. In practice, a cooling fluid is circulated inside the casting while a thermogram is made of the outside surface. Liquid freon 12 (having its boiling point at −21.6°F) is blown through the blade's cavities, as schematically shown in Figure 226. By controlling the nozzle dimensions it is possible to optimize the Freon expansion into gaseous state, so that the cooling effect will be localized in correspondence of the critical items to be checked. Two mirrors placed at a 45° angle at the two sides of the blade allow simultaneous viewing of its whole surface, as schematically shown in Figure 227.

Key to the success of this process is its speed: the temperature variations reach the outside surface of the blade in just a few milliseconds and

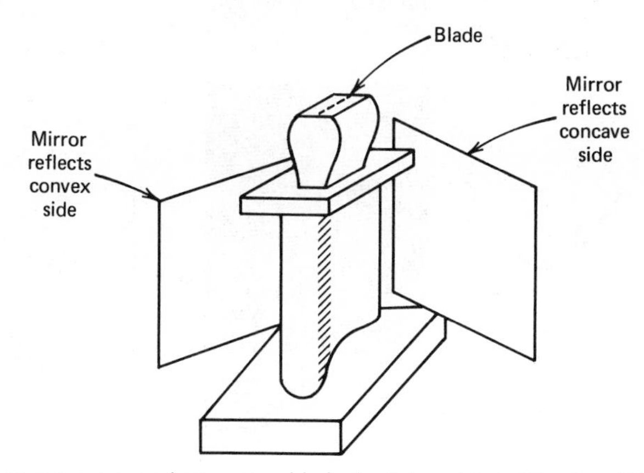

Figure 227. Viewing setup of jet engine blade for infrared test. (Courtesy W. T. Lawrence, Novatek, Inc.)

unless the whole target (blade and mirrors) is scanned very fast, lateral heat transfer will wipe out the detail of the features under investigation.

This is achieved by using the fastest scanning infrared system currently available: the AGA Thermovision, described in Chapter 3, under "surface Scanning Radiometer." The rapid scan rate of the system allows to complete one frame in 40 msec, so that the fast thermal transients can be detected and recorded on synchronized Polaroid pictures. By using a "negative" display (bright for cold and black for warm) and adjusting the threshold between black and white, it is possible to eliminate all information except for those areas where the wall thickness is below the design tolerance. These areas will appear as bright spots on a black background.

Figure 228 shows test results for two blades, together with a plot of the wall thickness for both the concave and convex sides of the blade. In the first unit, the convex side of the blade (left-hand side) is less than 0.017 in. thick, so a white area appears in the top line. This area grows larger

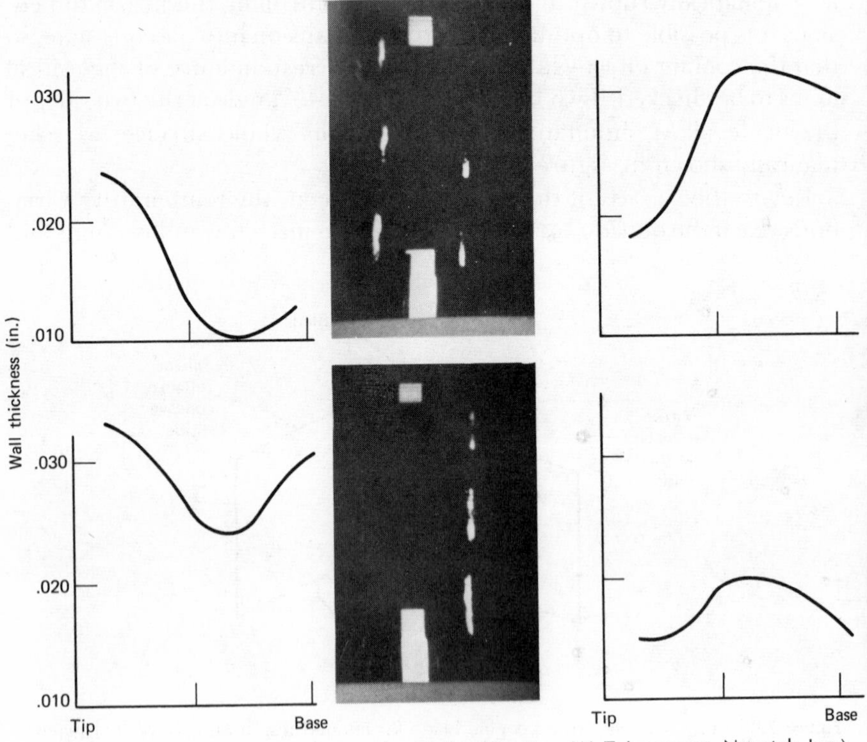

Figure 228. Wall thickness test of jet engine blades. (Courtesy W. T. Lawrence, Novatek, Inc.)

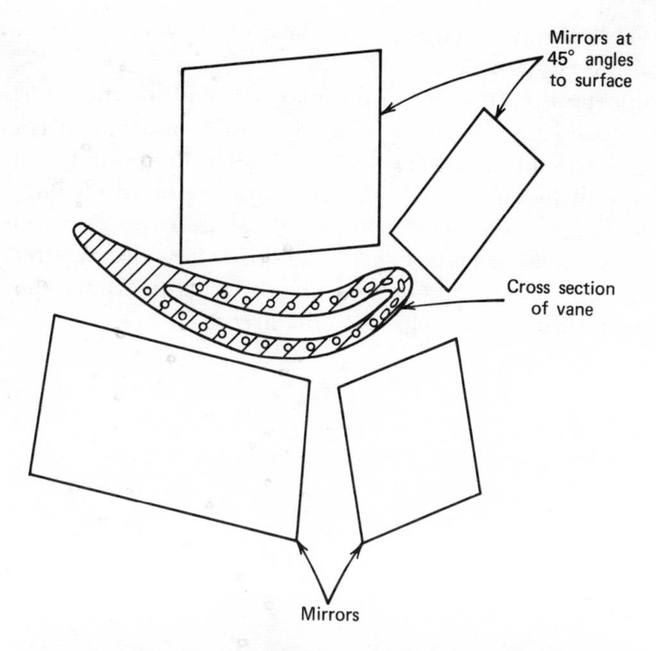

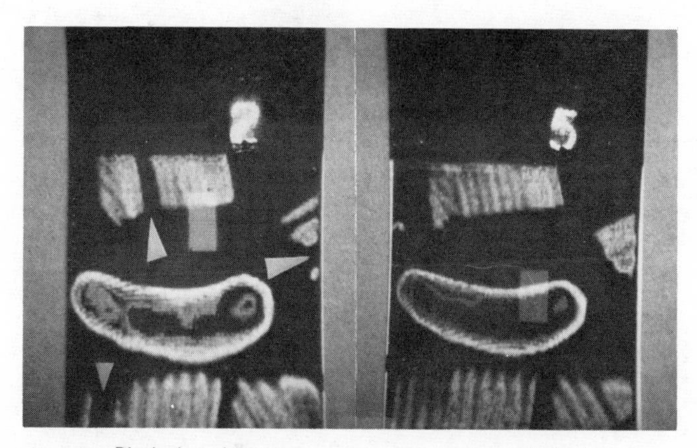

Figure 229. Infrared detection of blocked cooling channels in jet engine blades. (Courtesy W. T. Lawrence.)

in the lower mirrors. The concave side has a thin spot whose thickness is 0.018 in. It shows up first in the second mirror.

The second unit has a convex side whose wall thickness is everywhere greater than 0.021 in. Therefore, nothing is seen in the left-hand portion of the photograph. The concave side has a thickness less than

0.017 in. at both ends. These areas show up as white spots in the first mirror on the right.

One important feature of this test procedure is that the interpretation of the results does not require an analysis of grey areas. It is necessary only to look for the absence or presence of a temperature indication.

Besides wall thickness measurement, obstruction of cooling passages can be detected in the same way, with less exacting demands on the timing of the whole process. Figure 229 shows how clearly three blocked channels stand out in the CRT display of a setup where the blade is viewed head-on with the help of four mirrors.

Chapter 8 Fields of Implementation of Infrared Techniques for Electronics

Design and development and manufacture and maintenance are the principal technical phases along the life of a program connected with modern electromechanical hardware. Each of these phases has its problems, which must be solved before the program could be called successful.

In particular, design should be flawless, components failure-free, assembly faultless, and maintenance guided from verified facts instead of ESP.

The use of infrared techniques can help to come closer to this goal. Figure 230 shows in what areas infrared measurements can be usefully applied to achieve better design, higher reliability, more thorough quality control, and sensible field maintenance. The rest of this chapter discusses every area in detail, but first it is appropriate to expound the basic concept common to all these applications: "infrared signature."

INFRARED SIGNATURE

Every physical object, from the simplest to the most complex, from the smallest to the largest, has a visible shape and its surface exhibits visible colors. A color picture of it will unequivocally identify it; the only limitation is the impossibility of differentiating between identical copies of the object.

The same is true with infrared. Instead of visible radiation, infrared radiation is the parameter being monitored. As long as the object exhibits a different radiation level than the background, its "picture," "signature" or "pattern," or "profile" will define its surface thermal distribution, as affected by its surface emissivity, and will be consistently the same for identical thermal conditions.

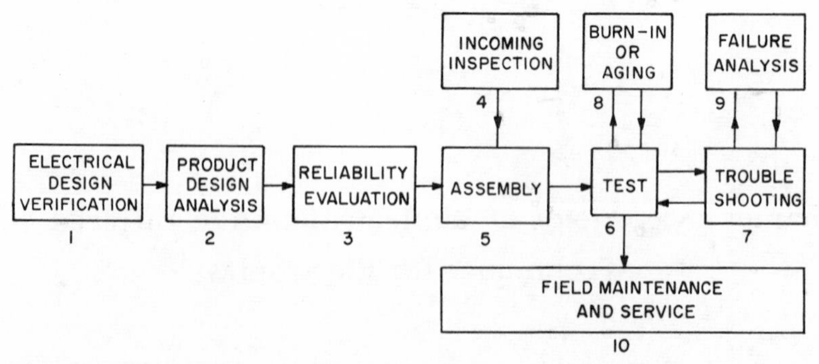

Figure 230. Areas of infrared techniques applications for electronics.

It makes no difference whether the thermal regime of the target is caused by heat injection from outside or by internal sources such as power dissipating elements. As long as the physical structure and the emissivity of the surface are constant, the infrared information emitted by the target will be consistent with its thermal distribution. This information, detected and displayed or recorded by infrared sensing equipment, is called the infrared picture, pattern, or profile or signature of a certain target. In a sense, it is an "infrared fingerprint" that can be matched only by identical units operating at identical thermal regime. Every deviation from it indicates a variation of some physical or thermal characteristics of the target.

In the case of electrically energized electronic assemblies, this concept of infrared signature can be readily put to use for failure identification purposes. When the assembly is operating correctly, in the desired mode, every component carries a certain amount of current and dissipates a certain amount of power, thus assuming a well-defined operating temperature that becomes stabilized at the end of the initial warm-up period. At this point, its infrared signature can be recorded and used as a standard, typical of that assembly during proper operating conditions. All identical units operating in the same mode will exhibit an infrared signature that matches the standard.

Now let us suppose that a failure occurs. One or more components will change their impedance, and the electrical current flow through the failed components will vary accordingly. Other elements of the circuitry might be affected too, so that the overall distribution of electric current through the circuitry will be different from the "sound" condition.

However, the current distribution related to the failure condition will be consistently the same every time that failure occurs on boards of the same type. After all, what is a failure? Just another mode of operation of an electronic assembly. An unwanted mode, for sure, but one that is typical of that failure, and one that can readily be identified by a unique infrared signature.

This is true for every failure. For every electronic assembly, each failure mode has its very own infrared signature that reflects the thermal regime of every component of the assembly. It could be called the "infrared standard" of the failure.

Besides identifying beyond doubt the operating mode of an assembly, the infrared signature yields a very valuable by-product—the measurement of the operating temperature of every component. As already mentioned, this measurement can be correlated with electrical power dissipation and with failure ratio, thus supplying design engineers and reliability engineers with all the data needed for a full evaluation of the design features and for reliable life expectancy calculations.

Furthermore, infrared signatures of failure modes will point out the secondary overstresses created by the primary failures. Generally ignored so far, these secondary overstresses can be the cause of early field failures otherwise difficult to explain.

Finally, infrared signatures can be the key to a novel maintenance approach: the infrared signature of a unit taken after a certain length of operating time can be compared with the original signature of that same unit, recorded the day it left the factory. Any deteriorating trend will stand out clearly, thus indicating when a replacement is needed.

Therefore, I would like to describe how the intelligent use of the infrared signature can be of substantial help in the following areas of application:

1. Electrical design verification.
2. Product design evaluation.
3. Component stress analysis.
4. Reliability calculation.
5. Inspection, test, and trouble-shooting.
6. Detection of hidden failure conditions.
7. Detection of secondary overstresses.
8. Novel maintenance policy.

ELECTRICAL DESIGN VERIFICATION

When I was a student (this was several wars ago) and first came in contact with the slide rule, I was offered a definition and a warning. The definition went: "The slide rule is a tool for which two multiplied by two makes . . . well . . . very close to four." The warning was to the effect that the slide rule knows no decimal point, whose location must be established by the operator.

Ever since I was a student, I had trouble in placing the decimal point. Oh, not serious trouble: I seldom miss its location by more than one digit, plus or minus. However, it annoys me quite a bit not to know whether the balance of my mortgage is $2000, or 20,000 or 200,000. Oh, well, it can't be 2 million, anyway.

However, am I glad that I am not a design engineer! I would never know whether a certain load resistor is supposed to dissipate 0.2, or 2, or 20 W. As by magic, design engineers always know where the decimal point belongs. Or, rather, *almost* always. Only seldom in my career of quality assurance engineer have I discovered a design mistake of a full decimal point.

Infrared scanning, of course, can pick out decimal-point design mistakes in no time flat. All that is required is an operating breadboard of the circuit and an infrared scanner. Not only will decimal-point mistakes be as conspicuous as sore thumbs but also smaller overstresses will stand out, disclosing conditions that may have been overlooked in the design notes.

Figure 231 is an example of engineering design verification carried out on a recently developed clamper amplifier unit. For every component, the infrared radiation level corresponding to its operation under full rated load was calculated and plotted as a white vertical bar. Then, next to it, the radiation level corresponding to the calculated derated stress was plotted as a black vertical bar. It can be seen that the designer used safe derating ratios; he did not want to take any chances. Finally, in the dashed vertical bars, the actual radiation levels of each component are shown, as measured by an infrared radiometer.

In examining the chart, a few interesting facts are observed. First, every component works at a thermal level higher than calculated. Second, one component, R49, is operating at a temperature above the level corresponding to its full rated load. Finally, two critical components, CR-1 and CR-2, are stressed at 65% of their full rating, while reliability considerations had induced the designer to use only 5% of the full rating. In view of the fact that the infrared measurements were carried out on an assembly operating in free air whose temperature was 21°C, the higher temperature measured on all components cannot be attributed to ambient conditions. On the contrary, in practical use the unit is enclosed inside a chassis closed with a metallic cover, so that the air inside this enclosure will be warmer than the outside air. This condition will just make the situation worse.

How much worse? Two approaches are available—the first is approximate, the second much more precise.

In the first approach, we must assume that an increase in temperature will not appreciably change the electrical characteristics of the components. This is not true, but we will assume it in first approximation, leaving the corrections for a later recalculation. In this hypothesis, an increase of 1° in the ambient temperature will cause an increase of 1° in the temperature of every component. In other words, the temperature profile "floats" above a baseline that corresponds to the temperature of the ambient.

When the temperature difference between inside and outside the enclosure is limited to a few degrees, we might be satisfied with this approximation. But when the difference is large, corrections will be necessary, due to the fact that semiconductors, for instance, are nonlinear devices whose electrical characteristics are heavily affected by temperature variations.

Under these conditions, the more precise infrared approach is preferable. The assembly is placed in an enclosure made totally or partially of infrared-transparent material. Scanning takes place after the components and the air inside the enclosure have reached thermal equilibrium. In this way, realistic operating conditions are simulated, and the infrared readings are indicative of the true temperature of the components when the unit is in actual use.

Now, these are the temperatures that the design engineer needs to know to verify the soundness of his calculations. If these are confirmed, he will be proud of himself. If they do not, he still has time to introduce design changes before the unit is released to production.

When this takes place, he knows that his unit is as good as it can be designed and is free from hidden weaknesses. He will be proud again.

PRODUCT DESIGN EVALUATION

After the electrical design engineer is through with his task, he turns the whole project to the product designer. In practice, he gives him a few basic dimensions, a schematic, and a list or a handful of components. From then on these two gentlemen never talk to each other again. The product designer will try to squeeze all those parts and the wiring onto a board that always looks too small, especially in view of the inflexible esthetic rules that must be observed, such as components must be either parallel or at right angle with each other; similar components must look

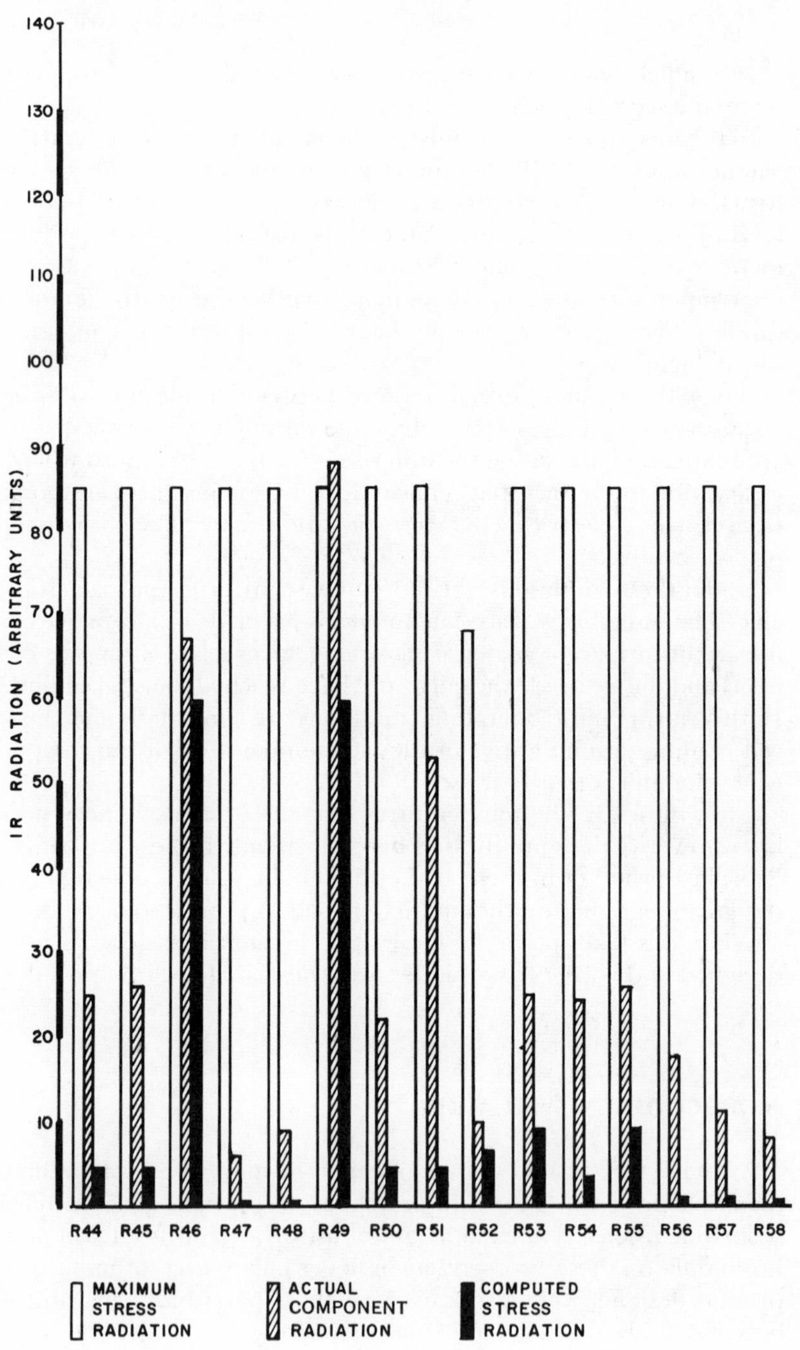

Figure 231. Engineering design evaluation of clamper amplifier.

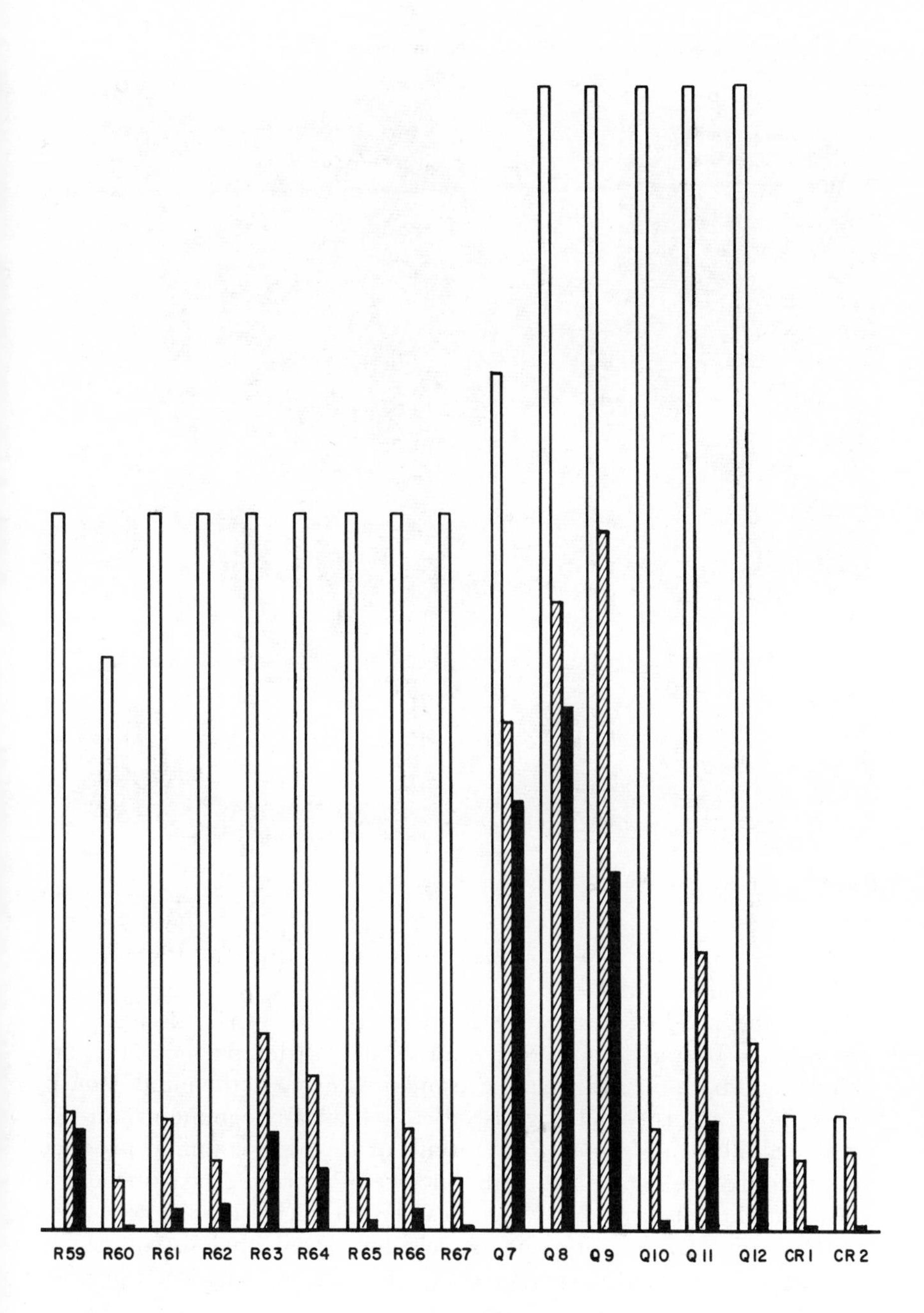
R59
R60
R61
R62
R63
R64
R65
R66
R67
Q7
Q8
Q9
Q10
Q11
Q12
CR1
CR2

Figure 232. Product design evaluation of power amplifier board.

nicer when grouped together; heat sinks can be dispensed with if the room gets scarce, and such.

When the product designer is finished with his work, the assembly usually looks beautiful. It gives an impression of order, balance, and harmony that is both pleasing to the eye and reassuring to the mind. How it works, however, is something else. Better than any comments, an example will illustrate some of the pitfalls of this very common product design approach. Figure 232 is the picture of a printed board where a power amplifier is assembled. Electrical design calculations showed adequate derating for all components; reliability stress analysis confirmed the absence of overstresses; thermal analysis did not disclose the existence of any part working at excessive temperature. Nevertheless, after a few hours of operation, the fiberglass material of which the printed board is made was turning brown. A quick infrared scan revealed temperatures exceeding 130°C in two areas of the assembly; in the center, where the heavy wattage resistors are tightly grouped, and in correspondence to the Zener diodes, whose heat sink was inadequate. Product design changes were introduced as a result, and an infrared check subsequently carried out confirmed the disappearance of the hot spots.

Before closing this section, I would like to mention that unsound thermal conditions are often not so blatantly conspicuous as illustrated in the previous example. One not infrequent occurrence is, for instance, the case of a thermally sensitive component (a diode, a transistor) wired directly to a heat dissipating element (a high wattage resistor) and unwittingly acting as a heat sink for it. Even with its temperature below that of the resistor, its electrical performance will be affected and its stress level may become unsafe.

Another item to beware of are the snap-on heat sinks designed to cool semiconductors. They usually have fins, and these are made black to raise the emissivity factor. But let us not forget Kirchhoff's law: a good radiator is a good absorber. Thus if the heat sink is located in an area near strong sources of infrared radiation, it might absorb heat instead of emitting it, thus contributing to increase the temperature of that very device that it was supposed to cool.

And finally, there is value engineering. As new products are being developed, chance is that some day a new component, a capacitor for, instance, will become available, smaller, cheaper, and electrically identical to a bulky capacitor until then used in a certain assembly. Who would not jump at the opportunity? So a change is implemented, and the man who made the suggestion gets his $10.00 award, and after a while field failures start piling up on that assembly. It takes quite a while to figure out the whole story.

It is really quite simple. The old bulky capacitor was acting as the heat sink of a hot resistor. Now that it is gone, the heat floods the closest semiconductors, which become thermally and electrically overstressed.

The morale of the story is that every time a change is made, one ought to perform a quick infrared check. Sometimes one will be surprised by the findings and may avoid a lot of field trouble.

COMPONENT STRESS ANALYSIS

On October 31, 1966, at the 26th National Conference of the Society for Nondestructive Testing, Kenneth E. Appley of the Raytheon Company presented a paper entitled: "Infrared Replaces Calculations for Realistic Stress Analysis of Electronic Modules."

Since the subject is totally pertinent to the contents of this chapter, I would like to quote here the first three paragraphs of this paper:

A technique used to enhance the reliability of electronic equipment is to reduce the stress under which the individual component parts are operated. Stress in electronic components is primarily a function of voltage, current, and temperature, but the majority of components fail due to overheating.

The method usually employed by design and reliability engineers in determining the operating stress levels of electronic parts is by calculating the internal operating temperature of parts, from given or computed ambient temperatures, and calculated operating currents and power. But these calculations can be grossly inaccurate, primarily because the precise degree of heat conduction, radiation, and convection of the part is often too difficult to determine or too time-consuming to calculate, and is, therefore, usually ignored in stress calculations. Of course, thermocouples could be and often are used to measure surface temperatures of parts. But in many applications, such as in high-voltage equipment, the use of thermocouples is either hazardous or impossible; and even where the use of thermocouples is feasible, the method has many limitations. Determining the optimum location for placing thermocouples is often extremely difficult. The more serious hot spots are often missed; at best, it offers spot measurements. It is also a slow, time-consuming process. For these and other reasons, the use of thermocouples is often limited to investigating known thermal problems.

Infrared scanning for detecting the precise surface temperature of electronic parts under actual operating conditions has none of the limitations mentioned for thermocouples. It can be used safely on high-voltage equipment. It is quick, accurate, and does not miss hot-spots. For these reasons, and primarily because of its accuracy in measuring actual surface temperatures, it is a powerful tool for stress analysis and provides accurate thermal indices for calculating part stresses in Reliability Engineering work.

What else remains for me to say? Perhaps just add another quote from the same paper:

It is estimated that analysis of the circuits tested by conventional calculations would require an average of four hours, with some taking up to eight hour to complete. . . . The infrared technique is economically competitive with calculated stress analysis at this time, but further reductions in test time can be achieved.

Further reductions have been achieved since then. At the present time, scan times of just a few seconds are met by several infrared scanners, some of which are capable of yielding a picture, a thermal map, or a printout sheet in real time.

For the benefit of the conservatives, who still prefer to cling to the conventional time-tested stress analysis system, I would like to recommend a paper[17] published by Gordon Cawood, Reliability Manager, Raytheon Company. In this paper, the author states that out of the first 30 different modules picked at random for infrared design evaluation in a large military communications program, 32 overstressed components were detected during a preliminary superficial examination of their infrared signatures.

RELIABILITY CALCULATION

As discussed in the first chapter of Part II, failure ratio charts based on component surface temperature make much more sense than those based on ambient temperature. At the present time NASA is planning to fund the development of these new charts, so that hopefully they will soon become available.

The infrared signature provides the reliability engineer with all the information necessary to obtain realistic failure ratio figures for every component, and consequently for whole assemblies. The availability of the new charts will make this task easy and fast, so that reliability calculations will require very little time expenditure, while the results will be much closer to reality.

Elements required to carry out this simplified, more exact reliability calculation system are: a sample of the electronics assembly to be evaluated; its infrared signature and the new failure ratio charts.

In the meantime, implementation of this new approach is feasible in first approximation by using the presently available failure-ratio charts. The only difference will be substituting in place of the "ambient temperature" parameter the value of the temperature of the thermal region around each of the critical components. Measurement of the temperature in these areas can be made by reading the infrared radiation of "passive" elements (i.e., elements not electrically energized or not dissipating electrical power) located in said areas, or by temporarily installing there small blackbodies that will acquire the temperature of the ambient at that particular location. This procedure might sound rather complex, but it is the best that can be done for the time being.

INSPECTION

"Inspect" as a verb means "to view closely and critically. *Examine.*"* The term seems appropriate to define what an infrared detector does: it actually views without physical contact, whatever target is in its field of view (just as the human eye does), with one added parameter, *temperature,*

* From Merriam-Webster Dictionary. June 1968 edition.

which the human eye cannot see or measure. According to the design of its optics and to the processing of its output signal, the infrared detector will view the target very closely and very critically, even pointing out any deviation that might exist between the unit under inspection and an established standard.

Incoming Inspection

The term is misleading: it covers only one portion of the operations that are carried out on the items purchased outside the plant to ascertain their compliance to the purchasing specifications. It should be called "incoming inspection and test."

Anyway, we have already seen that component parts, subassemblies, and assemblies have an infrared signature, and conventional inspection and test techniques are likely to miss "hidden anomalies" affecting them that will be disclosed in the signature comparison process. Therefore, an infrared scanner should find good use in an incoming inspection department to weed out incoming items affected by defects that are either difficult or impossible to detect with conventional means. For instance, resistors with poor mechanical connection of the end wires, overdissipating semiconductors, de-formed electrolytic capacitors, short-life fuses, overheating transformers, and such.

The major problem in incoming inspection is probably the need to process rapidly large number of presumably identical items; the limiting factor will be the time required to preheat the items to the desired level. Special conveyors equipped with the capability to electrically energize the items and synchronized with the scanning speed of the radiometer will be necessary to meet the requirements above.

In-Process Inspection

In general, the following items are subject to normal inspection during the manufacturing process—type of component used at every point of the assembly; orientation of it; proper connection; absence of damage; correct value according to print or color-code; quality of soldered connections; and possible absence of needed components, etc. It is a long, complex operation, and it is never foolproof: even the best inspectors will miss 2% of the existing faults, as verified by the author during a study in which inspectors were submitted units "prepared" with a variety of defects.

Infrared scan of electrically energized electronic assemblies can eliminate most of the visual inspection as it is performed today. The infrared signature contains information about all the items listed in the last paragraph, except for minor details such as superficial damage to a component, or appearance items. Actually, these details are usually the subject of the last visual inspection, which is called "final inspection."

Final Inspection

While conventional manufacturing processes are keeping a strong differentiation between in-process and final inspection, the separation almost disappears in infrared applications. As soon as the electronic assemblies can be electrically energized, infrared inspection will become possible, and in one single sweep of the scanner an infrared signature will be generated that will give us both inspection and test data, thus saving a major portion of the time and effort required by today's processes. With one difference: it will not miss 2% of the existing defects, or the "hidden anomalies" that account for most of the early field failures, at the surprising rate of 3% of the printed circuits tested out as good by conventional electronic test equipment.

TEST AND TROUBLE-SHOOTING

First of all, I would like to point out a basic difference between conventional and infrared testing.

In the conventional system, a certain number of electrical measurements are taken at preselected key points of the circuitry. If the readings are within tolerance, the unit is pronounced OK, and it is assumed that everything else between the test points works properly. This might not always be true, since the weak performance of a component might be hidden by the compensating effect of a "hot" component. Especially in pulse circuitry, where clipping stages always follow gain stages, conventional techniques do not always tell us how much we are clipping.

However, in the infrared approach, no stone is left unturned. The infrared signature contains the measurement of the power dissipated by each and every component of the assembly, and this is the reason why infrared testing can disclose actual or impending failures that conventional test equipment cannot detect.

Until now, test and trouble-shooting have been two distinct operations, separate from each other and from inspection. This is not true with infrared. All three operations are based on the infrared signature of a unit. Any deviation from the *standard* expected points out the effects of the discrepancy. Inspection, test, and trouble-shooting are consolidated in one single scan of an electronic assembly: all the information is there in the infrared signature and to understand it only a schematic of the circuit and some knowledge of how it works are required. Besides the standard signature, the signature of the failure also can be kept on file for identification purposes in case this should occur again in similar units. Or the deviations from the "good" standard can be kept on file for the same purpose. In the latest models of automated infrared scanning equipment, the "file" is a length of punched tape with the infrared signature permanently recorded on it, easy to handle by the computer performing the pattern recognition and comparison process.

A pilot infrared semiautomated trouble-shooting program took place in 1966 at the manufacturing plant of Raytheon located in North Dighton, Mass. Many hundreds of printed boards were tested, using a first generation infrared scanner called Compare, especially designed to perform infrared testing of printed circuit assemblies of a certain size, belonging to a military communications system. Figure 233 shows this system, which had the capability of processing one printed card every 100 sec (versus a $\frac{1}{2}$-hour average time for conventional trouble-shooting). To eliminate down time due to warm-up of the modules, the system was provided with a preheat rotating table where the modules could be plugged in in advance and then brought in the focal plane of the radiometer in time sequence, after they had reached thermal equilibrium.

The trouble-shooting cycle goes as follows: as soon as a module is brought in the focal plane by the rotation of the warm-up table, scanning begins. At a rate of 4 lines/sec, a 180-line raster is completed in 45 sec and stored in the magnetic core memory. During the remaining 55 sec of the cycle, the EDP compares the signature of the module with the signature on file. At the end of this process, a figure appears on the nixie lights display, representing the number of points of deviation (in absolute value) of the unknown signature from the standard stored in the file. If this number is zero, it means that the unit under test has a signature identical to the standard and therefore it is operating properly. Conversely if the number differs from zero, it means that deviations are present, indicative of failure(s). At this point, through the keyboard the operator can instruct the EDP to print out the deviations of the unknown signature versus the standard, so that the defective components might be isolated for quick pinpointing of the failure cause.

A typical trouble-shooting operation of a printed board whose "standard" signature was already on file is described in the following example:

Figure 233. The Compare system.

The output of the system was printed out in octal code figures, as shown in Figure 234. Column *A* shows the standard signature of a good module. The first three digits are the numerical designation of each component, while the last digits indicate the radiance level of each of them.

Column *B* shows the signature of a defective module. The meaning of the figures is the same as described for *A*.

Column *C* is the printout of the deviations of the radiance level for each component of *B* versus *A*. In this printout, the first three digits are the numerical designation of the component, while the last two digits are the deviation, complete with the indication whether it is positive or negative.

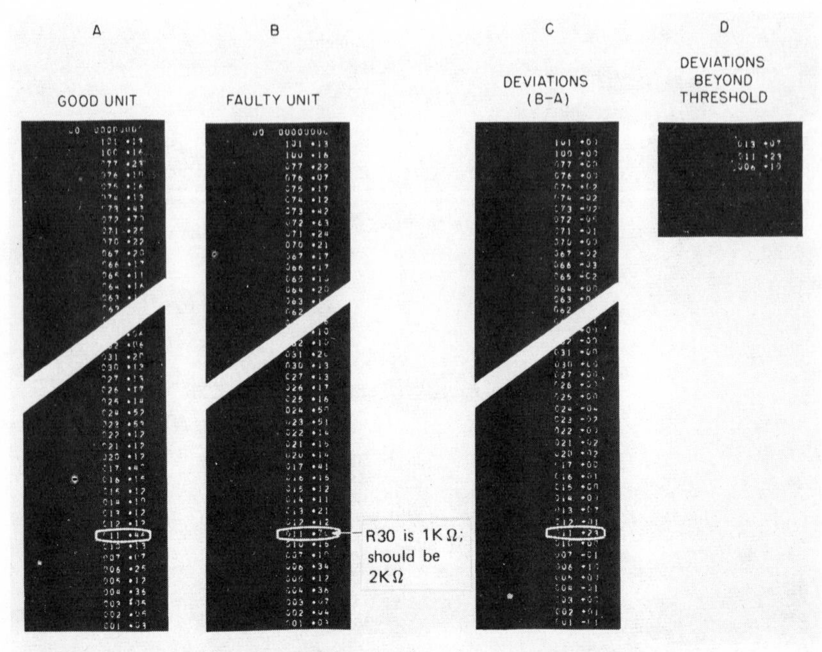

Figure 234. Infrared signatures of printed board.

Column D is the printout of only the major deviations of B versus A. Experience has shown that these alone are sufficient to achieve fast trouble-shooting with this system.

The first task to be performed during the pilot infrared trouble-shooting program was the establishment of the standard infrared signature of each type of printed board to be investigated.

Lack of existing knowledge or experience in this new field required a cautious approach, and repetitive, verified results before routine operating policies could be established.

For each type of board the plan called for scanning ten units that had been accepted as good at the electrical test, establishing their signatures, and for each component determining the average radiation value and the range of the deviations.

And right there, in the middle of the first group of "good" boards, one of them exhibited unexpected, high radiation peaks in correspondence to components that normally operate at moderate temperature level: resistor $R34$ was operating at 27.5°C above ambient, instead of the expected 4.7° Δt, and transistor $Q16$ was operating at about three times its standard Δt. What had happened?

A quick look at the deviating components pinpointed the trouble:

during the assembly operation, by mistake, a 510-Ω resistor was installed in place of the 5100 Ω required for *R*34, the collector resistor of transistor *Q*16. As a consequence, the current flowing through both components was much higher than expected, thus causing heavy thermal overstresses that would in a relatively short time bring about a failure of the transistor.

The surprising discovery was that conventional electrical test could not detect this anomaly. Had not it been for the infrared test, the unit would have reached the field in the precarious condition created by this hidden anomaly. What would have happened then? Evidently, after a few weeks of operation, the heavily overstressed transistor would have failed, and the unit would be sent to trouble-shooting. Eventually, a diagnosis would have been formulated: "*Q* 16 burnt-out. Replace." After repair, the unit would have passed test, and be returned to the field for another brief, overstressed life cycle. And so on.

Further printed board scanning disclosed that this was not an isolated case. Out of the first 1000 "good" units tested, almost 3% were found to contain anomalies that were not detected by conventional test equipment. These "hidden" anomalies cover a wide range of defects, such as shown in Table 8, which is a partial list of them.

And now a few words about troubleshooting. Conventional techniques are based on checking electric characteristics and signals at various points of the circuitry, slowly "closing in" toward the region where the faulty component is located. It is a deductive process: when a certain amount of incriminating evidence has been collected, the troubleshooter takes an educated guess and decides which component is most likely to be at fault. It is a hit-or-miss process, with more misses than hits. Years ago, I ran a survey of parts replaced during one month of troubleshooting, and I found that 80% of them were perfectly good. Of course, this 80% includes good components removed because they had been installed wrong (for value, orientation, location, etc.) but even after having made allowance for this, the number of good parts needlessly replaced is much larger than the number of faulty parts.

Table 8 List of Most Common "Hidden" Anomalies in Electronic Assemblies

Reversed polarity of electrolytic capacitors
Missing or open component of a pair wired in parallel
Wrong wattage resistor
Missing or open voltage clamping diode
Incorrect gain before limiting or clipping stages

Not so with infrared. To begin with, the all-encompassing ability of measuring the radiation emitted by each component is many times superior to the spotty conventional technique of taking a few measurements here and there. It is like having an infinite-impedance probe connected across every component: all the readings are there—available and clear. When compared with the standard "good" signature, the deviations stand out conspicuously. They spell out the trouble, with hardly any doubt, except in those instances where an "open link" is indicated, and there is more than one single element along this link.

The infrared anomalies related to a failure can be strictly localized and sometimes limited to a single component, or can affect a large number of components, when for instance the defective item is at the input of the circuit or in the power supply.

Let us see a few examples taken from the work recently performed by

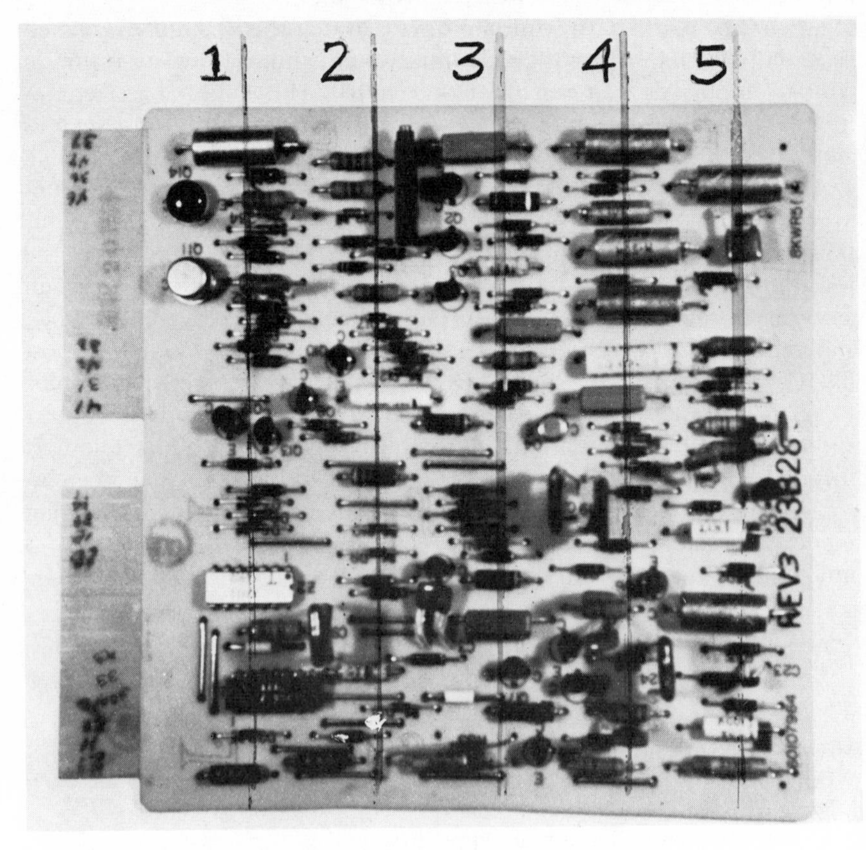

Figure 235. Printed circuit board of computer system.

the author, using the Inspect System described in Chapter 3 under "Computer Controlled Scanner", for instance, the printed circuit of Figure 235. It is a digital board used in a computer, and its schematic is shown in Figure 236. We can see that several distinct networks are assembled on the board. And now let us look at the printout of the deviations from the standard infrared signature, for two defective boards, designated as *A* and *B*. Figure 237 shows the printout of board *A*: only two deviations—*R*4 and *D*2—both with a minus sign. The diagnosis is easy: Zener diode *D*2 fires at the wrong voltage. However, for board *B* the situation appears quite complex: there are 17 deviations in the printout of Figure 238. Is it possible that all those components could be defective?

Of course not. In most instances, only one component is at fault, and to identify it quickly from among the 16 anomalies all that is needed is the schematic of the circuit and a red and blue pencil. We circle in red those components showing a positive deviation, and in blue those showing a negative deviation. In the case of Figure 238 we only have blue circles, actually shown as empty circles. Marking these circles takes most of the trouble-shooting time—perhaps a minute or two. But once they have been drawn, the diagnosis is self-evident; *Q*4 is off, and so is *Q*3, and *Q*2 and *Q*1. Since the signal travels from *Q*1 toward *Q*4, evidently something is wrong with *Q*1 (in the particular instance, the base wire of *Q*1 was not connected to the circuitry, so the whole amplifier was turned off).

In the case of printed boards carrying integrated circuits instead of discrete components, the infrared technique is basically the same. Each integrated circuit is treated as a single unit, which is expected to dissipate a certain amount of electrical power when operating in the desired mode. Any failure mode resulting in a variation of the power dissipation will be detected by infrared in exactly the same way as discussed for discrete components, provided it is within the sensitivity limits of the infrared test equipment being used. This was verified in several test programs carried out, among others, on printed boards of the type shown in Figure 239. In general, however, the surface temperatures of the flat-packs and of the other configurations containing the integrated circuits are relatively low when compared to the temperatures of most discrete components. This is due to the mass ratio chip/package, and also to the large number of terminal leads, through which heat can reach the sinks by conduction.

Therefore, very fine thermal resolution is a primary requisite for infrared scanners designed to test assemblies of integrated circuits on printed boards. Ability to measure Δts as small as 0.01°C is essential, but

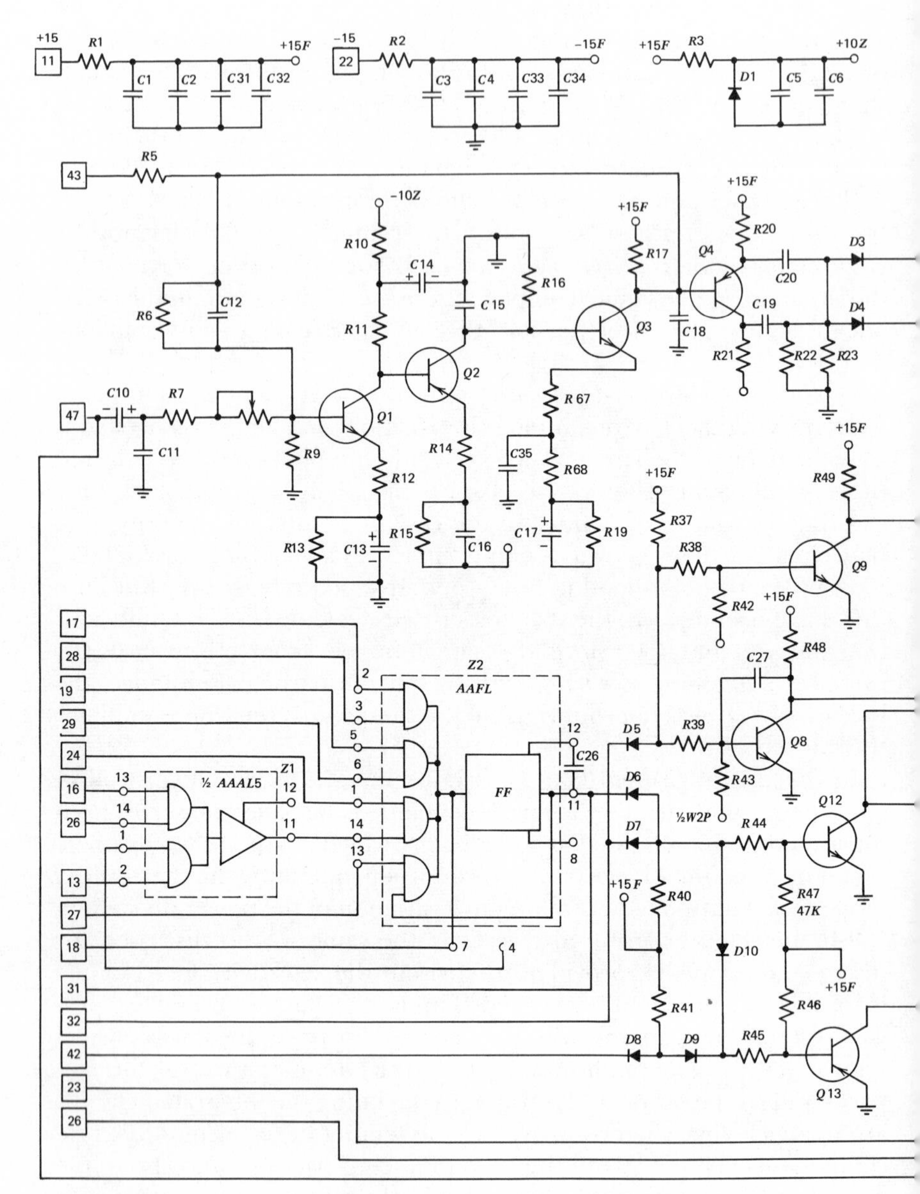

Figure 236. Schematic of computer board.

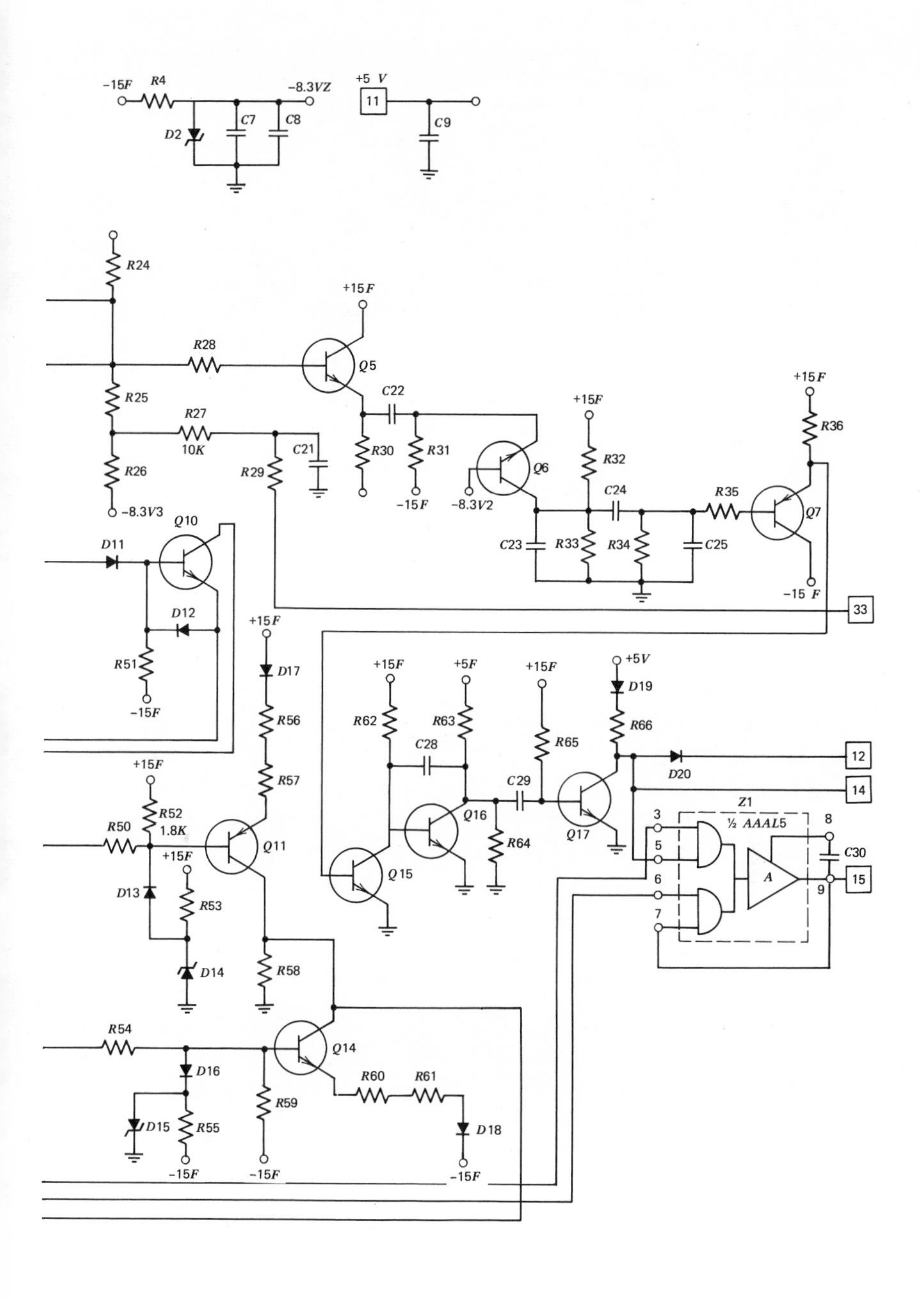

−15F
R4
−8.3VZ
D2
C7
C8
+5 V
11
C9
R24
+15F
R28
Q5
R25
C22
R27
10K
R30
R31
C21
R26
R29
Q6
+15F
R32
+15F
R36
−8.3V3
−15F
−8.3V2
C24
R35
Q7
Q10
D11
C23
R33
R34
C25
−15 F
D12
33
+15F
R51
−15F
D17
+15F
+5F
+15F
+5V
D19
R62
R63
R65
R66
R56
C28
12
D20
+15F
R57
C29
14
Z1
R52
R50
1.8K
Q16
Q17
3
½ AAAL5
8
+15F
Q11
R64
5
C30
D13
Q15
6
A
15
9
R53
7
D14
R58
R54
Q14
D16
R60
R61
R59
D15
R55
D18
−15F
−15F
−15F

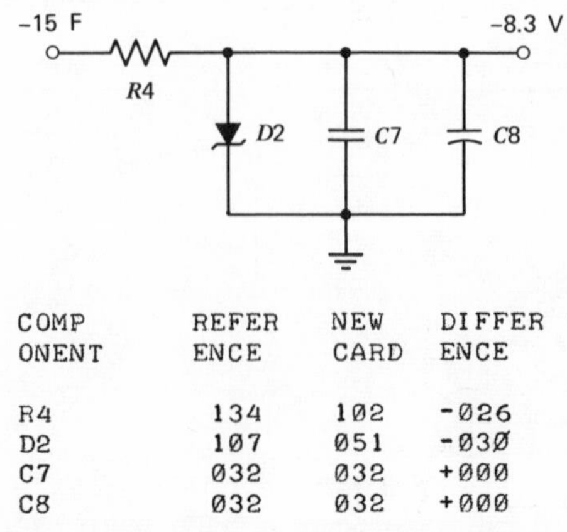

COMP ONENT	REFER ENCE	NEW CARD	DIFFER ENCE
R4	134	102	-026
D2	107	051	-030
C7	032	032	+000
C8	032	032	+000

Figure 237. Board *A* deviations printout.

an even finer resolution, close to 0.005°C is preferable. This can be translated in the ability to detect changes in power dissipation as small as 1 mW in integrated circuits encapsulated in flatpacks and similar packages.

The printouts of the board of Figure 239 are shown in Figure 240. On the left is the standard signature, and on the right the signature of a defective board. All the deviations are negative; that is, the components affected by the failure are underdissipating. The major deviations are listed in the column at the extreme right. They appear in base 10, while the printouts are in base 8, hence the apparent discrepancy (from 0 to 11 there are only 9 decimal points).

The temperature scale used in these printouts was 0.01°C per point. Consequently, an integrated circuit recorded at +050 octal points operates at a temperature of 4°C above ambient.

DETECTION OF SECONDARY OVERSTRESS

One additional bonus of infrared trouble-shooting is the identification of those components that have been overstressed during the time that the unit was operating under failure condition. Electronics engineers have been always concerned with what are usually called "secondary failures," that is, failures caused by the faulty operation of some other component. But very seldom has a thorough study of what could be called "secondary overstress" been carried out, even for high reliability programs. The reason for this lack of interest is probably the fact that, as

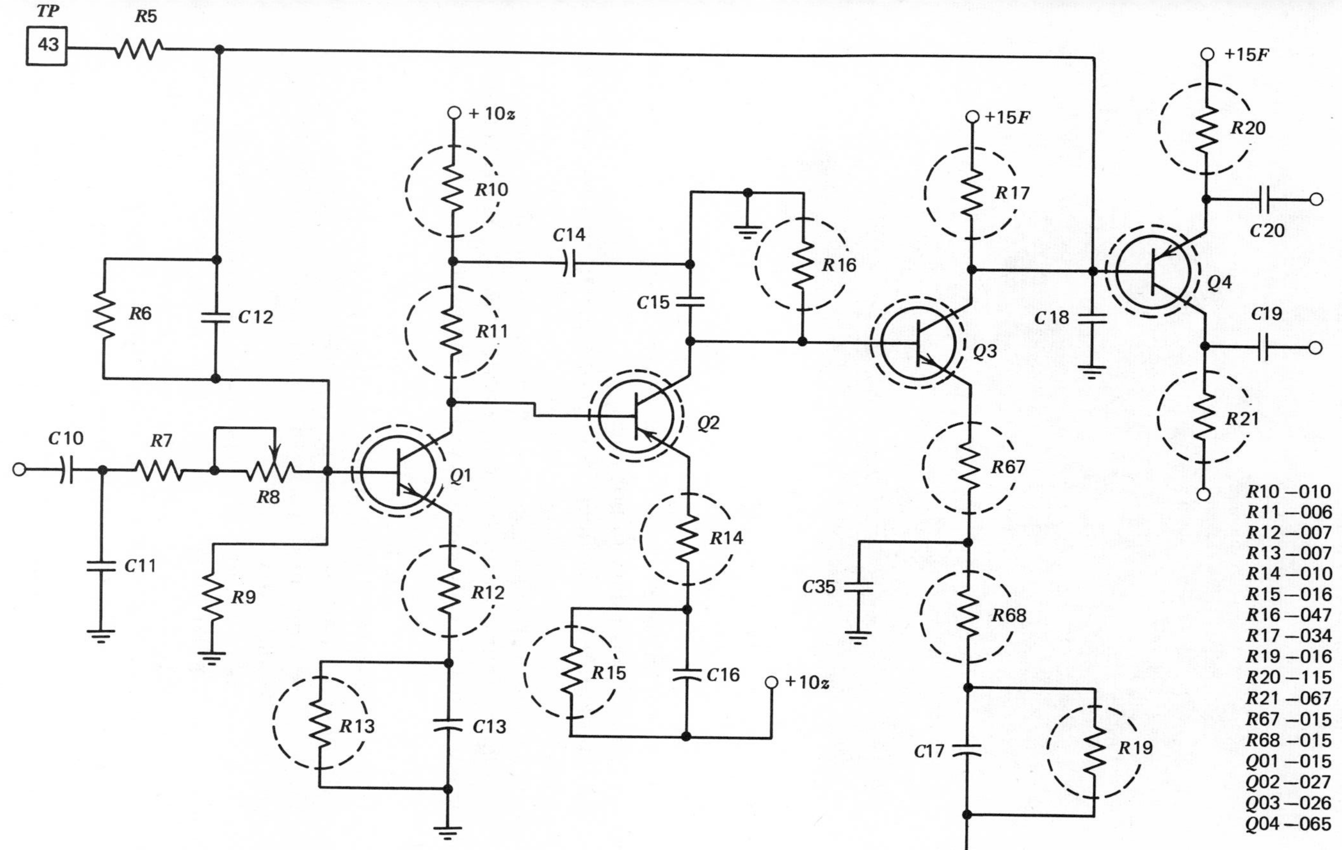

Figure 238. Board *B* deviations from standard.

Figure 239. Printed board with integrated circuits. (Courtesy Teradyne Corporation.)

long as the secondary overstress does not cause catastrophic failures, it is assumed to be of secondary importance.

In our opinion, the consequences of secondary overstress are more serious than those of the outright failures, because the latter must be corrected by replacement by new components, while the former are usually ignored, thus allowing components that have been weakened by overstress to remain in the assembly, whose life expectancy is therefore reduced.

The all-encompassing ability of infrared scanning to point out every component overstressed during a failure mode makes it possible to generate a list of those parts that must be replaced when a certain type of failure takes place. Only in this way the original reliability level can be restored for any failed unit.

The following is an example of a typical case of a secondary overstress condition that can take place in the magnetic amplifier of an azimuth-elevation driving power assembly. This unit is made of two networks powered by current equally divided between two balanced circuits; if a failure causes one of the two branches to fail by opening, the balanced condition disappears, and all the power is dissipated through the other branch. This condition raises the temperature of a 6.5-W wirewound resistor (*R*212) up to 300°C. The magnitude of the overstress is apparent in Figure 241 where two infrared profiles, (*a*) and (*b*), are shown for

Standard signature		Defective board signature		Difference
RE1	+2Ø6	RE1	+211	
R17	+Ø4Ø	R17	+Ø36	
DØ1	+Ø33	DØ1	+Ø26	
DØ2	+Ø33	DØ2	+Ø33	
RØ1	+Ø24	RØ1	+Ø21	
QØ1	+Ø1Ø	QØ1	+ØØ6	
RØ2	+Ø15	RØ2	+Ø12	
IØ1	+Ø44	IØ1	+Ø41	
IØ2	+Ø53	IØ2	+Ø52	
RØ3	+Ø1Ø	RØ3	+ØØ5	
QØ2	+ØØ6	QØ2	+ØØ1	
DØ3	+Ø25	DØ3	+Ø17	→ −6
DØ4	+Ø41	DØ4	+Ø35	
RØ4	+ØØ7	RØ4	+ØØ4	
DØ5	+Ø34	DØ5	+Ø16	→ −14
DØ6	+Ø44	DØ6	+Ø4Ø	
RØ5	+Ø11	RØ5	+ØØ7	
QØ3	+ØØ5	QØ3	+ØØ2	→ −3
IØ3	+Ø47	IØ3	+Ø17	→ −24
IØ4	+Ø55	IØ4	+Ø5Ø	
RØ6	+ØØ5	RØ6	+ØØ2	
RØ7	+ØØ3	RØ7	+ØØ2	
QØ4	+ØØ3	QØ4	−ØØØ	
DØ7	+Ø25	DØ7	+Ø14	→ −9
DØ8	+Ø44	DØ8	+Ø37	
RØ8	+ØØ2	RØ8	+ØØ1	
DØ9	+Ø35	DØ9	+Ø17	
D1Ø	+Ø45	D1Ø	+Ø4Ø	
RØ9	+ØØ7	RØ9	+ØØ4	
QØ5	+Ø11	QØ5	−ØØØ	→ −9
R1Ø	+ØØ6	R1Ø	+ØØ3	
IØ5	+Ø46	IØ5	+Ø22	→ −20
IØ6	+Ø6Ø	IØ6	+Ø47	→ −9
QØ6	+ØØ4	QØ6	−ØØØ	
R11	+ØØ4	R11	+ØØ2	
D11	+Ø32	D11	+Ø23	→ −7
R12	+ØØ6	R12	+ØØ3	
D12	+Ø43	D12	+Ø4Ø	
R13	+Ø12	R13	+ØØ7	
D13	+Ø37	D13	+Ø35	
D14	+Ø45	D14	+Ø43	
QØ7	+Ø11	QØ7	+ØØ4	
R14	+Ø1Ø	R14	+ØØ6	
IØ7	+Ø45	IØ7	+Ø44	
IØ8	+Ø54	IØ8	+Ø54	
QØ8	+ØØ1	QØ8	−ØØØ	
R15	+Ø1Ø	R15	+ØØ5	
D16	+Ø36	D16	+Ø35	
R16	+ØØ7	R16	+ØØ3	
D15	+Ø36	D15	+Ø32	
XØ1	+Ø16	XØ1	+ØØ7	
XØ2	+Ø12	XØ2	+ØØ5	
IØ9	+Ø5Ø	IØ9	+Ø53	
I1Ø	+Ø42	I1Ø	+Ø45	
ØØØ	+Ø11	ØØØ	+ØØ5	
R18	+324	R18	+273	→ −25
R19	+241	R19	+216	
TEST COMPLETE		TEST COMPLETE		

Figure 240. Inspect printouts of units of Figure 239.

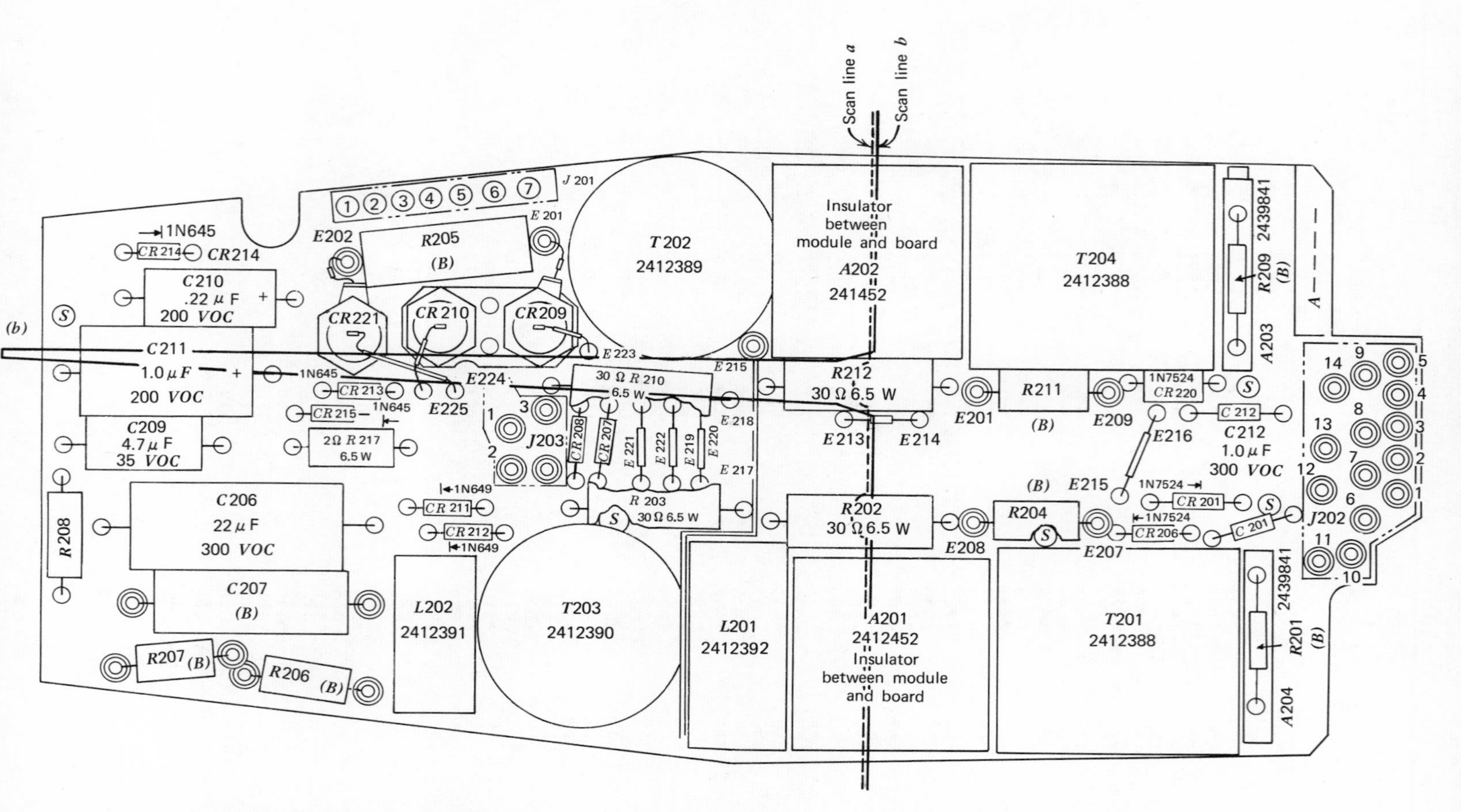

Figure 241. Detection of a secondary overstress condition.

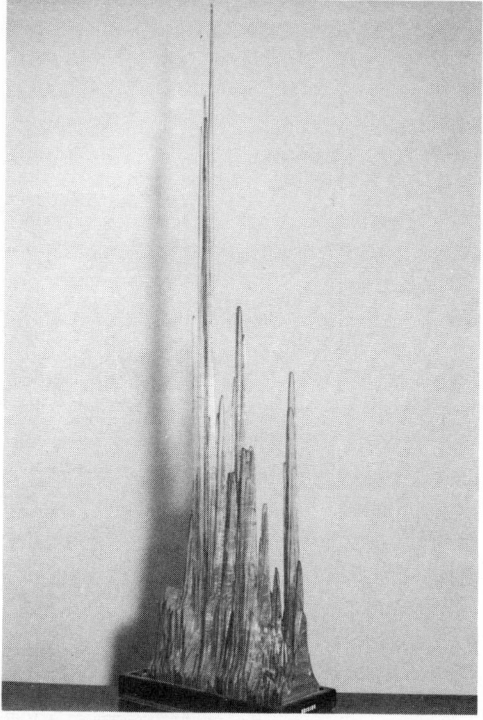

Figure 242. Three-dimensional representation of infrared radiation of unit in Figure 241.

comparison, plotted on a panel layout drawing of the unit. Both profiles are the oscilloscope traces of scan line 45, *a* representing the "balanced" operation (dotted line), and *b* the "failure" condition (solid line). We can see that in this case the temperature of *R*212 is way above the normal operating level.

A three-dimensional representation of the infrared radiation intensity for every point of the energized panel is shown in Figure 242. In the highest peak of this display we can easily recognize the excessively high infrared radiation emitted by *R*212 during the malfunction of the twin circuit.

In spite of the high level of overstress, conventional test and troubleshooting procedures never called for replacement of the overstressed resistor, which consequently used to reach the field in damaged condition.

Besides outright failures, secondary overstresses can be caused by undesired conditions that sometimes are not detected by conventional test equipment. One such case is illustrated in Figure 243 where heat sink

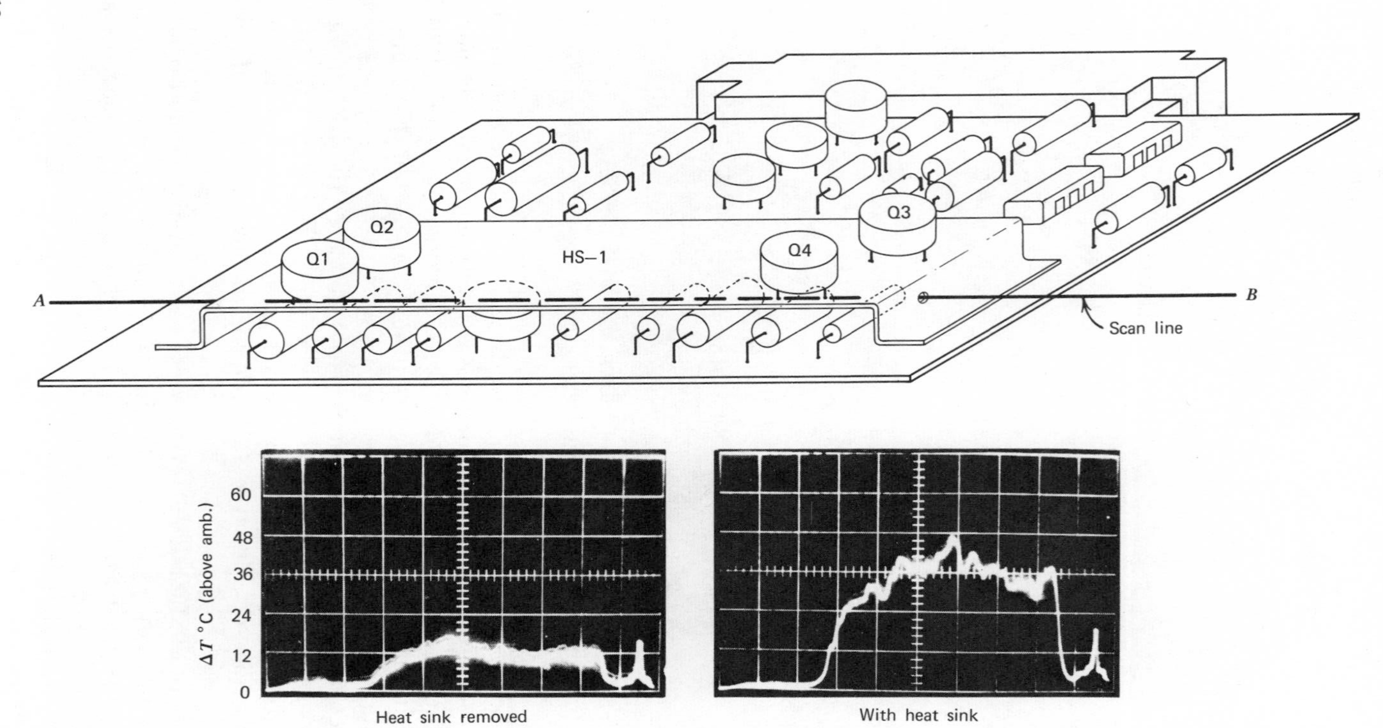

Figure 243. Effect of heat sink on nearby electronic components. The oscilloscope traces indicate the temperature of the components located under the heat sink HS-1, along scan line *A–B*.

HS-1 has the function of cooling the four power transistors $Q1$ to $Q4$. This is done at the expense of *all* the components located under it, which are subjected to an average increase in temperature of 24°C, as shown in the oscilloscope traces depicting their temperature first without and then with heat sink in place.

NOVEL MAINTENANCE POLICY

Current maintenance practices are based on two opposite philosophies: (*a*) "*a posteriori*" or after the fact: wait for a failure and then replace the failed part or assembly (we will not even discuss this primitive approach); (*b*) *Preventive* maintenance: based on statistical calculations of MTBF; according to the degree of failure-free operation that is desired, a schedule of replacements, both at the component-part and at the assembly level, is derived and implemented.

This approach has the drawbacks that are typical of any statistical system: only averages are taken into consideration and the exceptions are ignored. This leads us to the paradox that now we have the statistical certitude that in some instances the replacement used is actually less reliable than the part being replaced.

Comparison of infrared profiles of energized assemblies taken at predetermined intervals of time versus their "original" profile will tell us when it is time to replace a certain part or assembly and will also tell us whether the replacement is better than the part being discarded. In special instances when we cannot afford to let failure occur, we could carry out continuous infrared surveillance of electronic equipment so that every trend that might develop will appear immediately in the infrared pattern constantly being generated. Consequently, we will be able to determine from the magnitude of the trend how much operating life is left in the drifting component. This will enable us to replace it before catastrophic failure occurs.

If the procedure sounds complex and of difficult implementation, I would like to mention that the Inspect system was designed precisely with this infrared maintenance concept in mind. During final test at the manufacturing plant, all the signatures of the boards of a system are recorded and filed in a "system tape" that is then forwarded to the main depot in charge of maintenance. After its cycle of field service, the system is brought back for check and servicing to its main depot. At this point, all the system's boards are sequentially infrared tested by an Inspect system that measures the deviations between the present signature and the original one, automatically retrieved from the file. If there are no deviations, it means that no deteriorating trend is present, and the unit is as good as new, no matter how many hours of work have elapsed. If, on the other hand, the infrared signature indicates the existence of a degradation process, replacement of the drifting assembly or component can be planned in time to prevent reaching an out-of-spec condition, that will cause a catastrophic failure.

In this way, preventive maintenance is turned into a realistic, factual operation, based on component performance assessment instead of statistical probability.

I have just used the term "catastrophic failure." A word of caution might be in order. When we complain that "everything was operating beautifully, when, all of a sudden, a catastrophic failure occurred," we seem to imply that during all the time prior to the failure, everything was perfect.

In most of the cases, this is not true. The catastrophic failure was the last step down a long incline of progressive performance degradation. As long as this degradation was contained within operational tolerance limits (and these limits might be wider than those allowed by design tolerances), no inkling of failure was evident. But the fact that we do not notice the degradation does not mean that it is not there. It merely means that we are not aware of its presence. However, with infrared we will not miss it: it will stand out right there in the infrared signature. And we will know when it is time to replace an electronic assembly or a component of it.

Chapter 9 The Future

Although this is the last chapter, the final infrared word has not been spoken, the final work not performed, and the final goals not achieved. On the contrary, infrared as a new technique just starts here. The little progress done so far is almost nothing when compared with the future developments in this new field.

All we have achieved during this initial period of work has been to answer some basic questions on the capability of infrared as a new tool for more thorough evaluation of materials and electronics. Now that we did obtain a positive answer, we are faced with the task of implementing these techniques in all the areas where they can contribute to the technological progress.

There are areas, for instance, where feasibility has not even been tried, in spite of the fact that both theory and common sense tell us that the idea should work. One of these areas, for instance, is the detection of waveguide malfunction.

Radio frequency losses in microwave guides manifest themselves as heat sources. A study should be conducted to determine how to use infrared detecting equipment to locate the area of the radio frequency loss and its magnitude.

These techniques could be used both for the evaluation of new designs and for the control, test, and trouble-shooting of production equipment.

Effective shielding of the radio frequency detector against RFI is a prerequisite for proper infrared measurements in this application.

Another area of interest is the possibility to detect "hidden" stresses caused by vibration, in electronic or mechanical equipment. We are all aware that during vibration tests, resonances develop at different points, while the vibrating frequency is being varied, but many of these resonances go undetected, unless they are so conspicuous as to be visible, or so destructive as to cause a failure.

Surveillance of the equipment under vibration by an infrared scan system would disclose temperature increases in most places where resonances develop, thus eliminating the possibility that dangerous resonances might go undetected. The technical difficulty lies in the need to synchronize the scan speed with the vibration frequency of the target, so that the latter might appear stationary to the infrared detector. Since the vibration frequency is a variable, the scan speed must also vary in correspondence. As an additional bonus, this system would eliminate the need for testing to failure.

THE FUTURE OF FIBER OPTICS

The image-transfer capability of coherent fiber bundles, plus the large NA values, pose a serious challenge to the conventional radiometers designed for thermal measurement of close targets.

Conventional optical systems are based on the principle of reconstructing an image in a focal plane, after having thoroughly scrambled all the rays captured from the target. No wonder that so many aberrations of so many different types will affect the reconstructed image!

Besides the critical requirements for alignment, focusing, aiming, and the like, the conventional system can work only if the target can be directly viewed by the detector without obstacles standing in the way. This is quite a severe limitation, for instance, in the field of electronics. Components and assemblies are always contained in some sort of opaque protective envelope, and their performance when operating in their actual environment cannot be monitored by conventional infrared means.

The use of infrared-transmitting fibers will soon give us this capability. Figure 244 shows how.

Incoherent bundles of these fibers are attached to every component whose performance must be monitored, and their outputs are aligned on an outside surface of the module. After this, potting can take place, and all components and fibers are buried inside the protective compound, forever removed from view.

However, the infrared radiation of every component travels along the fibers, and emerges at the output "windows" where it can be measured by conventional means.

In the case illustrated in Figure 244, the fiber arrangement produces a welcome simplification: it reduces the scanning requirements from two dimensions (area scan) to one dimension (line-scan).

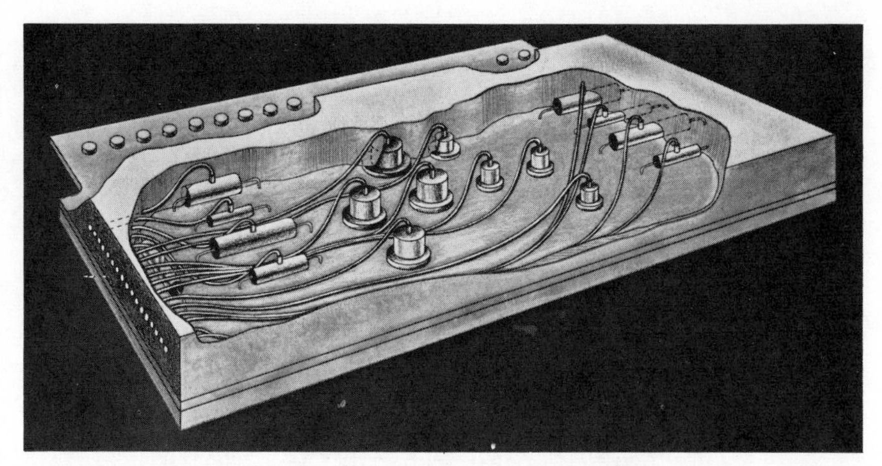

Figure 244. Use of infrared transmitting fibers to monitor electrical performance of component parts inside potted assembly.

The feasibility of this concept has already been demonstrated. Its practical implementation, however, is still awaiting a sponsor.

And if this is possible for a single module, why not for a number of them? For instance, it is possible for the 500 units assembled in the five-rack system shown in Figure 245. For each module, the infrared "windows" are located on the rear surface of the system, as illustrated. An infrared system facing such surface, constantly scans the target along scan lines carefully aligned over the rows of windows, picking up the infrared signal emitted by each of them.

When the electronic system is operating at thermal equilibrium, the detector will generate a definite output pattern, dictated by the sequence in which the windows' infrared signals are sampled. This output pattern will repeat over and over for every scan cycle of the scanner. Each pulse of the pattern measures the operating performance of the corresponding component part, which can be readily identified.

For electronic systems whose operation is of critical importance, this approach can allow continuous monitoring without physical contact, nor measurement interruption. As soon as a component's performance deteriorates, its temperature will change accordingly, and this will immediately be reflected in the infrared output from the corresponding window, and in the magnitude of the related pulse in the electrical pattern generated by the scanner. Furthermore, the speed of the drift can be computed, so that extrapolation can be made to figure out the time when replacement must be made.

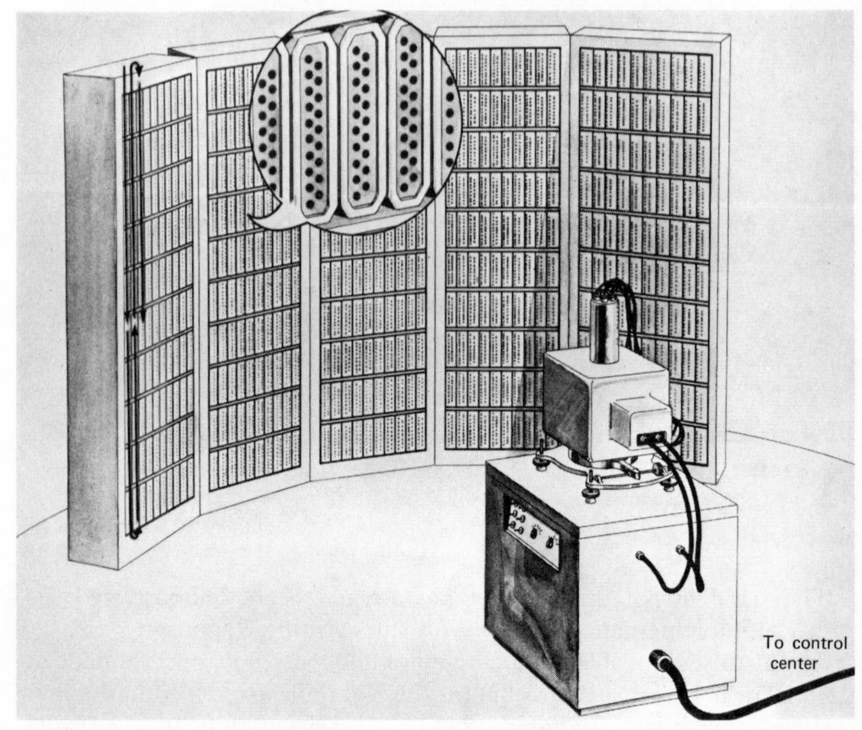

Figure 245. Infrared surveillance of five-rack electronic system. Dots indicate infrared windows; arrows indicate path of scan.

Eventually, however, an infrared vidicon tube having adequate sensitivity in the spectral region of the radiation emitted by electronic components will become available, and its video output will contain the electrical pattern indicative of the infrared output of each window.

And what about processing the detector output? A number of approaches and systems are available, their degree of complexity increasing with the level of tasks to be performed by them. For instance, if we want a system capable of checking drift, extrapolating to the point in time when an out-of-spec condition will be reached, and issuing the necessary instructions for component or module replacement, the complexity of the EDP system will be quite high. Quite high also is the speed at which modern computers work; therefore, this one, with only one radiometer output to process, will have a lot of idle time on its hands. Perhaps in the spare time, it could have a look at the information generated by another radiometer, or by ten or twenty more. They do not have to be located in the next room. They can be hundreds or thousands of miles away; their output can always reach the control center by cable or by radio. They can be located underground, such as in hard missile sites, at sea aboard ships of all kind, or even in outer space orbiting the earth. The computer will receive their information, compress it, store it temporarily, and then process it and flash back the diagnosis and the instructions to the interrogating party. It might say, translated into English: "module 387 as good as new. Leave it alone," or "module 718 drifting at *Q*8, you have a week to replace it, or a month, or one hour."

THE FUTURE OF INFRARED MICROSCOPY

The fast progress being made in microelectronics and the definite trend toward large scale integration will increase the popularity of infrared microscopy. Fast scanning infrared microscopes will be installed at the end of semiconductor production lines, and go-no-go pattern comparison systems will accept or reject on the basis of the infrared signature. Rapid conveyors equipped with synchronized energizing devices will allow processing hundreds or even thousands of units per hour.

Besides thermal radiation, microscopes will be able to isolate and detect recombination radiation, thus adding a modulation and pulse measuring capability anywhere there is a junction.

To allow these measurements to be taken at any time during the life of semiconductors, their encapsulation might be made of material transparent to infrared radiation. It could be silicon lids such as those shown in Figure 246, or arsenic-selenium-tellurium glass, or irtran, or some other material of those listed in Figure 13, Chapter I, Part I. The important requirement is adequate optical finish of the surfaces, so that transmission can take place with the least amount of losses.

Or perhaps the cover will be made of coherent fibers, so that the infrared image of the semiconductor chip will be brought out, onto the external surface of the cover. Perhaps these covers could be much longer than wide, so as to reach a display surface located at some distance from the devices where examination of the infrared pattern transmitted through the fibers could take place with all the comforts.

And the microscopes might do away with lenses and conventional optics. Coherent fiber bundles will replace these. Figure 247 is an example of an application of this concept, limited to the measurement of the infrared radiation picked up by one single fiber.

The optics are replaced by a coherent fiber assembly of which one

Figure 246. Flatpacks with silicon lids.

single fiber, located in the center of the bundle, is made of infrared-transparent material. An infrared detector of comparable size is deposited at the output end of the fiber, whose front end is brought directly in front of the point to be observed. Aiming is achieved by observation (through an optical microscope) of the visual semiconductor pattern appearing at the output end of the fiber bundle.

A setup of this type has been successfully used for the study of recombination radiation,[11] with the only difference being that all the fibers of the bundle were of the same material, a special glass transparent equally well in the visible and in the near infrared regions of the spectrum.

Since this setup works with one fiber, it will work with more. We could conceive an array of a large number of fibers, for instance, 50, placed on a straight line as shown in Figure 248. The front end would have the fiber centers 20 μ apart from each other, while the output end would

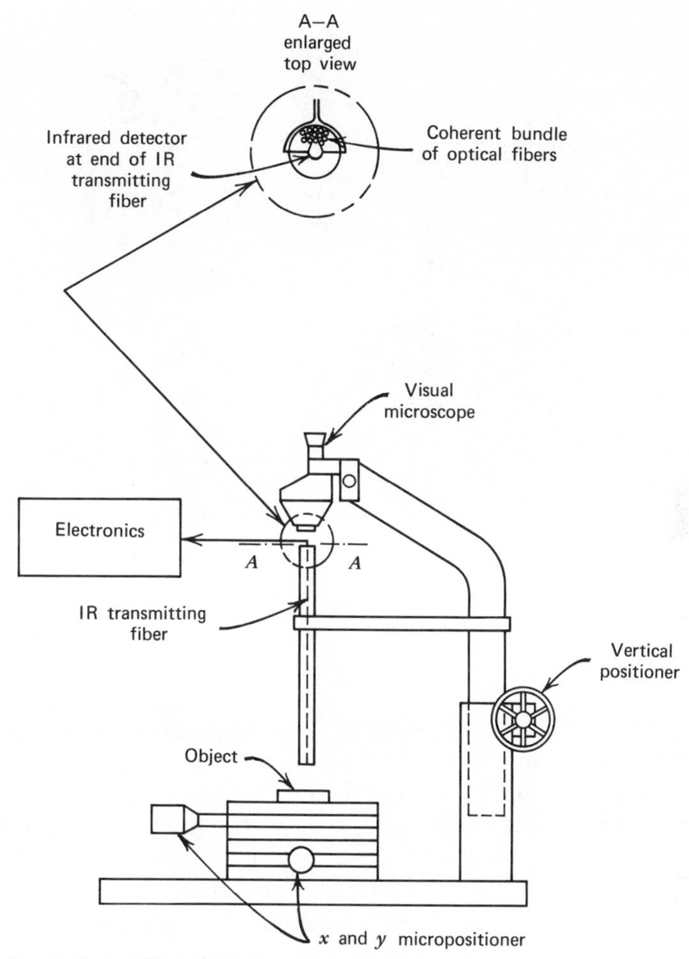

Figure 247. Coherent fiber optics pickup and viewing system.

open up in a fan-like fashion, to allow enough room for the detectors. Just one pass of this array would give 50 scan traces of an area measuring 1 mm on the side. If detector cooling were required, the output array could be enclosed inside a specially designed Dewar bottle.

In video tube technology, an infrared coherent fiber bundle could enter a vidicon tube and deposit the radiating energy directly onto the photoconductive layer where the image is conventionally formed. The use of fibers having a large aperture number would allow the transfer of large quantity of energy and thus increase the detectivity of the system.

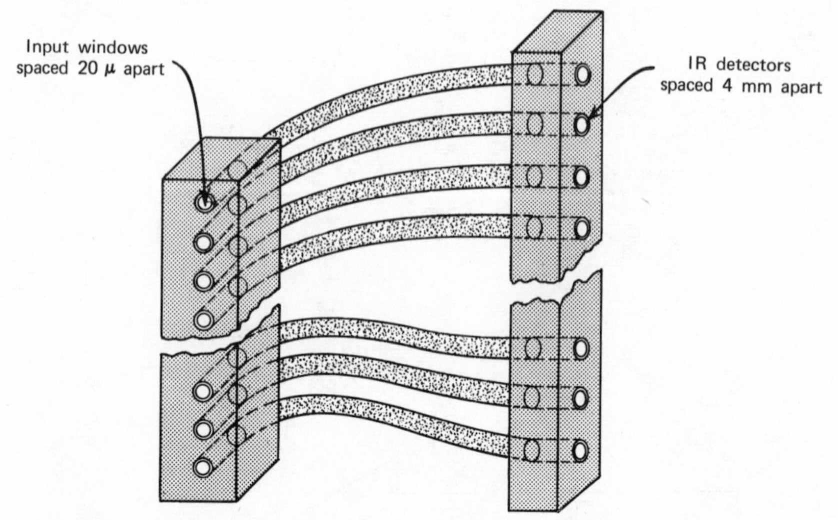

Figure 248. Fiber detector array.

NEW APPROACH

Thus far, we have been successful in tearing down the conventional, conservative concepts. First, we discovered that the present-day electrical testing cannot be fully trusted, and that infrared can detect many hidden anomalies and otherwise nondetectable defects. It took the greater part of this book to reach this point. But no sooner had this conclusion been formulated and infrared systems had risen over the horizon, carrying our best hopes with them, than some of their key elements began to be pushed away by new concepts. Out went the optics, under pressure from the fibers; out went the choppers, casualties of fast scan; even the signal processing was radically curtailed in favor of a centralized solution. The only thing that still stood firm was the infrared detector. After all, to measure infrared radiation we could not do without an infrared detector.

Or could we? Wouldn't it be possible to use some other kind of device, possibly one more readily available, possibly one that can be easily carried around, possibly one that comes already made by Mother Nature, possibly. . . . You guessed it! Possibly the Human Eye?

The fact that the human eye is not sensitive to infrared radiation does not matter in the least. All we have to do is to convert the frequency, and turn infrared into visible radiation.

In Part I, Chapter 2, we have already discussed this type of frequency conversion, describing the different electroluminescent and phosphor luminescent materials and devices at our disposal. Since here we are talking about the future, we will suppose that full development of them has been completed, and that infrared-to-visible converting paints are available. In such a case, the setup shown in Figure 249 will be rather easy to implement. In this application, the paint is applied directly onto the body of the electronic components, and the visible light emitted by them is picked up by incoherent bundles of optical fibers, and forwarded to the male connector of the board. In the socket part of the connector are placed matching optical fibers that will carry the visible light, indicative of the thermal condition of each component, to an observation window conveniently located for ease of viewing by an operator. Figure 250 shows this convenient observation window.

Since the human eye is almost worthless as a quantitative meter, the fibers must be arrayed in such a way as to take advantage of the eye's ex-

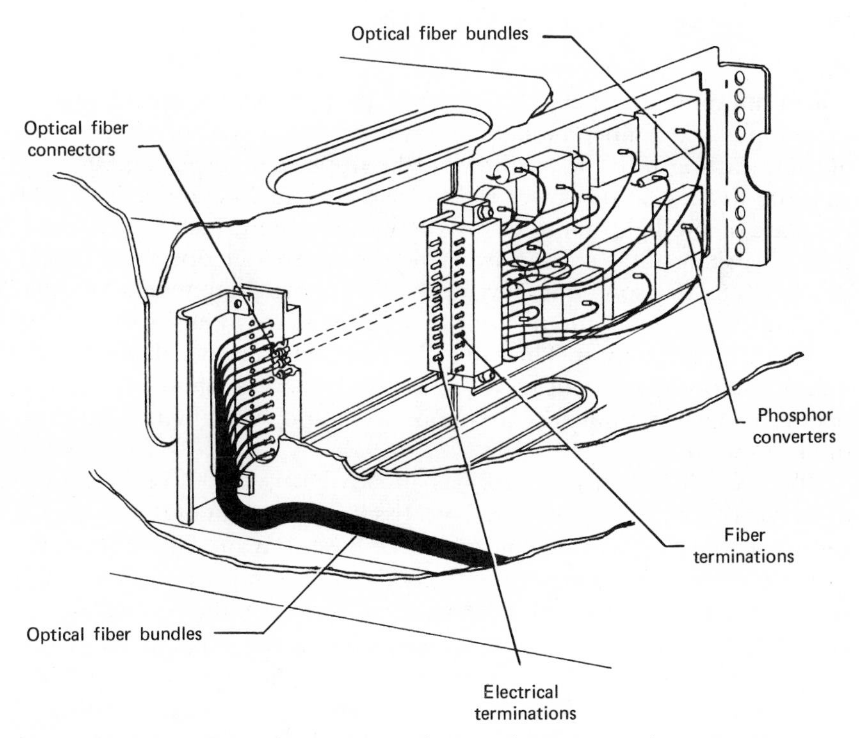

Figure 249. Thermal-to-visible conversion for optical monitoring electronic components performance.

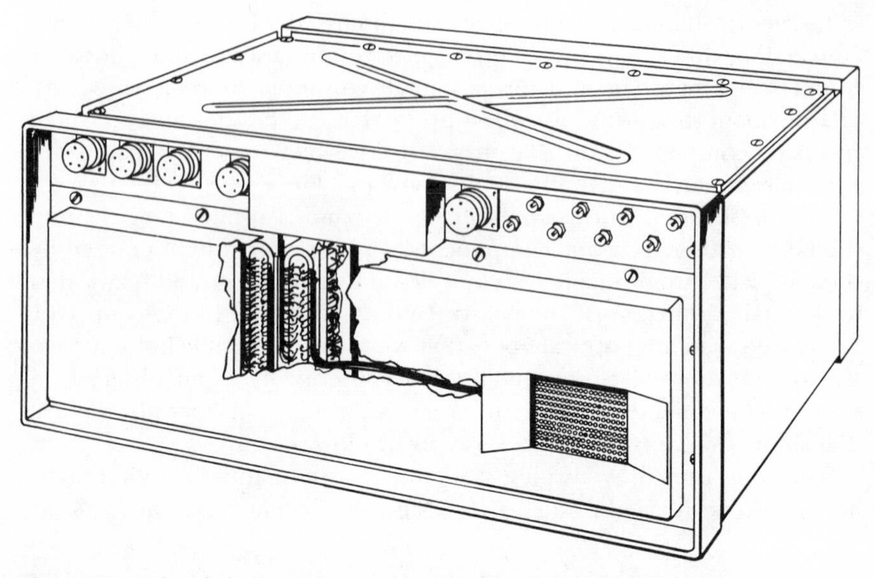

Figure 250. Infrared checkout of a portable unit with window display.

cellent discriminative power. Therefore, in the window display, the terminations of the fibers will be arrayed in sequential order of light output, starting with the brightest at the upper left corner and ending with the weakest at the lower right corner. In this fashion, a continual decrease of the light intensity will be the indication that every component is working at the correct level, and the system is in operating order.

When a failure condition develops, one or more components will vary their thermal regime, and this will be reflected in the intensity of the corresponding light output. The eye will discern, literally at a glance, which luminous dots appear out of sequence, and a service chart will tell the operator which components are out of order and which failure this condition represents.

The feasibility of this concept has already been verified in the course of an Air Force program directed by Miss Ruth Herman of the Aeropropulsion Lab., Wright Patterson Air Force Base, Dayton, Ohio.[13] Radelin phosphor 1807 was painted onto the body of six component parts of the board shown in Figure 251, where jagged squares indicate the components selected for this purpose: one resistor, three transistors, and two integrated circuits. In the laboratory setup of the experiment, optical fiber bundles shaped in Y form were used, to allow simultaneous ultraviolet energization and optical viewing of the phosphor converters, as illustrated in Figure 252. It was observed that the human eye had no difficulty in aligning the fiber terminations in decreasing or increasing

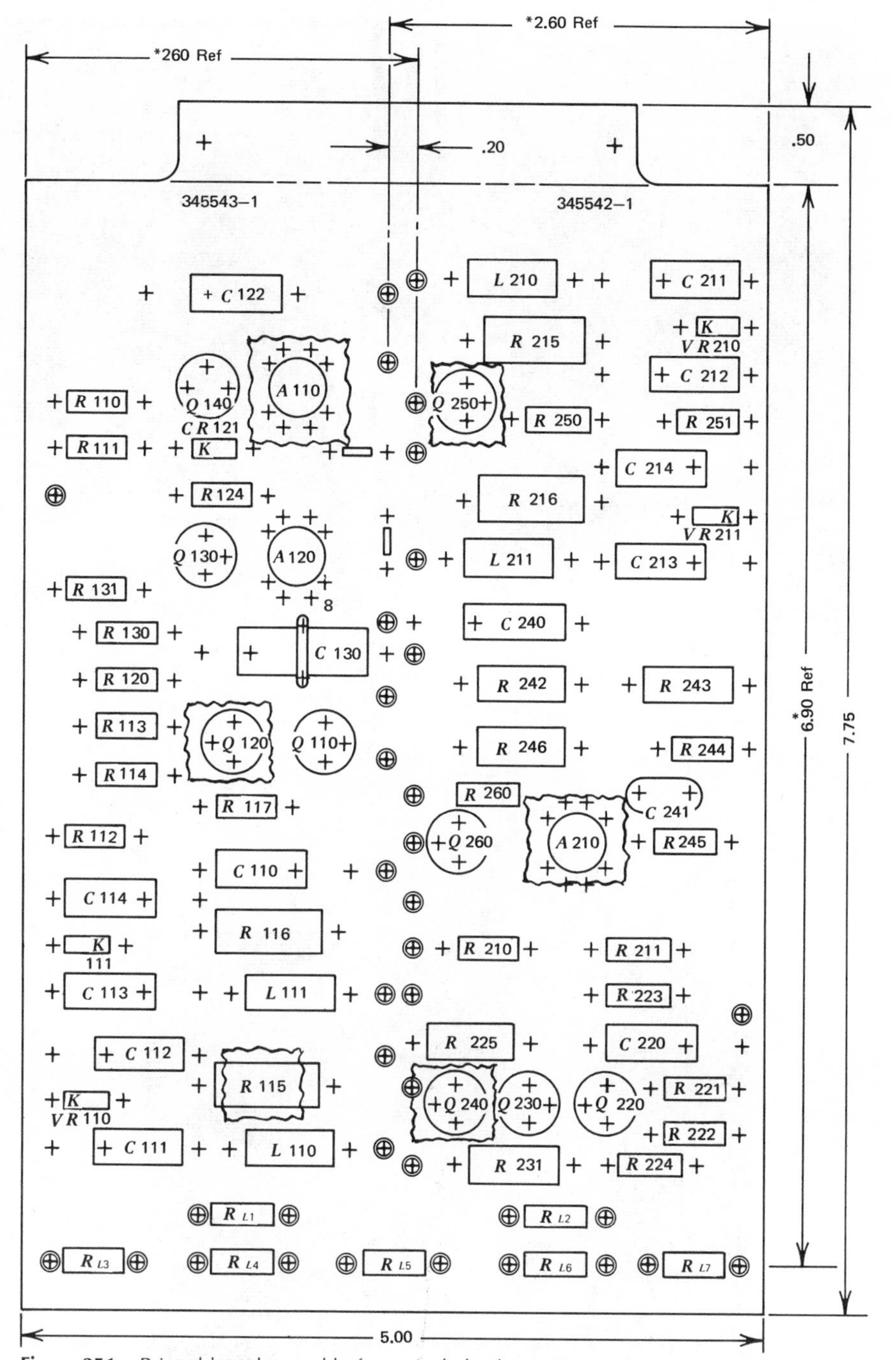

Figure 251. Printed-board assembly for optical checkout test.

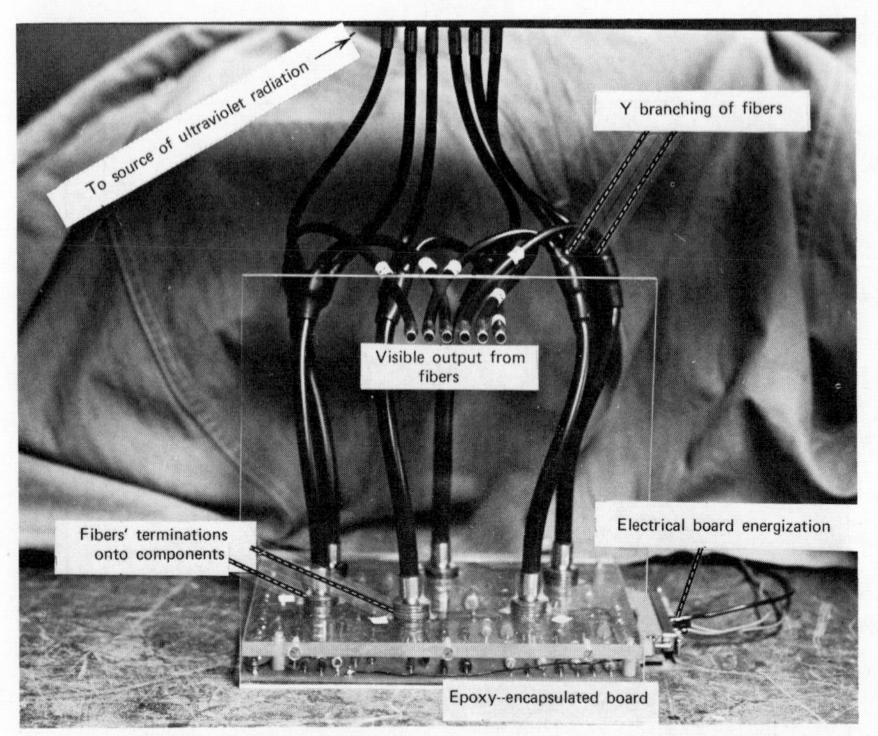

Figure 252. Fiber setup for optical checkout of printed board.

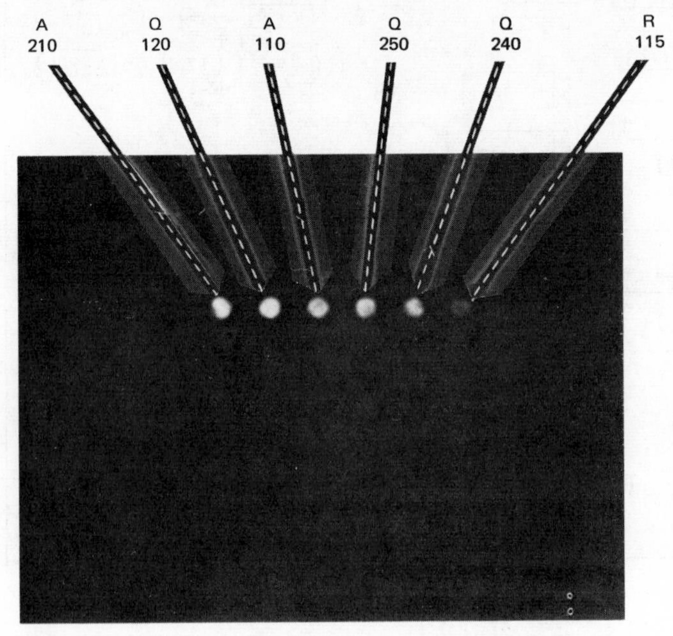

Figure 253. Light output from fibers in optical checkout test.

Figure 254. In-flight infrared checkout for aircraft blind landing.

order of light output, which in the particular type of phosphor used is inversely correlated with temperature. Figure 253 is a picture of the fiber terminations, as aligned according to their light output.

The applications could be legion. When a walkie-talkie does not talk, who can tell whether this unit or the other unit at the opposite end is at fault? The window display will tell. What shall one do when trying to obtain a working unit out of two that are out of order? Just look in the window and cannibalize accordingly. How can a pilot trust his automatic landing equipment when the airfield is engulfed in thick fog? If all the dots in the display are in correct, smooth sequence, he can trust his equipment (see Figure 254) and let it land the plane.

In outer space, things become even easier. Vacuum tubes do not need an envelope to operate; the vacuum is even better outside anyway. In a space-borne satellite, the fiber terminations could be brought to bear

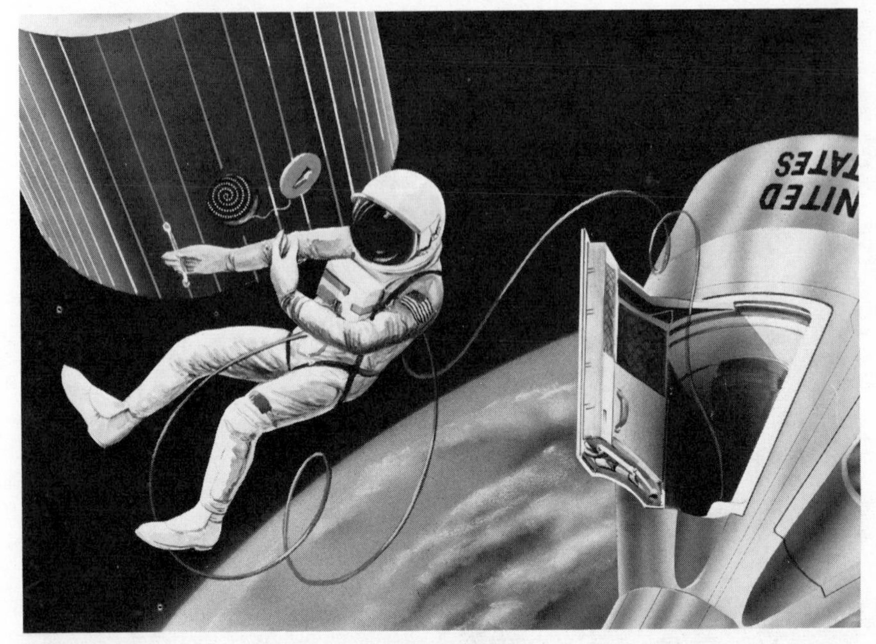

Figure 255. Space serviceman on troubleshooting duty.

right on the front plate of a vidicon tubeless tube, and the video information telemetered to earth to keep the service people informed on the status of the equipment and to point out if and when some part of it needs rest or replacement.

But what if the telemetering system breaks down? Well, in that instance, the service man will have to go out on a space call. Figure 255 shows him performing instant troubleshooting. He has just left his service truck in a parking orbit and is having a quick glance at the fiber display. In no time, he will have the satellite back on the air again, or should I say back on the vacuum?

And finally, there is our astronaut on a planet. He is through with his mission and wants to go home. All he has to do is press two buttons: the first starts the automatic take-off process; the second activates the radar controlling the rendezvous with the mother ship.

Before pressing the two buttons, he checks the operating conditions of the electronic equipment that will control his flight. Figure 256 shows the check-out device: a panel display of fiber optics. If the brightness and the color and the sequence are correct, he can confidently press the two buttons and successfully complete his mission.

Figure 256. Infrared optical checkout on a planet.

CONCLUSION

In this book, I attempted to discuss the several elements of a new emerging technology: the theoretical foundations, the physical laws, the feasibility studies, the current applications, the results, and the potential. It has not been easy, especially with all my shortcomings, for which I apologize. I am strongly biased in favor of infrared and I enjoy taking potshots at whatever appears to contrast, compete with, or combat my favorite subject.

But what a fascinating subject it is! Suddenly we realize that we have just set foot on a new world of undreamed vastness, that we hold the key to exciting, extraordinary discoveries. All things are broadcasting information in this strange new language. Already we recognize some words. Soon we will understand the full messages.

We are now at the very beginning. New roads are being opened, new concepts formulated, new techniques developed, new equipment fabricated, and new applications devised. The little work carried out so far is dwarfed by the magnitude of the effort still ahead of us.

Men of good will, let's join hands and push forward.

References

1. G. A. Rutgers, *Encyclopedia of Physics,* Vol. 26, S. Flugge (Ed.), Springer-Verlag, Berlin, 1958, p. 129.
2. H. Melchior and W. T. Lynch (Bell Laboratories, Murray Hill, N.J.), "Signal and Noise Response of High Speed Germanium Avalanche Photodiode," *IEEE Transactions of Electron Devices,* Vol. E.D. 13, December 1966, p. 829.
3. W. N. Shaunfield, J. R. Biard, and D. W. Boone (Texas Instruments, Inc., Dallas, Texas), "A Germanium Avalanche Photodetector for 1.06 μ," Presented at the Annual Conference of IEEE's Electron Devices Group, Washington, D.C., 1967.
4. Glenn H. Brown (Kent State University, Kent, Ohio), "Liquid Crystals," *Industrial Research Magazine,* May 1966.
5. W. E. Woodmansee (Boeing Company, Seattle, Wash.), "Cholesteric Liquid Crystals and Their Application to Thermal Nondestructive Testing," a report by IDEP, November 1965.
6. S. N. Bobo and A. H. Crowley (Raytheon Company, Waltham, Mass.), "Use of Contiguous Optical Fibers as a Means of Carrying Thermal Information from Welds," *Appl. Opt.,* Vol. 7, No. 9, Sept. 1968, p. 1839–1844.
7. A. J. Feduccia (R.A.D.C. Griffin Air Force Base, New York), "Reliability Screening Using IR Radiation," *Proceedings of the 3rd Annual Meeting of the ITEC (Infrared Techniques for Electronics Committee),* Huntsville, Alabama, published by the American Society for Nondestructive Testing, Evanston, Ill., Feb. 1964.
8. B. Selikson and J. DiMauro (Sylvania Electric Products, Inc., Woburn, Mass.), "A Development Study of the Reliability Screening of Large Numbers of Operating Transistors Using Infrared," *Current IR Papers,* ASNT, Evanston, Ill., 1966.
9. R. Vanzetti and A. S. Dostoomian (Raytheon Company, Wayland, Mass.), "Design and Develop Fast Scan IR Detection and Measuring Instrument," Report on NASA Contract NAS8-11604, Oct. 1969.
10. W. M. Berger (Philco Corporation, Lansdale, Pa.), "Application of Isothermal Mapping as a Reliability Tool," *Transactions of the IR Session,* ASNT Spring Convention, Los Angeles, 1965.
11. Vanzetti Infrared & Computer Systems, Inc., Dedham, Mass. Final Report on Contract NAS12-585 with NASA-ERC, Cambridge, Mass., July 1969; and on Contract NAS8-25862 with NASA-MSFC, Huntsville, Ala., March 1971.

12. A. A. Bergh and G. H. Schneer (Bell Telephone Laboratories, Murray Hill, N.J.), "The Effect of Ionic Contaminants on Silicon Transistor Reliability," *IEEE Transactions on Reliability,* Vol. R-18, May 1969.

13. J. F. Stoddard (Raytheon Company, Wayland, Mass.), Technical Report AFAPL-TR-68.84, "Infrared for Electronics Equipment Diagnosis," August 1968.

14. A. H. Crowley (Raytheon Company, Waltham Mass.), "Weld Quality as Seen by IR," *Transactions of the IR & T Session of ASNT 27th National Conference,* October 1967, Evanston, Ill.

15. R. S. Horne (Lockheed-Georgia Company, Marietta, Georgia), "IR Techniques for Component Derating of Printed Circuit Modules," *Current IR Papers,* ASNT, Evanston, Ill. 1966.

16. W. T. Lawrence (Novatek, Inc., Burlington, Mass.), "A Thermal Method for Measuring Wall Thickness and Detection of Incomplete Core Removal in Investment Castings," *Materials Evaluation,* May 1971, ASNT, Evanston, Ill.

17. G. Cawood (Raytheon Company, Norwood, Mass.) "'Infrared', The Last Word in Stress Analysis," *Transactions of IR & T Session of ASNT 27th National Conference,* October 1967, Evanston, Ill.

Bibliography

Infrared Techniques for Materials and Structures

Infrared, A Bibliography, Part II, Mauree W. Ayton, Thomas J. Darby, and others, Technical Information Division, Library of Congress, Washington, D.C. March 1957, 150 pp.

"Methods of Bond Testing," W. J. McGonnagle, *SNT Journal,* March–April 1955.

"Qualitative Analysis of Brazed Sandwich," Frank J. Filippi, Solar Aircraft Co., San Diego, Calif., *Nondestructive Testing,* January–February 1959.

"Summary of the SNT Symposium on Stainless Steel Brazed Honeycomb Structures," R. C. McMaster (Ohio State University), *SNT Journal,* September–October 1959, pp. 263–269.

Sensor State of the Art Survey, Vol. II, Dystron Inc., December 1959, 267 pp. (secret report). Contents: Gravity Measurement, Sonar, Infrared Radiation, Non-Acoustical Submarine Detection, and Inertial Navigation.

Instrumentation in Scientific Research K. Lion, McGraw-Hill Book Company, Inc., 1959, 324 pp.

"Evaluation of Brazed Honeycomb Structures," Robert C. McMaster, Anthony T. D'Annessa, and Henry W. Babel (Ohio State University Engineering Experiment Station), Columbus Report for January 1959–May 1960, on "The Chemistry and Physics of Materials" September 1960, March 8, 1961.

"Progress Report on the Technological Advances in Thermographic Inspection by Brazed Honeycomb" J. Borucki, *Proc. First National Symposium on NDT of Aircraft and Missile Components,* 1960.

"Proceedings of Infrared Information Symposia," (secret report), *Office of Naval Research,* Vol. 5, No. 1, January 1960, p. 525, AD 316 224.

Symposium on Nondestructive Testing Trends in the AEC Reactor Program. Held at Atomic Energy Commission Headquarters Bldg., Germantown, Md., May 20, 1960, TID-7600.

"Retardation and Diffraction Effects in the Conduction of Heat in Solids," F. E. Alzofon, Lockheed Missiles and Space Division 800-322, Technical Report, November 1960, U.S. contract No. ORD-17017.

IRIA Annotated Bibliography of Infrared Literature, W. Wolfe Vol. IV, No. 4 (confidential report) Institute of Science and Technology, January 1, December 31, 1960, February 1961.

Infrared Detection and Weld Defects Robert K. Lewis and John T. Norton (Advanced Metals Research Corp., Contract DA-19-020-ORD-5228), U.S. Army Materials Research Agency Technical Report, WAL TR 550/1, January 1961 and June 31, 1961 (final report).

Thermographic Profiles of Material Discontinuities, D. K. Wilburn, Report No. RR-37, March 1, 1961, Project No. 2210.4500.061.

"Feasibility Study of a Nondestructive Testing Infrared Inspection System for Bonding Flaw Detection," P. Yettito and R. Gorman (Perkin-Elmer Corp.), April 1961, p. 168.

"A Survey of Infrared Inspection and Measuring Techniques," D. K. Wilburn, *Materials Research and Standards,* Vol. 1, July 1961, pp. 528–531.

"Infrared Physics," H. L. Sach (Perkin-Elmer Corp.), Argonne National Laboratory Report No. 6516, Second Symposium on Physics and Nondestructive Testing, October 3, 4, and 5, 1961.

Feasibility Study of a Nondestructive Testing Infrared Inspection System for Bonding Flaw Detection, R. Gorman, H. L. Sachs, and P. Yettito (Perkin Elmer Corp.), December 1961.

"Applications of Infrared Technology in Nondestructive Testing," Frederick E. Alzofon, *Proceedings of Bureau of Naval Weapons, Missiles and Rockets Symposium,* 1961, Concord, Calif., pp. 247–251.

"Resolution of Flaws in Infrared Nondestructive Testing," F. E. Alzofon (Baylor University, Waco, Tex.), *American Rocket Society, Solid Propellant Rocket Conference,* January 24–26, 1962.

Development of Nondestructive Testing Techniques for Large Solid-Propellant Rocket Motors, Charles Harris (United Technology Corp., Sunnyvale, Calif.), Monthly Progress Report No. 2, April 1962, Contract AF 04(611)8018.

"Nondestructive Infrared Inspection of Three Layered Slabs," Ronald James Baschiere, Master's Thesis—I.I.T., June 1962.

Progress Report on the Infrared Nondestructive Testing Program, F. E. Alzofon (Lockheed Missiles & Space Company), July 1962.

"Infrared Nondestructive Testing of Glass Filament Wound Rocket Motor Cases," F. E. Alzofon, L. E. Florant, R. K. Ronarl, and M. J. Vann (Lockheed Missiles and Space Company), Paper No. 76, 1962.

Feasibility Study of a Nondestructive Testing Infrared Inspection System for Bonding Flaw Dection, P. Yettito, R. Gorman, and H. Sachs, Perkin-Elmer Corp., August 1962.

Development of an Infrared Inspection Technique, L. H. Caveny and W. P. Corley (Thiokol Chemical Corporation, Redstone Arsenal, Huntsville Ala.), ASD TR 62-7-842 September 1962.

Nondestructive Testing of Solid-Propellant Rocket Motors, R. P. Pasley and C. L. Seale (Battelle Memorial Institute), October 25, 1962.

The Feasibility of Utilizing Thermal Detecting Devices to Determine Fatigue Damage in Metals,

G. Stephen Tint and Marvin Herman (Franklin Institute, Laboratories for Research & Development, Philadelphia, Pa.), Report NAVAIRENGGENASL-1064, December 1962.

A National Survey on Activities in Infrared Nondestructive Testing, D. K. Wilburn, (Army Tank-Automotive Command), December 3, 1962.

Applied Research to Establish Infrared Detection Methods for Nondestructive Analysis of Metallic and Ceramic Structures, D R. Maley, H. T. Pinnick, and R. J. Barton (Automation Industries, Inc., Boulder, Colo.), Wright-Patterson AFB, Ohio; AF Mater. Lab., August 1963, 97 pp. Refs for part 1 see N64-10251 01-23, Technical Documentary Report, January 1962–February 1963, Contract AF 33(616)-7725 (ASD-TDR-62-385, Pt. 11: AD-605511).

"Sources of Error in Measuring Flame Temperature With Infrared Radiation," A. E. Kadishevich and V. A. Kokuchaeva, *Izvestiya Vuz-Chernaya Metallurgiya,* 1962, pp. 184–195; *SNT Journal,* May–June 1963.

Applied Research to Establish Infrared Detection Methods for Nondestructive Analysis of Metallic and Ceramic Structures, Aeronautical Systems Div., Metals & Ceramics Lab., Wright-Patterson AFB, Ohio, Report ASD-TDR-62-385 (final report), January 1963, 85 pp.

"Infrared – A New Approach in Industrial Quality Control," R. Vanzetti, *Industrial Quality Control Magazine,* April 1963.

Feasibility Study of a Nondestructive Testing Infrared Inspection System for Bonding Flaw Detection, Gorman, R., et al. (Perkin-Elmer Corp.), Final Report, PE-TR-7345, June 28, 1963.

Evaluation of Coated Refractory Metal Foils, V. S. Moore and A. R. Stetson (AD-423-672 Solar, San Diego, Calif.), Quarterly Report 5, July 1,–September 30, 1963.

"Interim Report on IR-NDT System Research," John L. Radsliff (Hercules Powder Company Bacchus Works, Magna, Utah), 28th Meeting of Polaris/Minuteman/Pershing NDT Testing Committee, Paper 28-p, July 9–11, 1963.

Status Report on the Infrared Nondestructive Test Program F. E. Alzofon (Hercules Powder Company Bacchus Works, Magna, Utah) 28th Polaris/Minuteman/Pershing Propulsion NDT Test Committee Meeting, July 9–10, 1963.

Infrared for Nondestructive Testing and Inspection, P. R. Stromer (Lockheed Missiles & Space Co., Sunnyvale, Calif.) (gov. limited rights), August 1963.

"Nondestructive Test Development for Polaris Wound Chambers," A. M. Granat and P. A. Sreinkritz (Lockheed Missiles and Space Co.), *Proceedings of the Fourth Annual Symposium on Nondestructive Testing of Aircraft and Missile Components,* San Antonio, Tex., 1963.

"A Thermal Infrared Inspection Procedure for Detection and Location of Flaws in Solid Propellant Rocket Cases," Harold L. Sachs, *Proceedings of the 4th Annual Symposium on Nondestructive Testing of Aircraft and Missile Components,* San Antonio, Tex., 1963, Southwest Research Institute.

"A Thermal Scanning Technique for Nondestructive Testing," D. R. Maley and G. J. Posakony, *Proceedings of the 4th Annual Symposium on NDT of Aircraft and Missile Components,* San Antonio, Tex., 1963.

Report No. 2 of the Aerospace Manufacturing Techniques Panel, Material Advisory Board of the Division of Engineering and Industrial Research National Academy of Sciences, National Research Council, Washington, D.C., Report MAG-139-M (AMT-2), October 1963, 396 pp.

"Factors Influencing the Detection of Flaws in Glass Filament Wound Rocket Motor Cases by IR-Scanning," F. E. Alzofon (Chemical Propellant Information Agency Applied Physics Laboratory, Johns Hopkins University, Howard Country, Md.), *Second Symposium on NDT, of Solid Propellant Rocket Motors,* September 30–October 1, 1963.

IR-Image System Program. Second Progress Report. Polaris, A-3 NDT, J. L. Radsliff (Quality Evaluation Laboratory, U.S. Naval Weapons Station, Concord, Calif.), November 1963, Vol. 1., revision 2.0.

"Detection of Propellant/Liner Separations by Infrared Scanning," F. E. Alzofon (Aeroject-General Corporation, Sacramento, Calif.), 30th Meeting, Polaris/Minuteman/Pershing NDT Testing Committee, Paper 30F, January 1964.

"Nondestructive Testing Information Retrieval Systems Here and Abroad," U.S. Army Materials Research Agency, Watertown Arsenal, Watertown, Massachusetts, NTN-641, Nondestructive Testing Newsletter, February 1964, 5–7 p.

"Hot Targets Flash Flaw Patterns," W. N. Redstreake, *The Iron Age,* March 5, 1964.

Applied Research to Establish Infrared Detection Methods for Nondestructive Analysis of Metallic and Ceramic Structures, Technical Documentary Report, February 1963–January 1964 D. R. Maley and S. W. Maley (Wright-Patterson AFB, Ohio), AF Mater. Lab., March 1964, 157 pp. (Contract AF33(616)-7725) (ASD-TDR-62-385, Pt. III: AD-605510).

"Inspection Becomes Resourceful, More Automatic, More Effective," W. N. Redstreake and Raymond Shah, *Iron Age Metalworking International,* Vol. 3, No. 5, May 1964, pp. 13–16.

Recent Accomplishments in Nondestructive Testing Research at Hanford Laboratories, G. F. Garlick (General Electric Co., Hanford Atomic Products Operation, Richland, Wash.), July 16, 1964, 40 pp.

Nuclear Fuels and Materials Development, William L. R. Rice, Ed., USAEC Report, Clearing house for Federal Scientific and Technical Information, National Bureau of Standards, U.S. Department of Commerce, Springfield, Va., July 1964.

"Radiographic Penetrometers as Infrared Standards," D. K. Wilburn, *Materials Evaluation,* Vol. 22 No. 10, October 1964.

"Preliminary Report on the Development of a Nondestructive Test for Bridgewires," Douglas, W. Ballard, L. J. Klamerus, and J. D. Stewart (Sandia Corporation, Albuquerque, N. Mex.), *Proceedings of the 3rd Annual Meeting of the Infrared Techniques for Electronics Committee,* February 19–21, 1964; published by the Society for Nondestructive Testing, Inc., Evanston, Ill., October 1964.

"Nondestructive Testing," Warren McGonnagle and Ford Park (Southwest Research Institute, San Antonio, Tex.), *International Science and Technology Journal,* July 1964; also *Materials Evaluation,* December 1964, p. 561.

Infrared/Thermal Nondestructive Testing Abstracts, D. K. Wilburn (U.S. Army Tank-Automotive Center, Warren, Mich.), December 1964.

"The Inspector's Role in Onstream Inspection," E. F. Ehmke, *American Petroleum Institute, Proceedings,* Section III, Refining, Vol. 45, No. 3, 1965, pp. 163–168.

"Fundamentals of Infrared Radiation," R. Vanzetti, presented at the 24th National Convention of the Society for Nondestructive Testing, Philadelphia, Pa., October 23, 1964; published in *Materials Evaluation Magazine* (An Official Journal of SNT), January 1965.

"Infrared NDT of Solid Propellant Missile Motors," Emil M. Bergh, Spring Convention of SNT, February, 22–26 1965, Los Angeles, Calif.

"Emissivity Independent Infrared Thermal Testing Methods," D. R. Green, Spring Convention of SNT, February, 22–26, 1965, Los Angeles, Calif.

"Evaluation of the Resistance Microwelding Process by Infrared Energy Measurements," Sheldon Leonard and James K. Lee, Spring Convention of SNT, February, 22–26, 1965, Los Angeles, Calif.

"Application of Infrared Imaging to Failure Prevention on High Voltage Power Lines," Stephen N. Bobo, Spring Convention of SNT, February, 22–26, 1965, Los Angeles, Calif.

"The Relative Contributions of Emissivity and Thermal Conductivity in Infrared Nondestructive Testing," F. E. Alzofon, Spring Convention of SNT, February, 22–26 1965, Los Angeles, Calif.

"Two Thermal Nondestructive Testing Techniques," D. R. Maley, Spring Convention of SNT, February, 22–26 1965, Los Angeles, Calif.

"Nondestructive Testing Applications of Military Infrared Technology," Otto Renius, and David Wilburn, Spring Convention of SNT, February, 22–26 1965, Los Angeles, Calif.

"Thermal Evaluation of Bond Integrity in Welded, Brazed and Adhesive Joints," W. R. Apple and D. R. Maley (Automation Industries, Inc. Boulder, Colo.), *Proceedings of the Fifth Annual Symposium on Nondestructive Evaluation of Aerospace and Weapons Systems, Components, and Materials,* April 20–22, 1965, San Antonio, Tex.

"Survey of Infrared Techniques for Nondestructive Evaluation," Burton Bernard and H. E. Williams (Infrared Industries Inc., Santa Barbara, Calif.), *Proceedings of the Fifth Annual Symposium on Nondestructive Evaluation of Aerospace and Weapons Systems, Components, and Materials,* April 20–22, 1965, San Antonio, Tex.

"Infrared Scans for Inner Defects," T. H. Malim, Associate Editor, *Iron Age,* May 27, 1965.

Large Motor Case Technology Evaluation, The Boeing Company, Aero-Space Division, Seattle, Washington. Wright-Patterson Air Force Base Contract No. AF 33(615)-1623, Annual Progress Report, Vol. II, June 1964–June 1965.

Investigation of Nondestructive Method for the Evaluation of Graphite Materials, G. E. Lockyer (Research and Advanced Development Division AVCO Corp., Wilmington, Mass.), Wright-Patterson Air Force Base, Ohio. Technical Report AFML-TR-65-113, June 1965.

"A Thermal Imaging Technique of Nondestructive Testing," D. K. Wilburn, U.S. Army Tank Automotive Center Report 8883, June 1965.

"A Review of Nondestructive Testing for Plastics: Methods and Applications," Nicholas T. Baldanza (Plastics Technical Evaluation Center (PLASTEC) Picatinny Arsenal, Dover, N.J.), Plastec Report 22, August 1965, 182 pp, includes references and bibliography.

"Reducing Emissivity Errors in Infrared Detecting Systems," F. R. Bareham, *Sixth Annual Symposium on Physics and Nondestructive Testing* September 28–30, 1965, Dayton, Ohio.

Elastic Constants of Small Sintered Ceramic Specimens, Orson L. Anderson and Naohiro Soga (Lamont Geological Observatory, Columbia University, Palisades, N.Y.), Wright Patterson Air Force Base, AFML-TR-65-202, September 1965.

"Optics and Infrared Nondestructive Testing," F. E. Alzofon, *Sixth Annual Symposium on Physics and Nondestructive Testing,* September 28–30, 1965, Dayton, Ohio.

Development of Nondestructive Methods for the Evaluation of Organic Nonmetallic Materials, J. R. Zurbrick and C. K. Chiklis (Research and Advanced Development Division, AVCO Corporation Wilmington, Mass.), Wright-Patterson Air Force Base, Contract AF 33(615)-1705; TR AFML-TR65-267, October 1965.

"Fundamentals of Infrared Technology for Infrared Nondestructive Testing," William L. Wolfe, 25th National Convention, Society for Nondestructive Testing, October 18–22, 1965, Detroit, Mich.

"Thermal Imaging for Material Integrity," David K. Wilburn, 25th National Convention, Society for Nondestructive Testing October 18–22, 1965, Detroit, Mich.

"The Application of an Infrared Radiometer to Measure Plastic Flow in Metals," Norman C. Small, 25th National Convention, Society for Nondestructive Testing October 18–22, 1965, Detroit, Mich.

Infrared Temperature Mapping of Weapons, P. Shajenko and C. L. Nickerson (Springfield Armory, Springfield, Mass.), Report SA-TR-20-2601, November 26, 1965.

Investigation of Nondestructive Methods for the Evaluation of Graphite Materials, G. E. Lockyer, E. M. Lenoe, and A. W. Schultz (AVCO Corp., Space Systems Divisions, Lowell, Mass.), Wright-Patterson Air Force Base Contract No. AF 33(615)-1601 AFML-TR-66-101, July 1966.

"Detection of Flaws in Adhesive Bonded Metallic Honeycomb," F. E. Alzofon, Spring Convention, Society for Nondestructive Testing March 7–10, 1966, Los Angeles, Calif.

"Inspection of Reinforced Plastics by Thermal Nondestructive Techniques," G. J. Posakony et al., Testing Techniques for Filament Reinforced Plastics, sponsored by AFML and ASTM, September 1966, AFML-TR-66-274, pp. 577–604.

"An Infrared Method of Rocket Motor Inspection," J. C. St. Clair (Thiokol Chemical Corporation, Huntsville, Ala.) *Materials Evaluation,* August 1966.

Development of Nondestructive Methods for Evaluating Diffusion-Formed Coating on Metallic Substrates, R. C. Stinebring and T. Sturiale (Avco Corp., Lowell, Mass.), Tech Report AFML-TR-221, September 1966.

Development of Nondestructive Dynamic Monitoring Instrumentation for Resin Impregnated Glass Roving (submitted to U.S. Naval Applied Science Laboratory, Brooklyn, N.Y., by Ferro Corp., Cordo Division, Mobile, Ala.), 1st Quarterly Progress Report, April 30, 1966.

"New NDT Methods Probe: The Look, Sound, and Smell of Quality," *Steel,* February 14, 1966, pp. S-1-S-8.

A Case History for Infrared Nondestructive Testing, D. K. Wilburn, University of Wisconsin Extension Institute for Nondestructive Testing, March 1966.

Study on Development of Techniques for Resistance Welding, W. R. Hutchinson et al. (George C. Marshall Space Flight Center, Huntsville, Ala.), Martin Marietta Corp., Orlando, Fla. OR-8741, Phases I & II, February 1967.

"NDT on Adhesive Joints May Open up New Jobs," *Steel,* March 27, 1967 pp. 59–62.

"Infrared for Product Evaluation," R. Vanzetti, presented at the 5th Conference on New Horizons in Quality and Reliability, Binghamton, N.H., March 18, 1967.

"An Analytical Approach to Infrared Nondestructive Testing," A. W. Schultz, Fifth International Conference on NDT, Montreal, Canada, May 1967.

"Infrared Techniques for NDT, Past, Present and Future," R. Vanzetti, 5th International Conference on NDT, Montreal, Canada, May 1967.

"New NDT Techniques for Aerospace Materials and Structures," E. J. Kubiak (General American Transportation, Niles, Ill.), Symposium on Correlation of Material Characteristics with Systems Performance, USAF Conference Facility, Orlando AFB, Fla., May 10–12, 1967.

"Eddy Current and Infrared Inspection of Graphite," C. V. Dodd, Paper 61, ASTM 70th Annual Meeting (June 1967). Published in symposium transactions.

"Infrared Testing of Bonds Between Graphite and Protective Coatings," D. R. Green and C K. Day, Paper 58, ASTM 70th Annual Meeting (June 1967). Published in symposium transactions.

Investigation of Nondestructive Methods for the Evaluation of Graphite Materials, G. E. Lockyer et al. (AVCO Corp., R&D Division, Lowell, Mass.), Wright-Patterson AFB, Ohio 45433, Technical Report AFML-TR-67-128, June 1967.

"Infrared Evaluation of Microweld Quality," F. E. Alzofon and A. D. McDonald, *Materials Evaluation,* Vol. XXV, No. 8, August 1967, pp. 183–184.

"Infrared and Thermal NDT," R. Vanzetti, presented at the 22nd International Conference of the Instrument Society of America, Chicago, Ill. September 13, 1967.

"Armor Plate Bond Evaluation With Infrared," Paul E. J. Vogel, SNT, 27th National Fall Conference, October 16–19, 1967, Cleveland, Ohio.

"Analysis of Three Modes of Cooling in Infrared Nondestructive Testing," F. E. Alzofon, SNT, 27th National Fall Conference, October 16–19, 1967, Cleveland, Ohio.

"Measuring Microbond Integrity with an Infrared Microradiometer," D. H. Schumacher, SNT, 27th National Fall Conference, October 16–19, 1967, Cleveland, Ohio.

"Evaluation of Helicopter Propeller Blades Bonding," W. B. Allen, SNT, 27th National Fall Conference, October 16–19, 1967, Cleveland, Ohio.

"Infrared Techniques," R. Vanzetti, presented at the Merrimack Valley Section of the American Society for Quality Control, Nashua, N.H., October 5, 1967

"Infrared Evaluation of Multilayer Boards, R. W. Jones and L. W. White, SNT, 27th National Fall Conference, October 16–19, 1967, Cleveland, Ohio.

Investigation of Nondestructive Methods for the Evaluation of Graphite Materials, G. E. Lockyer et al., AVCO SSD, Lowell, Mass. AFML-TR-67-128, 1968 Contract AF33(615)-1601.

"Planned Nondestructive Testing as Preventive Maintenance for Steel Plants," W. H. Tait, *Materials Evaluation,* April 1968, pp. 54–58.

"Thermal and Infrared Methods for Nondestructive Testing of Adhesive-Bonded Structures," E. W. Kutzscher (Lockheed Calif. Company, Burbank, Calif.), *Materials Evaluation,* July 1968.

"Sensing the Invisible World," Eric M. Wormser, *Applied Optics,* Vol. 7, No. 9, September 1968.

"Evaluation of Bonds in Armor Plate and Other Materials Using Infrared Nondestructive Testing Techniques," Paul E. J. Vogel, *Applied Optics,* Vol. 7, No. 9, September 1968.

"Electro-Optical Remote Sensing Methods as Nondestructive Testing and Measuring Techniques in Agriculture," Victor I. Myers and William A. Allen, *Applied Optics,* Vol. 7, No. 9, September 1968.

"Infrared Reflectography: A Method for the Examination of Paintings," J. R. J. Van Asperen De Boer, *Applied Optics,* Vol. 7, No. 9, September 1968.

"An Infrared Transient Method for Determining the Thermal Inertia Conductivity and Diffusivity of Solids," Arnold W. Shultz, *Applied Optics,* Vol. 7, No. 9, September 1968.

"Principles and Applications of Emittance—Independent Infrared Nondestructive Testing," Donald R. Green, *Applied Optics,* Vol. 7, No. 9, September 1968.

"Infrared Applications to Nondestructive Testing," W. R. Apple, ASNT, 1968 Fall Conference, October 14–17, Detroit, Mich.

Investigation of Nondestructive Methods for the Evaluation of Graphite Materials, R. C. Stinebring, A. W. Orner (AVCO Space Systems Division, Lowell, Mass. 01887), Tech. Report AFML-TR-68-128, Part II, February 1969.

"Use of IR-NDT Techniques for Evaluating Materials," R. Vanzetti, presented at the February 12, 1969 Meeting of the Boston Section of ASNT.

"Nondestructive Test Methods for Reinforced Plastic/Composite Materials," G. Epstein (Aerospace Corp., El Segundo, Calif.), Aerospace Rept. No. TR-0200(4250-20)-4; Air Force Rept. No. SAMSO-TR-69-78, 3, February 1969.

"Fundamentals and Applications for IR-NDT Techniques," R. Vanzetti, presented at the March 30, 1969 Meeting of the Albany Section of ASQC.

"National Survey of Current IR-NDT Systems," Paul E. J. Vogel, ASNT, 1969 Spring Conference, March 10–13, Los Angeles, Calif.

"Nondestructive Inspection of an Advanced Geometry Composite Blade," R. D. Whealy and A. Intrieri (Boeing Co., Vertol Div., P.O. Box 16858, Phila., Pa. 19142), presented at Conference on NDT Plastic/Composite Structures March 1969, Dayton, Ohio.

"Infrared Temperature Measurements Using a Radiation Calibrator," A. J. Intrieri and N. W. James, ASNT 1969 Spring Conference, March 10–13, Los Angeles, Calif.

"Measurement of Anodic Coating on Aluminum by Emissivity," A. E. Lawler and N. Pirzadeh (Quality Assurance Engineering, The Boeing Co., Seattle, Wash.) *Materials Evaluation,* May 1969, p. 118.

"A Thermal Method of Measuring Wall Thickness and Detection of Incomplete Core Removal in Investment Castings," W. T. Lawrence, ASNT, 1969 Fall Conference, October 13–16, Philadelphia, Pa.

"Infrared Nondestructive Testing in Medicine," D. H. Whittier, ASNT, 1969 Fall Conference, October 13–16, Philadelphia, Pa.

"Infrared Sensor for Evaluation of Electric Welding Modules," S. N. Bobo and A. Crowley, ASNT, 1969 Fall Conference, October 13–16, Philadelphia, Pa.

"Infrared Method for Detection of Inhomogeneities in Titanium," F. J. Vicki, ASNT, 1969 Fall Conference, October 13–16, Philadelphia, Pa.

"Infrared Detection of Invisible Shorts in Electrolytic Copper Refining Tanks," Bergstrom, Leif, IEEE International Convention, NYC Coliseum, March 23–26, 1970.

"Infrared Bond Inspection of Multilayer Laminates," Richard Leftwich and G. B. Ordway,

IEEE International Convention, NYC Coliseum, March 23–26, 1970.

"Infrared Techniques for Surveillance and Crime Deterrence," J. S. Jackson and R. L. Cosgriff, IEEE International Convention, NYC Coliseum, March 23–26, 1970.

"Economic Effectiveness of IR-NDT," H. M. Hedgpeth, ASNT, 1970 Spring Conference, March 9–13, Los Angeles, Cal.

"IR Evaluation of Rubber Tires," P. E. J. Vogel, ASNT, 1970 Spring Conference, March 9–13, Los Angeles, Cal.

"IR and Thermal Evaluation of a Phased Array Antenna," M. Pitasi and R. J. Joachim, ASNT, 1970 Spring Conference, March 9–13, Los Angeles, Cal.

"Electro-Thermal Method for Nondestructively Testing Welds and Complex Metal Parts," L. D. McCullough (Battelle-Northwest Lab. Richland, Wash.) and D. R. Green (WADCO, Hanford Eng. Dev. Lab., Richland, Wash.), ASNT 30th Natl. Fall Conference, October 1970, Cleveland, Ohio.

"A Capability and Limitation Study of Structual Thermography," R. H. Yoshimura (Bernard Stiefeld, Sandia Laboratories, Albuquerque, N.M.), ASNT 30th Natl. Fall Conference, October 1970, Cleveland, Ohio.

"Thermal and Infrared Nondestructive Testing of Composites and Ceramics," Donald R. Green (WADCO, Hanford Engineering Dev. Lab. Richland, Wash.), ASNT 30th Natl. Fall Conference, October 1970, Cleveland, Ohio.

"Practical Problems Related to the Thermal Infrared NDT of a Bonded Structure," J. W. Sneering, R. J. Roehrs, and K. P. Hacke (McDonnell Aircraft Co., St. Louis, Mo.), ASNT 30th Natl. Fall Conference, October 1970, Cleveland, Ohio.

"Application and Evaluation of a Practical Infrared System for Nondestructively Predicting Tire Failure," George H. Halsey (Halsey Industrial Systems, Inc. Indiana, Pa.), ASNT 30th Natl. Fall Conference, October 1970, Cleveland, Ohio.

"Latest Applications of Infrared Techniques for Industrial Process Control," Riccardo Vanzetti (Vanzetti Infrared & Computer Systems, Inc. Dedham, Mass.), IEEE-IGA Annual Meeting, Cleveland, Ohio, October 1971.

"Computer-Based Display of Nondestructive Evaluation Data," B. Stiefeld (Sandia Laboratories, Albuquerque, N.M.), Sandia Laboratories Development Report, May 1971.

A Test Method for Nondestructive Testing of Fuel Filtration Equipment Using Thermography, Anthony P. Pontello (Naval Air Propulsion Center, Naval Base, Philadelphia, Pa.) Technical Report NAPTC-AED-1963, December 1971.

"A Capability and Limitation Study of Thermography of Carbon-Carbon Cones," B. Stiefeld and R. H. Yoshimura (Sandia Laboratories, Albuqerque, N.M.), *Sandia Lab. Develop. Rep.,* January 1971.

"Analytical Predictions of the Thermal Signatures of Honeycomb Panel Bond Defects," R. M. Ashley (Martin Marietta Corp., Denver, Colorado), *ASNT Nat. Spring Conf.,* Los Angeles, Calif., March 1972.

"Recent Developments in the Use of Infrared for Nondestructive Evaluation of Automotive Tires," Stephen N. Bobo (U.S. Department of Transportation, Cambridge, Mass.), *ASNT Nat. Spring Conf.,* Los Angeles, Calif., March 1972.

"Application of Infrared Radiometric Scanning to the Detection of Flexible Package Seal

Defects," Dr. R. A. Lampi, Norman Roberts, and Frederick A. Costanza (U.S. Army Natick Labs., Natick, Mass.), *ASNT Nat. Spring Conf.,* Los Angeles, Calif., March 1972.

"A Survey of Applications of Infrared and Thermal Techniques to NDT," Paul E. J. Vogel (Army Materials and Mechanics Research Center, Watertown, Mass.), *ASNT Nat. Spring Conf.,* Los Angeles, Calif., March 1972.

Infrared Techniques for Electronic Circuits and Components

"Infrared: A Tool for Reliability Engineering, Design Analysis and Quality Control," R. Vanzetti, presented at the Seminar on Utilization of Infrared Techniques for Electronics Equipment Reliability Improvement, Dedham, Mass. May 10, 1962, published in the *ITEC Meeting Transactions.*

"Use of an I-8A Camera to Test Electronic Equipment," R. Vanzetti and George A. Solli, presented at Space Systems Reliability Committee Meeting, March 6, 1962 Lockheed Missiles and Space Division, Sunnyvale, Calif. published in the *Minutes of the Meeting.*

"Fundamentals of Infrared Radiation," R. Vanzetti, presented at the Seminar on Utilization of Infrared Techniques for Electronics Equipment Reliability Improvement, Dedham, Mass., May 10, 1962, published in the *ITEC Meeting Transactions.*

"Infrared Techniques for Electronics Progress Report," R. Vanzetti, presented at the First Meeting of Infrared Techniques for Electronics Committee (ITEC), Dedham, Mass. September 25, 1962, published in the *ITEC Meeting Transactions.*

"Infrared: A New Tool for Reliability Improvement," R. Vanzetti, presented at Northeast Electronics Research and Engineering Meeting, Boston, Mass., Novemember 6, 1962; published in *NEREM Record,* 1962.

"Infrared Techniques Enhance Electronic Reliability," R. Vanzetti presented at the Ninth

"Component Failures Predicted by Infrared," P. J. Klass (reporting on work by R. Vanzetti), Aviation Week and Space Technology Magazine, December 1962.

"Infrared for Circuit Checkout," R. A. Herman, *Electrical Engineering,* January 1963. National Symposium on Reliability and Quality Control, San Francisco, Calif., January 23, 1963; published in the *Proceedings of the Symposium.*

"New Applications for Infrared Techniques," R. Vanzetti, presented at the 19th Annual Quality Control Conference of the Rochester Section ASQC, University of Rochester, N.Y., March 26, 1963; published in the *RSQC Conference Transactions,* 1963.

"Report on Raytheon Progress in Utilization of Infrared Techniques," R. Vanzetti, presented at the second Meeting of the ITEC, May 1, 1963, Boston, Mass; published in the *ITEC Meeting Transactions.*

"Infrared Radiation: A New Dimension for Production Reliability and Maintainability," R. Vanzetti, presented at the Joint Aerospace Reliability and Maintainability Conference, May 7, 1963, Washington, D.C.; published in the *Conference Proceedings.*

"Infrared—A New Dimension for Electronics Reliability," R. Vanzetti, published in May–June 1963 issue of *Electronic Progress.*

"Infrared: A New Approach for Higher Reliability," R. Vanzetti, presented at the International Conference and Exhibit on Aerospace Support, August 7, 1963, Washington, D.C.; published in the *Conference Proceedings.*

"Infrared Techniques Enhance Electronic Reliability," R. Vanzetti, *Solid State Design Magazine,* August 1963.

"Infrared Testing of Electronic Components and Circuits," R. Vanzetti, 23rd National Convention, Society for Nondestructive Testing, October 21–25, 1963, Cleveland, Ohio.

"Infrared, Another Medium of Evaluation," R. Vanzetti, *Quality Assurance Magazine* December 1963 and January 1964.

"Infrared Nondestructive Testing for Improvement of Integral Electronic Circuits," H. S. Kleiman and J. D. Reese, 3rd Annual Meeting of the Infrared Techniques for Electronics Committee, February 19–21, 1964, Huntsville, Ala.

"Infrared Technique for Electronics Testing and a Plan for Its Implementation," L. R. Judd, and J. F. Pina, 3rd Annual Meeting of the Techniques for Electronics Committee, February 19–21, 1964, Huntsville, Ala.

"Infrared Emissivity Equalization Via Use of Special Coatings," Neville Burrowes, 3rd Annual Meeting of the Infrared Techniques for Electronics Committee, February 19–21, 1964, Huntsville, Ala.

"IR Techniques for Electronics: Latest Progress in R&D and Applications," R. Vanzetti, 3rd Annual Meeting of the Infrared Techniques for Electronics Committee, February 19–21, Huntsville, Ala.

"Reliability Screening Using Infrared Radiation," Anthony J. Feduccia, 3rd Annual Meeting of the Infrared Techniques for Electronics Committee, February 19–21, 1964, Huntsville, Ala.

Nondestructive Testing of Soldered Joints and Electronic Circuit Board Components, H. Heffan (Quality Evaluation Laboratory, Naval Weapons Station, Concord, Calif.), Progress Report QE/CO Report 64–14, March 2, 1964.

"Infrared Looks into Integration," *Electronic Design,* Vol. 12, May 11, 1964; *Reliability Abstracts and Technical Reviews,* Vol. 5 No. 1, January 1965, NASA Ser 1696.

"Nondestructive Testing for Microelectronics: An Appraisal," H. S. Kleinman, 24th National Convention, Society for Nondestructive Testing, October 1964, Philadelphia, Pa.

Detection of Solder Joint Imperfections, Autonetics, Anaheim, Calif., Final Report, January 12, 1965, 16 pp. report MPP26; C5 19 33 Contract AF04 694 402.

"Tolerance Studies for Infrared Production Testing of Electronics," L. R. Judd and T. J. Magee, Spring Convention of SNT, February 22–26, 1965, Los Angeles, Calif.

"A High Speed Infrared Mapping System for Reliability Assessment of Miniature Electronic Circuits and Examples of Its Use," Harvey F. Dean and Robert M. Fraser, Spring Convention of SNT, February 22–26, 1965, Los Angeles, Calif.

"Infrared Fiber Optics and Infrared Nondestructive Testing," R. J. Simms, Spring Convention of SNT, February 22–26, 1965, Los Angles, Calif.

"Infrared Radiation Measurement for Reliability and Quality Control," David Seltzer and W. R. Randle, Spring Convention of SNT, February 22–26, 1965, Los Angeles, Calif.

"Application of Infrared Technology in Development of Thin Film Resistors," P. R. Young, Spring Convention of SNT, February 22–26, 1965, Los Angeles, Calif.

"Evaluation of IR & Excess Noise Measurements for Integrated Circuit Reliability

Improvement by Detection and Rejection of Potential Early Failures," David J. Collins, Spring Convention of SNT, February 22–26, 1965, Los Angeles, Calif.

"A Review of RADC Research Efforts in Infrared Radiation," Anthony J. Feduccia and William E. DuLac, Spring Convention of SNT, February 22–26, 1965, Los Angeles, Calif.

"Emissivity Equalization by Thermosetting Coatings," Neville R. Burrowes, Spring Convention of SNT, February 22–26, 1965, Los Angeles, Calif.

"Progress Report on IR Techniques for Electronics NDT," R. Vanzetti, Spring Convention of SNT, February 22–26, 1965, Los Angeles, Calif.

"Isothermal Mapping as a Reliability Tool," W. M. Berger, Spring Convention of the Society for Nondestructive Testing, February 22–26, 1965, Los Angeles, Calif.

"A State-of-the-Art Evaluation of Infrared in Heat Transfer Engineering," G. W. Carter (IBM, Federal Systems Division, Owego, N.Y.), IBM No. 65-825-1434, April 1965.

"Infrared for Electronics Reliability," R. Vanzetti, published in the July 1965 issue of *EDN Magazine.*

"Preindications of Failure in Electronic Components," J. W. Klapheke, B. C. Spradlin, and J. L. Easterday (Battelle Memorial Institute, Columbus, Ohio) Contract DA-01-021-AMC-11706(z) Redstone Scientific Information Center, U.S. Army Missile Command, Redstone Arsenal, Ala., RS1C-445, July 31, 1965.

"Infrared Examination of Soldered Electrical Cable Connections: An Annotated Bibliography," D. W. Hill (Lockheed Missile and Space Company) Rept. LS-65-17, September 1965 (government limited rights).

"Thermomicrography—Techniques and Applications," J. Richard Yoder, 25th National Convention, ASNT Detroit, Mich., October 1965.

"Infrared Techniques for Component De-rating of Printed Circuit Modules," R. S. Horne, 25th National Convention, Society for Nondestructive Testing, Detroit, Mich., October 18–22, 1965.

"IR Nondestructive Testing Applications: State of the Art," R. Vanzetti, 25th National Convention, Society for Nondestructive Testing, Detroit, Mich., October 18–22, 1965.

"Some Concepts Which May be Useful in Infrared Evaluation of Electronic Circuits," J. Fred Stoddard, 25th National Convention, Society for Nondestructive Testing, Detroit, Mich., October 18–22, 1965.

"Infrared Thermal Measurements on Semiconductor Microcircuits," Walter Koste, 25th National Convention, Society for Nondestructive Testing, Detroit, Mich., October 18–22, 1965.

"A Developmental Study of the Reliability Screening of Large Numbers of Operating Transistors Using Infrared," Bernard Selikson and Joseph DiMauro, 25th National Convention, ASNT, Detroit, Mich., October 1965.

"Infrared Testing for Electronic Hardware," R. Vanzetti, (co-author S. N. Bobo), presented at the Ninth Annual Symposium on Quality Control, Villanova University, Philadelphia, Pa., November 13, 1965; published in the *Symposium's Transactions.*

Manufacturing In-Process Control and Measuring Techniques For Integral Electronics (*Structural Defects in Single Crystal Semiconductors*), Pei Wang, F. Tausch, and R. Wolfson, Sylvania Electric Prods. Inc., Woburn, Mass. AFML-TR-65-293, December 1965.

"Infrared: New Approach to Thermal Measurement for Reliability," R. Vanzetti, (Raytheon Company), and M. Mark (Northeastern University, Boston, Mass.), presented at 1966 IEEE International Convention, Session 73: Reliability, March 25, 1966, New York City; published in Part 9 of the *Convention Record.*

Thin Film Microcircuit Interconnections, H. M. Greenhouse et al. (The Bendix Corporation, Bendix Radio Division Baltimore, Md.), March 1966, ECOM Contract DA-28-043-01482(E).

"Planned Infrared Applications in Raytheon Guidance Electronics Program," H. F. Sweitzer and Berssenbrygge, Spring Conference, SNT, March 7–10, 1966, Los Angeles, Calif.

"Infrared Detection of Microcircuit Metalization Falure Mechanisms," William Berger, Spring Convention, SNT, March 7–10, 1966, Los Angeles, Calif.

"Progress Report on Infrared Techniques for Electronics," R. Vanzetti, Spring Convention, SNT, March 9, 1966, Los Angeles, Calif.; published in the *IR Session's Transactions.*

"The Use of Infrared Techniques for Nondestructive Testing in Modern Aircraft," J. C. Jenkins, Spring Conference, SNT, March 7–10, 1966, Los Angeles, Calif.

"Survey of IR-NDT Applications for Electronics 1965–1966," David Seltzer, Spring Convention, SNT, March 7–10, 1966, Los Angeles, Calif.

Investigation of Infrared Radiation for Checkout Purposes, R. Vannzetti, S. N. Bobo, and J. F. Stoddard (Raytheon Company, Wayland, Mass.), Final Technical Report, March 1966, Wright-Patterson Air Force Base, Contract No. AF33(615)-2430.

"Infrared: New Approach to Thermal Measurement for Reliability," R. Vanzetti, presented at the Boston Section of IEEE 1966 Spring Seminar "Reliability Techniques Today and Tomorrow," Bedford, Mass., April 14, 1966; published in the *Proceedings of the Seminar.*

"Infrared Thermography for Diagnostic Evaluation of Electronic Modules," Newton N. Chapnick and Leonard L. Fagin, 26th National Convention, SNT, October 31–November 3, 1966, Chicago, Ill.

"Infrared Replaces Calculations for Realistic Stress Analysis of Electronic Modules," Kenneth E. Appley, 26th National Convention, SNT, October 31–November 3, 1966, Chicago, Ill.

"Weld Quality as Seen by IR." Arnold H. Crowley, 26th National Convention, SNT, October 31–November 3, 1966, Chicago, Ill.

"Vendor Selection Through Infrared Evaluation of Transistors," James F. Mallin, 26th National Convention, SNT, October 31–November 3, 1966, Chicago, Ill.

Infrared Testing of Electronic Components, Final Report OR8347, June 1966, prepared for G. C. Marshall Space Flight Center, NASA, Huntsville, Ala., by Martin Marietta Corp., Orlando, Fla.

"Aspects of Using Infrared for Electronic Equipment Diagnosis," Ruth A. Herman, 26th National Convention, SNT, October 31–November 3, 1966, Chicago, Ill.

"Automated Troubleshooter Works on Infrared Signatures," Riccardo Vanzetti and J. Fred Stoddard (Raytheon Company) IEEE International Convention, March 23, 1967, New York City.

"Infrared Pinpoints Second Breakdown Before Failure," M. F. Nowakowski and F. A. Laracuente, IEEE International Convention March 20–24, 1967, New York City.

"Infrared Exposes Hidden Circuit Flaws," R. Vanzetti, *Electronics Magazine,* Vol. 40, No. 7, April 3, 1967.

Reliability Screening and Step-Stress Testing of Digital-Type Microcircuits, H. F. Dean and K. F. Harper, Naval Electronics Laboratory Center for Command Control and Communications, San Diego, Calif. 92152 Report 1512, September 1, 1967.

"Infrared—The Last Word in Stress Analysis," W. Gordon Cawood, 27th National Fall Conference, SNT, October 16–19, 1967, Cleveland, Ohio.

"Programming Support for Infrared Thermal Plotters," A. D. Levit, 27th National Fall Conference, SNT, October 16–19, 1967, Cleveland, Ohio.

"Infrared Microradiometry (IRMR)—Precision and Accuracy Considerations Applicable to Microcircuit Temperature Measurements," D. D. Griffin, 27th National Fall Conference, SNT, October 16–19, 1967, Cleveland, Ohio.

Infrared Evaluation of Microweld Quality, F. E. Alzofon and A. D. McDonald, Lockheed Missiles and Space Company, Sunnyvale, Calif., August 1967.

"Idealized Versus Operational Reliability of RF Power Transistors as Determined by Infrared Scanning Techniques," E. B. Hakim, Bernard Reich, and G. J. Malinowski, IEEE Convention 1968, March 18–21, New York City.

"IR Analysis of Second Breakdown Modes in Power Transistors," M. F. Nowakowski and A. S. Dostoomian, IEEE Convention 1968, March 18–21, New York City.

"Transient IR Radiation: Promising New Probe of Semiconductor Performance," Riccardo Vanzetti, IEEE Convention 1968, March 18–21, New York City.

"Infrarot—Mikroskopie: Eine Neue Method Der Mikroelektronik," R. Vanzetti, *Radio Mentor Electronik* (Germany), two-part article, April and May 1968 issues.

"Infrared Evaluation of Multilayer Printed Boards," R. Vanzetti, presented at the 1968 NEPCON EAST Conference, NYC, June 6, 1968; published in the *Conference Kit,* distributed to attendees.

"Infrared for Electronics Equipment Diagnosis," J. F. Stoddard (Raytheon Co., Advanced Dev. Lab., Wayland Labs. Wayland, Mass.), Techn, Rept. AFAPL-TR-68-84, August 1968.

"Infrared Techniques for Measuring Temperature and Related Phenomena of Microcircuits," D. D. Griffin, *Applied Optics,* Vol. 7, No. 9, September 1968.

"The Application of Infrared Measurement Techniques to Electronic Design and Testing," William R. Randle, *Applied Optics,* Vol. 7, No. 9, September 1968.

"Infrared Profiling: A Novel Approach to Electronics Maintainability," R. Vanzetti, presented at 36th Meeting of the EIA Committee on Maintainability, Wash., September 16, 1968.

"Infrared Analysis of Telemetry Amplifiers and Discriminators," J. F. Pina and J. W. Oppenheim, ASNT, 1968 Fall Conference, October 14–17, Detroit, Mich.

"What Infrared scanning can do," R. Vanzetti, *Electronic Packaging and Production Magazine,* December 1968, Section 2.

"Infrared Data Analysis," Ruth A. Herman, ASNT, 1969 Spring Conference, March 10–13, Los Angeles, Cal.

"Enhancing Electronics Reliability & Maintenance Through Infrared Techniques," R. Vanzetti, *Minutes of U.S. Naval Materiel Command, Systems Performance/Effectiveness Steering Committee Meeting,* December 18, 1969.

"Infrared Process Control of Semiconductor Thermo-Compression Bonding," R. Resta and A. S. Dostoomian, IEEE International Convention, NYC Coliseum, March 23–26, 1970.

"NDT Applications of IR Fast Scanning," F. J. Stoddard, F. A. Orabona, and R. Dorval, ASNT, 1970 Spring Conference, March 9–13, Los Angeles, Cal.

"Redesign Recommendations for Space Telemetry Transmitters Based on Infrared Studies," J. F. Pina and J. W. Oppenhiem, ASNT, 1969 Spring Conference, March 10–13, Los Angeles, Cal.

"Introduction to the Infrared Applications Session," R. Vanzetti, (Vanzetti Infrared & Computer Systems, Inc., Dedham, Mass.), *IEEE Transactions,* Vol. IECI-18, No. 2, May 1971.

"Infrared Testing of Solar Cell Arrays," Robert R. Prudhomme, David Waddington, and Harry L. Over (Martin-Marietta Corp., Denver Division, Colorado), ASNT Fall Conference, Detroit, October 1971.

Phosphors and Liquid Crystals

Bibliography Phosphorescent Sulfides N/A, E. I. DuPont De Nemours and Company, Pigments Departments, Newport, Del., Special Report, February 16, 1948, Contract W36-039-sc-32256.

"The Surface-Tension Method of Visually Inspecting Honeycomb-Core Sandwich Plates," Samuel Katzoff (National Aeronautics and Space Administration, Langley Field, Va.) *Nondestructive Testing,* March–April 1960.

"Fluorescence Thermography," W. H. Byler and F. R. Hays, *SNT Journal,* May–June 1961, pp. 177–180.

About a Certain Method of Recording Weak Infrared Radiation, S. M. Kozel, Russian Book, Moskovskii Fiziko – Technicheskii Institute Issledovaniya Po Fizike i Radiolekhnike No. 8, 1962, pp. 73–76, FTD-TT 63-180/1 + 2.

"Liquid Crystals Applied to Thermal Nondestructive Testing," E. E. Bauer, 25th National Convention, SNT, Detroit, Mich., October 18–22, 1965.

"Testing Blossoms Out Into Color," T. H. Malim, *The Iron Age,* January 27, 1966.

"Cholesteric Liquid Crystals and Their Application to Thermal Nondestructive Testing," Wayne E. Woodmansee, (The Boeing Company, Seattle, Wash.), *Materials Evaluation,* October 1966.

"Thermal Nondestructive Testing with Cholesteric Liquid Crystals," W. E. Woodmansee, and H. L. Southworth, Fifth International Conference on NDT, Montreal, Canada, May 1967.

"The Chameleon Chemical," *Life,* January 12, 1968, pp. 40–45.

"Liquid Crystals for Thermal Mapping of Semiconductors and Related Hardware," Frederick Davis and Michael Lauriente, IEEE Convention 1968, March 18–21 (paper includes a film).

"Detection of Material Discontinuities with Liquid Crystals," W. E. Woodmansee and H. L. Southworth. (ASNT journal) *Materials Evaluation,* August 1968, pp. 145–154.

"Aerospace Thermal Mapping Applications of Liquid Crystals," W. E. Woodmansee, *Applied Optics,* Vol. 7, No. 9, September 1968.

"Liquid Crystals in Nondestructive Testing," James L. Fergason, *Applied Optics,* Vol. 7, No. 9, September 1968.

"High Resolution Thermal Mapping with Liquid Crystals," Michael Lauriente and John F. Farhood, ASNT, 1968 Fall Conference, October 14–17, Detroit, Mich.

"A Memory-Type Liquid Crystal for Permanent Image Flaw Detection," Frederick Davis and Betty Partain, ASNT, 1968 Fall Conference, October 14–17, Detroit, Mich.

"Detection of Fracture Initiation Sites by Liquid Crystal Coatings," Lawrence J. Broutman, ASNT, 1969 Spring Conference, March 10–13, Los Angeles, Cal.

Infrared Test Systems and Equipment

"Properties of Infrared Radiation," "Detectors for Infrared Radiation," "Infrared Radiation—Photodetectors," "Infrared Radiation—Materials," C. R. Betz, *Electronics Equipment Engineering,* February, March, April, and May 1959.

A Thermal Comparator for Estimating the Wall Thickness of Hollow Turbine Blades, W. Thompson and A. T. Josling, Ministry of Supply, Aeronautical Inspection Directorate AID/NDT/2007, September 1961, ASTIA 280427.

"An Instrument for Nondestructive Testing Fuel Core-to-Cladding Heat Transfer," D. R. Green, *Nuclear Science and Engineering,* Vol. 12, 1962, pp.271–275.

"High Speed Instrument for Quantitative Mapping," Donald R. Green, *Review of Scientific Instruments,* Vol. 33, 1962.

"Hot Bearing Detector Circuits," R. Stapelfeldt, U.S. Patent Office No. 3,076,090, *SNT Journal,* May–June 1963.

"Thermal Detection Method and Apparatus," E. J. Sternglass, U.S. Patent No. 3,072,819, *SNT Journal,* May–June 1963.

"Multiple Element Infrared Detector," H. S. Jones, U.S. Patent No. 3,073,957, *SNT Journal,* May–June 1963.

"Radiation Detector Systems," F. Schwarz, U.S. Patent No. 3,081,399, *SNT Journal,* July–August 1963.

"Radiation Measuring Apparatus," E. M. Speyer, U.S. Patent No. 3,082,325, *SNT Journal,* July–August 1963.

"Continuously Self-Calibrating Differential Detection System," T. F. McHenry and R. W. Astheimer, U.S. Patent No. 3,084,253, *SNT Journal,* July–August 1963.

Infrared Bond Defect Detection System, O. R. Gericke and P. E. J. Vogel, AMRA, *TR63–14, September 1963, 12 pp.

"Infrared Instrument for Nondestructive Heat Transfer Testing," D. R. Green, *Proceedings of the Fourth Annual Symposium on NDT of Aircraft and Missile Components,* San Antonio, Tex. 1963.

Investigation of Secondary Phenomena for Use in Checkout, Gilbert S. H. Hwang, Air Force Aero Propulsion Lab, Wright Patterson Air Force Base Contract No. AF 33(657)-9913 Systems Research Laboratories, Inc., APL-TDR 64-4, January 1964.

"The Thermal Plotter and Its Uses in Microcircuity Analysis and Testing," B. G. Marks, G. Revesz, and M. Walker, Third Annual Meeting of the Infrared Techniques for Electronics Committee, February 19–21, 1964, Huntsville, Ala.

"An Aperture for Infrared Temperature Measurement of Small Areas," F. J. Arnold, Third Annual Meeting of the Infrared Techniques for Electronics Committee, February 19–21, 1964, Huntsville, Ala.

"Infrared Fiber Optics," R. J. Simms (Optics Technology, Inc., Belmont, Calif.) Proceedings of the 3rd Annual Meeting of the Infrared Techniques for Electronics Committee, February 19–21, 1964; published by SNT, October 1964.

"Recent Development in Infrared Thermography," Richard Yoder (Barnes Engineering Co., Stamford, Conn.), Proceedings of the Third Annual Meeting of the Infrared Techniques for Electronic Committee, February 19–21, 1964; published by SNT, Evanston, Ill., October 1964.

"Process for Gaging Dimensions by Means of Radiation," Morris Weiss, Patent No. 3,131,306, *Materials Evaluation,* October 1964.

"Cryogenic Methods Applicable to the Use of Long Wavelength Photoconductive Infrared Detectors," George Giggey (Raytheon Company, Infrared and Optical Research Laboratory, Burlington, Mass.), *Proceedings of the Third Annual Meeting of the Infrared Techniques for Electronics Committee,* February 19–21, 1964: published by the Society for Nondestructive Testing, Inc., Evanston, October 1964.

"The Testing of Electrical Components and Systems Using Thermal Plots," H. Revesz and B. G. Marks, 24th National Convention, SNT, October 1964, Philadelphia, Pa.

"New High Speed Infrared Scanner," Charles Butter, L. McGlauchlin, and C. Motchenbacher, Spring Convention, SNT, February 22–26, 1965, Los Angeles, Calif.

"Bent Light Opens Up Dark Places," T. H. Malim, Associate Editor, *Iron Age,* May 20, 1965.

"A Scanning Infrared Radiation Pyrometer," Ernest W. Bivan, Spring Convention, SNT, February 22–26, 1965, Los Angeles, Calif.

"The Design and Development of a High Speed Line Scanner for Infrared Nondestructive Testing," J. W. Patterson and R. A. Wallner, Spring Convention, SNT, February 22–26, 1965, Los Angeles, Calif.

"An Infrared System for Detecting Flaws in Metal," Gerald L. Schmitz, Spring Convention, SNT, February 22–26, 1965, Los Angeles, Calif.

"Emissivity Independent Infrared Thermal Testing Method," Donald R. Green, *Materials Evaluation,* Vol. XXIII, No. 2, February 1965.

A High Speed Infrared Mapping System for Reliability Assessment of Minature Electronic Circuits, H. F. Dean and R. M. Fraser U.S. Navy Electronics Laboratory, San Diego, Calif., March 15, 1965, NEL 1272, AD 615018.

"An Infrared Nondestructive Testing System," Gerald L. Schmitz, (General American Transportation Corp., 7501 North Natchez, Niles, Ill.), Proceedings of the Fifth Annual Symposium on Nondestructive Evaluation of Aerospace and Weapons System Components and Materials, April 20–22, 1965, San Antonio, Tex.

Intrinsic Infrared Detector Development D. Long, Honeywell Research Center, Hopkins, Minn., April 1965, 71 pp.

"Infrared Finds the Hot Spots" (Barnes Engineering Co., Stamford, Conn.), *Iron Age,* Editorial, April 8, 1965.

"The Microthermograph: An Instrument for Obtaining Thermal Pictures of Thin-Film and Integrated Circuits," Robert B. McIntosh, Jr. and John R. Yonder, 25th National Convention, SNT, October 18–22, 1965, Detroit, Mich.

"Infrared Testing of Microcircuits," B. G. Marks, G. Revesz, and M. Walker (Philco Corporation, Willow Grove, Pa.), *Electrotechnology,* October 1965.

"Infrared Measurements in Real Time," R. I. Brown and J. E. Harris, 25th Convention, SNT, October 18–22, 1965, Detroit, Mich.

"The Fast-Scanning Infrared Microscope, A New Dimension in Infrared Nondestructive Testing," Leon C. Hamiter, Jr., 25th National Convention, SNT, October 18–22, 1965, Detroit, Mich.

"Pair of Helical Annuli As a 99 Percent Duty-Cycle Scanning System," R. Vanzetti, S. N. Bobo and M. Hinkle, (Raytheon Company, Wayland, Mass.) presented at the 1965 Annual Meeting of the Optical Society of America, Phila., Pa., October 5–8, 1965.

"An Infrared Nondestructive Testing System for Rocket Motors," F. E. Alzofon (Lockheed Missiles and Space Company, Sunnyvale, Calif.), *Materials Evaluation,* November 1965.

"An Infrared Radiometric Microscope for Nondestructive Testing of Integrated Circuits" (english) R. B. McIntosh, Jr., *Chem Rund,* No. 23, November 11, 1965, pp. 725–727.

"Fast Scan Infrared Microscope for Improving Microelectronic Device Reliability," L. C. Hamiter, Jr. *Research Achievements Review,* Vol. II, Rept. No. 5 NASA TM X-53602, 1966 R&D Operations, MSFC, Huntsville, Ala., also NASA SP 5082, pp. 99–111.

"Non-Destructive Testing Using a Fast-Response Infra-Red Scanner," W. B. Allen, Current Infrared Papers, distributed by SNT ASNT, 1966 Spring Convention, Los Angeles, Calif., March 7–10.

"Fast Scanning Infrared Microscope for Semiconductor Evaluation," R. Vanzetti and Leon Hamiter, NASA, Huntsville, Ala., presented at 1966 IEEE International Convention, Session 39: Novel Sensing, March 23, 1966, New York; published in Part 10 of the *Convention Record.*

"Characteristics of an Infrared Vidicon Television System," Harold Berger and I. R. Kraska (Argonne National Laboratory, Argonne, Ill.), *Materials Evaluation,* April 1966.

"Thermal (Infrared) Radiometers As Instruments For Nondestructive Reliability Testing," R. M. Fraser, NEL/Report 1377, May 10, 1966, U.S. Navy Electronics Laboratory, San Diego, Calif.

"Multiple IR Heads for Complete Control of Temperature and Growth Thickness in Epitaxial Reactors," W. C. Hamlin (Huggins Laboratories, Inc.), IEEE International Convention March 20–24, 1967, New York.

"A Novel Sensor for Microelectronic Weld Quality Evaluation," Stephen N. Bobo, Jr., (Raytheon Company) IEEE Convention, March 20–24, 1967, New York City.

"Infrared Radiometry of Semiconductor Devices," David A. Peterman and Wilton Workman (Texas Instruments Incorporated), IEEE International Convention, March 20–24, 1967, New York City.

"Instrumentation for Detecting Failure Mechanisms in Integrated Circuits," R. Vanzetti, presented at the Boston Section of IEEE Section Lecture Series on Failure Mechanisms in Microelectronics, M.I.T., Cambridge, Mass., May 4, 1967.

"Dynamic Infrared Detection of Fatigue Cracks," E. J. Kubiak, B. A. Johnson, and R. C. Taylor, Fifth International Conference on NDT, Montreal, Canada, May 1967.

"Temperature Measurement System Utilizing Infrared Radiation Sensing Techniques," (Infrared Industries, Inc., Santa Barbara, Calif.), Goddard Space Flight Center, Greenbelt, Md. Final Project Report, August 14, 1967.

"Scanning Infrared Inspection Systems (SIRIS) Applied to NDT of Aerospace Materials," Karl H. Zimmermann and Edgar W. Kutzscher, SNT, 27th National Fall Conference, October 16–19, 1967, Cleveland, Ohio.

"Development of Laboratory Model Fatigue Crack Detection Device Based on Infrared Techniques," E. J. Kubiak and L. M. Frank, (General American Corp., Niles, Ill.). Technical Report AFFDL-TR-67-39, October 1967.

"A Scanning Infrared Inspection System Applied to Nondestructive Testing of Bonded Aerospace Structures," Edgar W. Kutzcher and Karl H. Zimmermann, *Applied Optics,* Vol. 7, No. 9, September 1968.

"Conversion of Infrared Images to Visible in Color," L. W. Nichols and J. Lamar, *Applied Optics,* Vol. 7, No. 9, September 1968.

"Wide-Angle Infrared Camera for Industry and Medicine," Erik Sundstrom, *Applied Optics,* Vol. 7, No. 9, September 1968.

"Use of Contiguous Optical Fibers as a Means of Carrying Thermal Information from Welds," S. N. Bobo and A. H. Crowley, *Applied Optics,* Vol. 7, No. 9, September 1968.

"Thermal Imaging With Real Time Picture Presentation," Borg Sven-Bertil *Applied Optics,* Vol. 7, No. 9, September 1968.

"Temperature Measurement With an Infrared Microscope," John R. Yoder, *Applied Optics,* Vol. 7, No. 9, September 1968.

"Thermal Imaging Using Pyroelectric Detectors," R. W. Astheimer and F. Schwarz, *Applied Optics,* Vol. 7, No. 9, September 1968.

"Infrared Detection of Fatigue Cracks and Other Near-Surface Defects," Edward J. Kubiak, *Applied Optics,* Vol. 7, No. 9, September 1968.

"An IR-NDT System for Rotor Blade Honeycomb Assemblies," A. J. Intrieri, ASNT, 1968 Fall Conference, October 14–17, Detroit, Mich.

"Infrared Detection, Isolation, and Prediction of Electronic Equipment Malfunctions," Ruth A. Herman, 1968 AF Science and Engineering Symposium, November 1, 1968, Colorado Springs, Colo.; *Research and Technology Briefs,* February 1969.

"An Infrared Microscope System for the Detection of Internal Flaws in Solids," W. A. Simpson, C. C. Cheng, C. V. Dodd, and W. E. Deeds, ASNT, 1969 Spring Conference, March 10–13.

"Infrared to Visible Image Translation Devices," R. W. Astheimer, *Photographic Science and Engineering,* Vol. 13, No. 3, May–June 1969, p. 127.

"Nondestructive Testing By High-Speed Thermography," L. Bergstrom and D. Baeu, *Materials Evaluation,* May 1969, p. 25A.

Development of Thermal Test Methods, J. A. Halloway and D. R. Maley (Automation Industries, Inc., Boulder, Colo.), Wright Patterson AFB Contract AF 33(616)-7725, AF 33(615)-1531.

Survey of the State of the Art of Thermal Type Infrared Detectors, E. M. Wormser, Barnes Engineering Company, Stamford, Conn.

"Enhancement of Infrared NDT Data Presentation," B. Stiefeld and D. W. Ballard (Sandia Laboratories, Albuquerque, N.M.), *ASNT Nat. Spring Conf.,* Los Angeles, Calif., March 1972.

Index